2022 15th International Conference on Computer and Electrical Engineering (ICCEE 2022)

Journal of Physics: Conference Series Volume 2320

Online
24-26 June 2022

ISBN: 978-1-7138-6040-2
ISSN: 1742-6588

Printed from e-media with permission by:

Curran Associates, Inc.
57 Morehouse Lane
Red Hook, NY 12571

Some format issues inherent in the e-media version may also appear in this print version.

Institute of Physics
Dirac House, Temple Back
Bristol BS1 6BE UK

Phone: 44 1 17 929 7481
Fax: 44 1 17 920 0979

techtracking@iop.org

Additional copies of this publication are available from:

Curran Associates, Inc.
57 Morehouse Lane
Red Hook, NY 12571 USA
Phone: 845-758-0400
Fax: 845-758-2633
Email: curran@proceedings.com
Web: www.proceedings.com

TABLE OF CONTENTS

Journal of Physics: Conference Series **2320** (2022) 011001 doi:10.1088/1742-6596/2320/1/011001

Preface

We are glad to welcome you to the proceedings of 2022 The 15th International Conference on Computer and Electrical Engineering (ICCEE 2022), which was held on June 24-26, 2022. Due to the cases increasing recently in China, in addition to the travel restriction and in respect of the willingness of all participants, ICCEE 2022 was entirely transferred as online conference instead of in Shenzhen University, China as scheduled.

ICCEE was held annually and covered a broad range of topics related to Computer and Electrical Engineering. This year, presenters have presented and discussed topics in their respective research areas. The conference was held for 3 days including the online test sessions on ZOOM on the first day, and then followed with a wonderful array of 6 guest speeches along with 5 technical sessions during June 25-26. The authors' presentations cover topics on Computer Science and Information Engineering; Computer and Control Automation; Machine Vision and Image Processing; Modern Electronic Information Technology and Application; Electrical Engineering and Automation. 15 minutes for each presentation, including 2 minutes for the Q&A. Delegates had a wide range of sessions to choose from and had enough time in deciding which sessions to attend.

After the peer-review of all submissions, 40 full papers were accepted and included in the proceedings. All papers in these proceedings were peer-reviewed by conference technical committee members and international reviewers. We'd like to recognize the hard work of the entire committee for their enormous contribution. Our special thanks go to our speakers as well as all the authors for sharing their latest research results. We hope these talks interest and inspire all of you.

Computer and Electrical Engineering have had a successful history. This field has pushed the frontiers of fundamental research, led to the emergence of entirely new disciplines, and revolutionized our daily lives. It extended to all walks of life, from the design of a switch to the research of aircraft. We do hope researchers in this field can create innovative, practical solutions as well as invent, design and build technologies and products that matter in the future.

The proceedings will be successful only if these papers could motivate some of you. We look forward to meeting all of you next year at the conference in person.

Best regards,

Conference Program Chair

Chengwen Luo, Shenzhen University, China

COMMITTEES

Conference Chair
Jianqiang Li, Shenzhen University, China
Ying Tan, Peking University, China

Program Chairs
Chengwen Luo, Shenzhen University, China
Letian Huang, University of Electronic Science and Technology of China, China

Student Program Chair
Feng Li, Shandong University, China

Award Chair
Yiran Shen, Shandong University, China

Regional Chair
Weitao Xu, City University of Hong Kong, Hong Kong, China

Technical Committee Members
Ana Reyes-Menendez, Rey Juan Carlos University, Spain
Assiya Haddout, Ibn Tofail University, Morocco
Chun-An Cheng, I-Shou University, Taiwan, China
Dezhi Chen, Shenyang University of Technology, China
En-Chih Chang, I-Shou University, Taiwan, China
F Yudi Limpraptono, National Institute of Technology Malang, Indonesia
Fredy Hernan Martinez, Universidad Distrital Francisco José de Caldas, Colombia
Gokhan Erdemir, Sabahattin Zaim University, Turkey
Harrizki Arie Pradana, Harrizki Arie Pradana, Indonesia
Levent Seyfi, Konya Technical University, Turkey
LIEW HUI FANG, Universiti Malaysia Perlis, Malaysia
Lili He, San Jose State University, The United States
Lin Wu, Chiang Mai University, Thailand
Mario Chauca Saavedra, Universidad Nacional Tecnológica de Lima Sur, Perú
Mau Luen Tham, Univeristi Tunku Abdul Rahman, Malaysia
Nadia Abd-Alsabour, Cairo University, Egypt
Ping Zhao, Donghua University, China
Po-Chun Huang, Taipei Tech, Taiwan, China
Pua Chang Hong, Univeristi Tunku Abdul Rahman, Malaysia
Recep Özdağ, Yuzuncu Yil University, Turkey
Rishab Mehra, The University of Texas at Austin, The United States
S. Kannadhasan, Cheran College of Engineering, India
Shruti Goel, Navis LLC, California, USA
Tsung-Mou Huang, National Sun Yat-sen University, Taiwan, China
Weizhong Qian, University of Electronic Science and Technology of China, China

Journal of Physics: Conference Series **2320** (2022) 011001 doi:10.1088/1742-6596/2320/1/011001

STATEMENT OF PEER REVIEW

All papers published in this volume of *Journal of Physics: Conference Series* have been peer reviewed through processes administered by the proceedings Editors. Reviews were conducted by expert referees to the professional and scientific standards expected of a proceedings journal published by IOP Publishing.

Peer Review Statement

All papers published in this volume have been reviewed through processes administered by the Editors. Reviews were conducted by expert referees to the professional and scientific standards expected of a proceedings journal published by IOP Publishing Publishing.

- **Type of peer review:** Double Anonymous
- **Conference submission management system:** easychair
- **Number of submissions received:** 72
- **Number of submissions sent for review:** 72
- **Number of submissions accepted:** 40
- **Acceptance Rate (Submissions Accepted / Submissions Received × 100):** 55.6
- **Average number of reviews per paper:** 1
- **Total number of reviewers involved:** 23
- **Contact person for queries:**
  Name: Letian Huang
  Email: huanglt_uestc@outlook.com
  Affiliation: UESTC

Design and Implementation of AGV Calling System Based on Android

Ouwen Zhu[1] and Yuli Zhao[1]

[1] Software College, Northeastern University, Shenyang, China

2171389@stu.neu.edu.cn, zhaoyl_neu@163.com

Abstract. This paper designs and implements an AGV calling application based on Android. In the actual working environment, administrators need to call automatic guided vehicles to perform tasks and monitor them anytime and anywhere, so the AGV calling application based on Android came into being. Five business function modules of the system are realized: calling module, map monitoring module, task list module, node configuration module and communication configuration module. The application software is developed using Android Studio development tool, with the use of Vue framework and front-end HTML5+CSS+JavaScript technology to achieve interface rendering. TCP and HTTP protocols are used in the connection between the system and the central control system.

1. Introduction

With the rapid development of Internet technology and computer technology, a lot of work in modern industrial production and logistics warehouse can be completed by automatic guided vehicles, such as picking goods, transportation, quality control and other work[1]. AGV is a system integrating environment perception, dynamic planning, dynamic decision making, action execution and control. Its technology includes sensors, electronic engineering, software engineering, automation and other technologies. It is one of the most active fields of science and technology development[2]. Nowadays, with the continuous improvement of robot technology, the application scope of automatic guided vehicle is gradually expanding, among which, robot technology is particularly important in the industrial field[3]. In industrial production, if automatic guided vehicles are to be used to improve production efficiency, usually technical personnel will install and debug the management system, central control system and AGV calling application before putting them into practical use. After obtaining the requirements, the staff responsible for software development will develop the automatic guided vehicle system. However, in the testing stage, the system cannot be installed in the actual working environment to test the algorithm and software functions, so the AGV simulation system is essential. And the customer also needs a AGV simulation system to test the operation of the robot and get familiar with the function[4].

AGV calling application software is an essential key link. Only when users call automatic guided vehicles through mobile devices and release tasks, can the following series of robot related behaviors occur. Therefore, we need a calling APP for mobile devices to make it more convenient for workers to call the AGV, issue tasks and other operations, display all the status of the AGV, and call the AGV to the designated place at any time, and then let it perform the specified task. The design and implementation of AGV calling application will improve the efficiency of industrial production and make AGV serve us more efficiently[5].

2. Design of AGV calling application

AGV calling application is an application software developed based on Android. It runs on Android tablet. The mobile application makes it more convenient for workers to deliver tasks and monitor the operation of robots, which can be operated on site. Material calling managers use Android tablet to input the material taking position and discharging position to call. There will be a robot to perform tasks, its running track will be displayed in real time on the map, the administrator can clearly view the running situation. The administrator can also manage tasks in a unified manner, and manage nodes for taking and discharging materials.

The application includes calling module, map monitoring module, task list module, node configuration module and communication configuration module. The calling module is responsible for calling the robot to perform the specified tasks, the map monitoring module is responsible for viewing the running status of the robot and the overall status of the task, the task list module is responsible for viewing the information of all tasks and the execution of a single task, and the node configuration module is responsible for configuring the node information of the robot. The communication configuration module is responsible for connecting the calling APP to the AGV central control system. The application uses C/S architecture, and the interface is rendered using Android interface rendering technology. The backend of Android is written using Java language, and the map is loaded using Vue framework and front-end HTML5+CSS+JavaScript technology.

Server-client, that is, client-server (C/S) structure. The C/S structure usually takes two layers. The server is responsible for data management, and the client is responsible for completing the task of interaction with the user. The client is connected with the server through the LAN, accepts the user's request, and puts forward the request to the server through the network, carries on the operation to the data. The client of this system is the calling application of Android tablet, and the server is the central control system. The corresponding functions are realized through the LAN connection in the factory[6]. The structure of this application software is divided into three layers, which are presentation layer, logic layer and persistence layer. The main task undertaken by the presentation layer is the interaction between the user and the system, the rendering of the interface, and all the data presented to the user are completed in this layer. The logical layer takes data from the persistence layer and processes it logically, then returns the results to the presentation layer and presents them to the user. The persistence layer completes the interaction between the system and persistent data, obtains data and writes data. The persistence layer of this application software is the data in Android files, including the IP address port number of the data source and task source of the calling system, node configuration information and task list information[7]. The system hierarchy diagram for this application is shown in Figure 1.

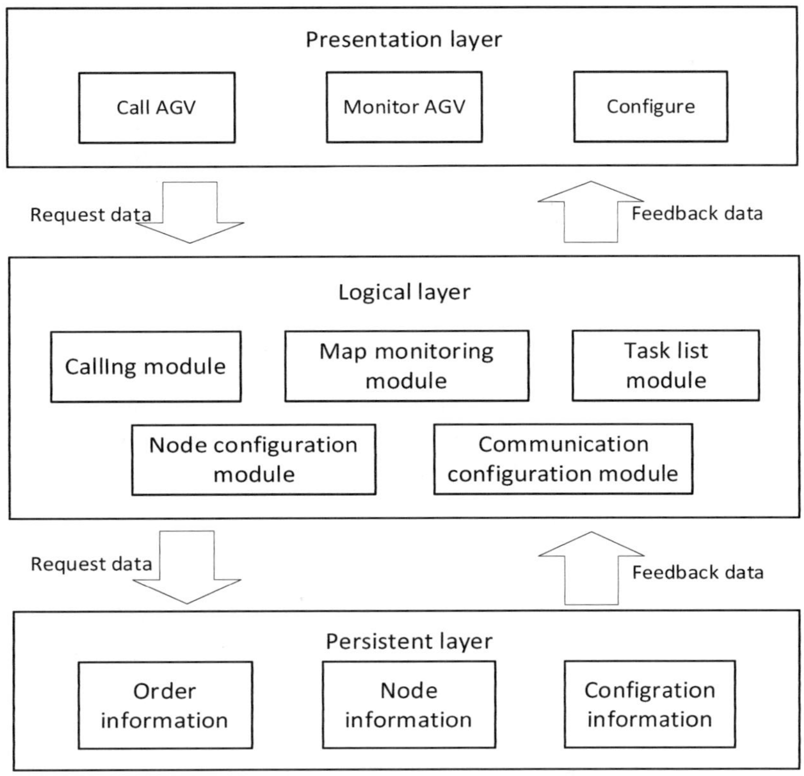

Figure 1. System hierarchy diagram

The calling module is a functional module for calling application to send tasks. The material manager enters the call module, selects the material picking node and discharging node, and calls the robot to perform tasks. Material calling managers can monitor the trajectory of the robot and the task status visually through the map monitoring module, keeping the map and work site one-to-one correspondence. The managers can also see the number of free robots, the amount of undistributed task of the robot, the number of the execution of task and the number of pending tasks. And they can be displayed in the form of percentage through the pie chart. The task list module is responsible for storing the task execution list sent by the central control system in the Android file and displaying it on the interface. The material calling managers can view, search and declare the task by the task list. Enter the single task interface, you can view the running track and task execution of the robot on the map, and monitor the current robot power, speed, etc. The communication configuration module has high permission and can be used only after the administrator logs in. Its function is to configure the IP address and port number of the data source and task source and save it in the Android file. If the IP address and port number of the central control system change, the calling application should also be configured accordingly[8].

AGV calling application uses TCP protocol for network communication. Therefore, TCP network request and reception need to be encapsulated into a TCP client module to process TCP network requests in a unified manner, including connection, disconnection, reconnection, data receiving, data sending, and heartbeat detection. The functions of the TCP client module include receiving and sending data, starting the client, disconnecting the client, and heartbeat detection. The data receiving function is responsible for processing the data transmitted by the central control system. First, it determines one byte of the received byte array. If the response is made according to the request of calling robot, it communicates with the calling module through Handler and resolves the task sent by the calling application into a certain format and writes the data into the system file. The task list module is called to update the task list in the public variable if it is responding to a request to update the execution of the task sent by the task list module. Data sending sends calling task requests or real-time system update requests sent by the material calling manager. Other functions are responsible for long network connections and exception handling.

3. Implementation of AGV calling application

HttpUtil is a low-level network request tool used by the system to make HTTP requests. It does not involve any system logic or data and only sends the incoming JSON data to the specified server through GET or POST requests. Cookie is carried when sending a request and saved when receiving a response from the server to realize the browser's cookie saving function.

Network request based on TCP/IP requires the system and the server to maintain a long connection, so the required method is relatively complex. Methods include starting the client, receiving data, sending data, disconnecting from the client, and detecting heartbeat. TCP network requests need to convert all transmitted data to byte array, and BytesToString function is used to convert array of bytes to string[9]. Calling APP needs calling interface and searching interface to cooperate with each other to achieve, the calling interface has feeding position and feeding position to input, which needs to be designed to enter the searching interface by clicking the feeding position, that is, display all the configured feeding position nodes; Click the feeding position to enter the display of all feeding position nodes, so as to ensure the reliability of the system. Calling module calls TCP module to send TCP request to central control system for calling robot operation. The calling module flow chart is shown in Figure 2.

The core technology of map monitoring module is map loading. The mapping tool provided by the company is used for map drawing, which can draw the dot plot map corresponding to the terrain of the factory. Next, export the txt file generated by the dot plot, which is used for loading the map on the front interface. The front-end VUE project converts dots and lines into corresponding path labels by reading txt files and presenting them on the web page. The system displays the map and the real-time running of the robot through the loading web page of Android.

Task list module is a unified management module for all tasks, and the content of each task including description, task code, task number, vehicle number and execution status should be clearly displayed. The module is mainly achieved through the coordination between the TCP client module. The main thread and the sub-thread coordinate with each other through the Handler, in order to achieve the effect of real-time update data[10].

Communication configuration module is the foundation of the whole application program. The calling function and monitoring function need to be connected through LAN and central control system to achieve, and the IP address and port number of data source and task source need to be correctly configured in this module. At the same time, the module is also responsible for saving the configured data source and task source information in the Android system file, which will be loaded when the system starts.

 2320 (2022) 012001 doi:10.1088/1742-6596/2320/1/012001

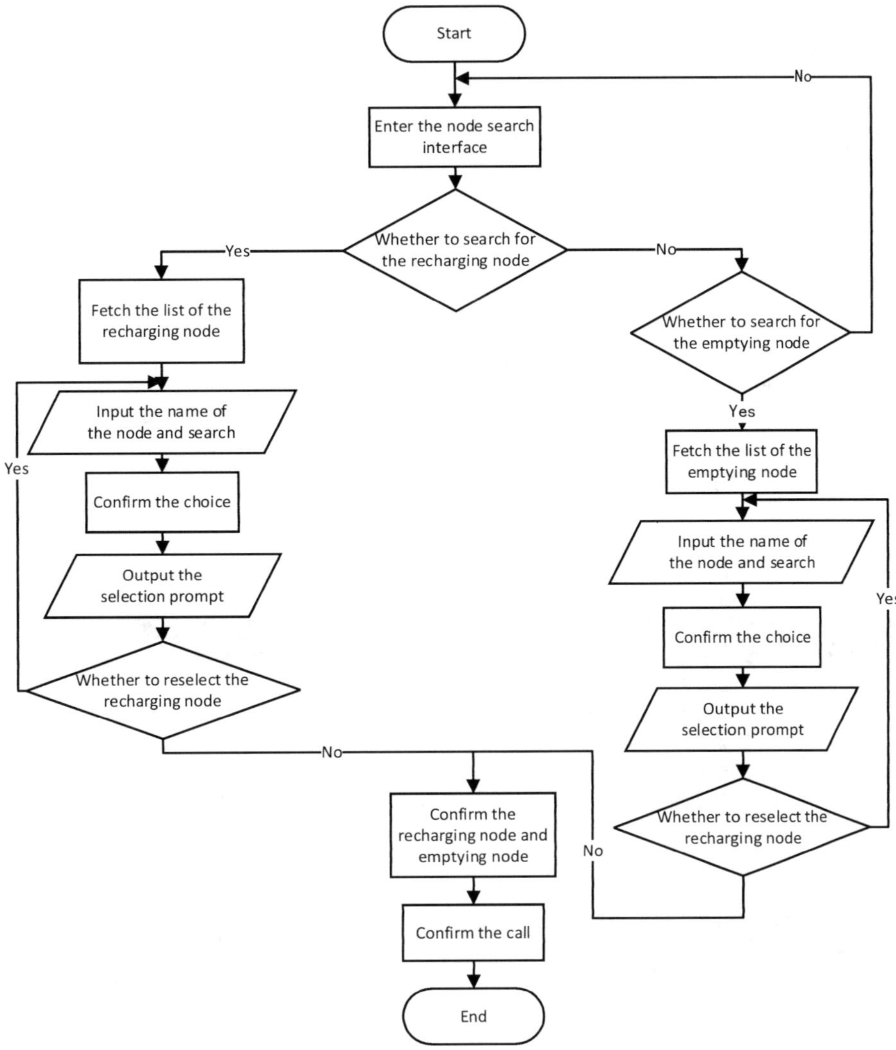

Figure 2. Calling module flow chart

The application development tool is Android Studio. First, click New Project under File to create the Project, input the Project name, select Java language development, and build the Project through Gradle Project automatic construction tool[11]. Select the Android version you want to develop and download the Android development kit SDK to develop Android projects. Next, select the Emulator of Android mobile device that needs to be debugged in real time. In this project, select the emulator of the corresponding size of Android tablet for debugging[12].

Java code is stored in Java packages automatically created by the system, including activity, adapter, bean, fragment, http, tcp, utils. The activity and fragment store the corresponding application components of each module, which provide the function of interacting with the screen. The adapter stores the intermediaries needed to map data to the list when the interface presents the data list. Http and tcp are the packages responsible for network requests. Utils is the low-level toolkit that works for the system. The res package for Android contains resource files such as layout files, images, common values, and system icons[13].

4. Simulation result

First of all, while entering the interface, initialize the controls and render the interface, and re-read the configuration information and node information in the system file. Next, add click monitoring event for input material taking position and discharging position button, and jump to search interface by carrying flag of material taking position and discharging position identifier through Bundle. Enter the search interface, load the node list in the public variable into the module, and obtain the identification flag from the Bundle in the interface before the jump, so as to determine whether it is necessary to display the node of material taking or discharging position and display the list. Finally, set the click listening event for the exit button to return to the call screen. Click the listening event for each setting in the list to bring the node information of the motor back to the call interface and display it on the button. The calling module interface is shown in Figure 3.

Initialize the Android control to display the currently configured data source and task source by default. The signal mark in the top bar is realized by sending information to MainActivity after the TCP connection between the TCP module and the central control system. If the connection is successful, the Message with the what attribute as 1 is sent to the MainHandler in the MainActivity, and the success icon is displayed in the top bar. If the connection fails, the MainHandler periodically reconnects and sends a Message with the what attribute as 2 to the MainHandler. The failure icon is displayed in the top bar. During the configuration, enter the IP address and port number in the input box and click Set. The FileReadAndWrite module will be called to write the configuration information into the system file in a certain format. The communication configuration module interface is shown in Figure 4.

Figure 3. Calling module

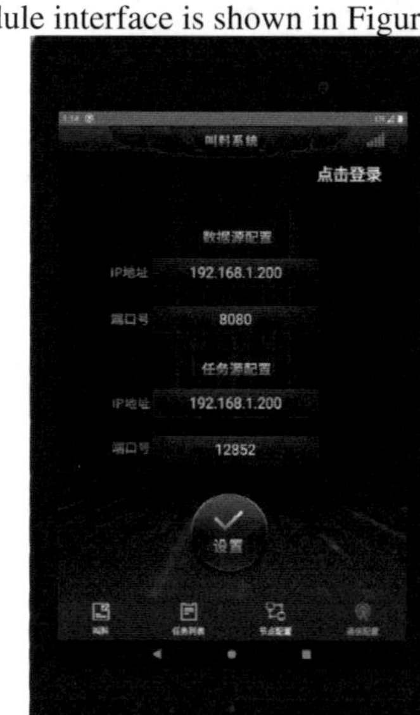

Figure 4. Communication configuration module

5. Conclusion

This application is an AGV calling APP developed according to the development status of industrial robots in production and combined with the needs of users and actual needs. Through the whole robot simulation system, before the robot is actually on-line, the software part of the system is debugged, the concurrent pressure is tested, and the number of robots and the distribution of tasks are systematically analyzed and tested according to the factory map. This application is designed for the factory material calling manager to call the robot to perform tasks and monitor it anytime and anywhere. It has strong practicability and matches the whole system. When the system is put into the actual work scene, workers will work more efficiently.

This application adopts C/S structure and it is designed and developed based on Android. The interface is rendered using Android interface rendering technology. The application is developed strictly in accordance with the software life cycle process. The APP has experienced the analysis of the current development status of industrial robots at home and abroad, an in-depth investigation of the needs of users and industrial workshops, the functional division and detailed design of the application software, and the realization and testing of the function. Each module of this application realizes dynamic display and effective interface interaction.

6. References

[1] Yin Niansheng. Research on Current Situation and Future Development of Industrial Robots in China[J]. *China New Telecommunication*, 2020, **22**(20): 135-136.
[2] Zheng Wenying. Current Situation and Prospect of Mobile Robot Technology[J]. *Information Recording Material*, 2020, **21**(10): 24-25.
[3] Li Xiaoxiao et al. Simultaneous Obstacle Avoidance and Target Tracking of Multiple Wheeled Mobile Robots With Certified Safety.[J]. *IEEE transactions on cybernetics*, 2021, **PP**.
[4] Hao Kun et al. The Application of an Adaptive Genetic Algorithm Based on Collision Detection in Path Planning of Mobile Robots[J]. *Computational Intelligence and Neuroscience*, 2021, **2021**.
[5] Xu Bingbing, Hao Rongfei. Status and Development of Robot Path Planning Technology[J]. *Electronic technology and Software Engineering*, 2018(24): 81.
[6] Walls C. Spring Boot in Action[J]. 2016.
[7] Ouyang Shen. *Android Studio Development of Actual Combat from Zero to App Online*[M]. Beijing Tsinghua University Press, 2018: 1-2.
[8] Li Zuobin. *Android Development Introduction and Actual Combat Experience*[M]. Beijing China Machine Press, 2011: 5-6.
[9] Kevin R. Fall, W. Richard Stevens. *TCP/IP Illustrated Volume 1 The Protocols 2nd Edition*[M]. Addison-Wesley Professional, 2011.
[10] Comer D, Stevens D L. *Internetworking with TCP/IP*[M]. Posts and Telecom Press, 2002.
[11] Hao Qiang. Overview of Android Platform Software Development Methods[J]. *Software Engineering*, 2020, **23**(08): 13-15.
[12] Atefeh Nirumand, Bahman Zamani, Behrouz Tork Ladani. VAnDroid: A framework for vulnerability analysis of Android applications using a model-driven reverse engineering technique[J]. *Software: Practice and Experience*, 2019(1): 10-12.
[13] Android Developer Website [EB/OL]. https://developer.android.google.cn/.

Acknowledgement

This work was financially supported by the Project of China's Ministry of Education and supported by Alibaba in 2020 (Grant No. 202002001063). Heartfelt thanks also to the reviewers for their constructive comments to the paper.

The Determination of Reward Function in AGV Motion Control Based on DQN

Yubin Chen[1], Dancheng Li[1], Huagang Zhong[2], Ouwen Zhu[1] and Ziqi Zhao[1]

[1] Software College, Northeastern University, Shenyang, China
[2] Four Leaves Clover(Suzhou) Intelligent Technology Co.,Ltd, Suzhou, China

Ldc@mail.neu.edu.cn, 2071267@stu.neu.edu.cn

Abstract. Motion control is a very important part in the field of AGV(Automated Guided Vehicle). A good motion control method can make the movement of AGV more stable. Network models of reinforcement learning is one of the methods to solve the problem of AGV in motion control. This paper introduces the Markov Decision Process and the role of reward function. Besides, it studies and analyzes several classic reinforcement learning cases. DQN(Deep Q-Learning Network) which belongs to deep reinforcement learning network model is adopted. We set up several sets of comparative experiments with different reward functions by using sparse reward setting method, formalized reward setting method and reward coefficient variation reward setting method. Also we adjust the time of training. Through comparison, a reward function suitable for solving the problem of AGV motion control is obtained. In the field of AGV motion control, reinforcement learning model can converge faster and make more correct decisions. The reward function is verified in the simulation environment built by ROS(Robot Operating System).

1. Introduction

PID(Proportion Integration Differentiation) algorithm is a basic and important part of AGV control algorithm. AGV can be controlled by using PID parameters. Each AGV produced is unique. Although the manufacturing process is fine, the assembly and slight error of assembly will cause differences in AGV. In addition, the working environment of AGV has a variety of different influencing factors such as ground material, friction, etc. These factors will affect the movement of AGV, which may make AGV unable to run smoothly and drive according to the specified path [1]. Using reinforcement learning can realize the interaction between AGV agent and environment. We take the movement of AGV on a straight path as the evaluation standard of whether AGV runs smoothly. We establish a Markov Decision Process for AGV in motion control. The initial state is set to random PID. Through the reinforcement learning model, the adjustment decision of PID is obtained. Through continuous exploration and adjustment, it is considered that the optimal PID is adjusted and AGV can run smoothly when the position deviation of AGV running on the straight path is less than the set threshold. Reinforcement learning model training needs a large amount of data and the training process takes a long time. The set reward function can affect the convergence speed and degree of the model. Appropriate reward function can improve the training speed of reinforcement learning model so that each AGV can quickly get the best PID through the model and run smoothly. This paper introduces Markov Decision Process, reward function and deep Q-Learning network. It also analyzes the classic cases of reinforcement learning. In the experiment part, the simulation platform is introduced. We set up four groups of experiments and obtain an appropriate reward function. Finally, the experimental results are summarized.

2. Reward function of Markov Decision Process

Reinforcement learning tasks are usually described by Markov Decision Process. Some technicians use reinforcement learning to solve the navigation and path planning problems of AGV [2]. Similarly, it can also solve the problems of motion control. Markov Decision Process is composed of < s, a, P, R, γ>. S is a finite state set. A is a finite behavior set. P is the state transition probability matrix in the set. R is a reward function. γ is an attenuation factor. Markov Decision Process introduces behavior. Individuals have the right to choose a behavior from the behavior set according to their own understanding of the current state. After an individual chooses a certain behavior, its subsequent state is determined by the dynamics of the environment. The basis for an individual to choose a behavior from the behavior set in a given state is called strategy π. Strategy π is a probability distribution based on the behavior set in a certain state [3]. In the Markov Decision Process, the strategy can produce an individual's behavior only by relying on the current state. The policy is only related to the current state, not the historical state. For different states, individuals may also produce different behaviors according to the same strategy. For the same state, individuals may also produce different behaviors according to the same strategy. Strategy which is considered to be static describes that the mechanism of individual behavior does not change with the change of state.

The goal of the reward function is to maximize the reward. This function explains how much reward can be obtained after each decision. The purpose of reinforcement learning is to enable the agent to find the best sequence of steps to solve the problem through continuous learning. This measurement standard is the cumulative reward obtained by the agent after performing a series of actions. Typically, there is a small amount of negative reward for each step to encourage a quick solution. In the final state, there will be a greater positive reward if the task succeeds or negative penalties if it fails.

There are many ways to set the reward function, including sparse reward, formal reward, distributed reward, reward coefficient change reward, intermediate difficulty starting point reward and reward shaping [4, 5]. Sparse reward is the most common kind of reward. It only gives rewards in the target state and there are no rewards in other states. The formal reward determines the reward value according to the gap between the current state and the target state. The closer to the target, the higher the reward value given by the system. The idea of distributed reward comes from the distribution in probability theory. The reward is usually set according to the distribution mode such as Gaussian distribution. The reward coefficient changes with other variables. The middle difficulty starting point reward avoids two situations. One is that all learning at the beginning is negative reward values and learning is very slow. The other is that all are positive reward values so that the system can't learn the bad side. Potential based reward shaping gives each state a potential energy. It gives positive reward value from places with low potential energy to places with high potential energy. The negative reward is given from the place with high potential energy to the place with low potential energy. The problem of repeatedly brushing points between two states is avoided.

3. Deep Q-Learning Network

DQN is a common network model in reinforcement learning. DQN can be regarded as an advanced version of Q-Learning [6]. The core of Q-Learning is the Q form. Provide guidance for action by establishing a Q form. This is applicable to the case where the state and action space are discrete and the dimension is not high. When the state and action space are high-dimensional and continuous, the Q table will become very large. It is unrealistic to maintain Q table and search and it takes up a lot of space [7]. DQN combines Q-Learning and neural network. It uses experience pool [8]. DQN takes advantage of the off-policy feature of Q-Learning. Experience gained through exploration or action will be stored in the experience pool. Randomly extract an experience from the experience pool to update the network. DQN uses the experience pool [9] of experience relay which has two advantages.

One is random selection which can cut off the correlation of experience. Another is that each experience can be learned repeatedly which improves the utilization rate of experience. DQN uses two networks with the same structure but different parameters [10]. One is used to predict Q estimation. Another is used to predict the Q target. The target network is used when calculating the target Q value. This reduces the correlation between the target calculation and the current value. After training for a period of time each time, copying the parameter values of the current Q network to the target Q network [11]. It can make the algorithm update more stable [12]. Therefore, the experiment in this paper is based on DQN.

4. Classic cases analysis

OpenAI Gym is an open source toolkit for researching and comparing reinforcement learning related models [13]. It contains many classic examples, simulation environments and various data. This article uses OpenAI Gym to analyze CartPole, Pendulum and MountainCar. Then we compare and summarize the settings of the reward function and the results of model training.

4.1. CartPole-v0

CartPole-v0 has a cart in the environment with a pole standing on it. After each reset, the initial state is different. The cart needs to move left and right to keep the pole vertical. The moving range x is [-2.4, +2.4]. The condition for ending is that the pole inclination angle θ exceeds ±12° or the cart exceeds the moving range x or the number of moves exceeds 200 times.

In the Gym open source toolkit, the reward function adopts a sparse reward method. Get a reward for each step performed. The reward function is as follows.

$$r = 1, \text{ every step} \tag{1}$$

The way of sparse rewards makes the algorithm converge very slowly. On this basis, some technicians improve it. A formal reward is proposed combining the current position of the cart and the angle of the pole. The reward function is as follows.

$$r1 = e^{(x\_threshold - |x|)} / x\_threshold - 0.8 \tag{2}$$

$$r2 = e^{(theta\_threshold - |theta|)} / theta\_threshold - 0.5 \tag{3}$$

$$r = r1 + r2 \tag{4}$$

Among them, x_threshold is the maximum range for the cart to move. x is the current cart position. theta_threshold is the maximum range of pole angle. theta is the angle of the current pole from the vertical position. This reward function allows CartPole case to converge within 1000 episodes.

The settings of the above two reward functions. The first one is suitable for scenarios where the execution time of each episode is as long as possible, that is, changing the state multiple times will not end the loop. Because every step you perform will get a positive reward. The more steps you perform, the more reward you accumulate. In the second way, the discrete range of reward values is narrowed. The reward value for each step is determined based on the current position and the angle of the pole. The model can better learn what decisions lead to higher reward values.

4.2. Pendulum-v1

Pendulum-v1 starts with a random position. The goal is to swing it up so that it stays upright. The state of the pendulum consists of the angle and the angular velocity. Each step applies a series of moments. The moment is measured as the action space [14]. The termination condition is that 200 steps have been executed. The reward function of the model's reward coefficient variation is as follows.

$$r = -\theta^2 - 0.1 * \dot{\theta}^2 - 0.001 * \tau^2 \tag{5}$$

θ is the angle of the current pendulum from the vertical position. The range is [-π, +π]. $\dot{\theta}$ is the angular velocity. τ is the moment, that is, the force arm multiplied by the force. The range is [-2, +2].

This reward coefficient variation reward function can combine multiple factors that affect the reward value. The reward value obtained for each step is more accurate. The difference in coefficients determines the degree of influence of different factors. But how to determine the value of the coefficient requires many experiments.

4.3. MoutainCar-v0

In the environment of MountainCar-v0, the car runs within a certain range. Viewed horizontally, the cart position is in the range [-1.2, 0.6] at any one time. The speed range is [-0.07, 0.07]. At each moment, the agent can apply one of three actions to the car. The three actions are left force, no force, and right force. The force exerted by the agent and the horizontal position of the car will jointly determine the speed of the car at the next moment. When the horizontal position of the car is greater than 0.5 at a certain moment, the control target is successfully achieved.

The model has two ways of setting the distributed reward function. One way is to get a negative reward value for each step performed and get a positive reward value when the end point is reached. The reward function is as follows.

$$r = \begin{cases} -1, & \text{not reach target point}, \; x < 0.5 \\ 0.5, & \text{reach target point}, \quad x \geq 0.5 \end{cases} \tag{6}$$

Another way is to divide according to the location of the car. When the horizontal position of the cart is less than -0.4, the reward value is negative. When reaching the target point, the reward value is 100. When x∈(-0.4, 0.5), the reward value is obtained according to the position of x. The closer to the target point, the higher the reward value will obtain. The reward function is as follows.

$$r = \begin{cases} -0.1 & , \quad x \leq -0.4 \\ 10 * (x + 0.4)^3 & , \quad -0.4 < x < 0.5 \\ 100 & , \quad 0.5 \leq x \end{cases} \tag{7}$$

The above two reward functions. The first setting method applies to all scenarios. Only when the goal is reached, a positive reward value is given. Other cases give a negative reward value. This approach is similar to how sparse reward are set in CartPole. The difference is that this setup expects to end the loop in fewer executions. The second way of setting distributed rewards requires setting different reward values in different ranges. The closer the task is to the target point, the higher the reward value it will obtain.

5. Experiments

5.1. Simulation environment

ROS is a highly flexible software architecture for writing robot software programs [15]. It can realize the construction of AGV model and the construction of simulation environment. Topic communication can also be implemented [16]. ROS works by specifying robot properties in a URDF file. URDF supports XML and Xacro. Compared with XML, Xacro is easier to implement and maintain. Xacro is also more readable [17].

The componentized toolkit of ROS is very rich, including the 3D visualization tool rviz, the physical simulation environment gazebo [18, 19], the data recording tool rosbag and the Qt toolbox rqt_*. This paper uses the physical simulation platform toolkit to realize the experimental environment. Simulate the operation of the AGV in the actual scene. Since the gazebo program cannot reset the AGV model after running, it can make the route of the AGV run in a straight line and the running distance is equal. If the AGV is kept running in one direction, it will cause the AGV to leave the displayed interface. You need to zoom out the scene or move the coordinates to see the AGV again. Such a situation affects observability and also increases unnecessary operations. Therefore, a square loop scheme is designed which can make the route of each AGV run in a straight line and the running distance is equal. Rotate after each task execution is completed. Then it can run on the next straight line. Finally through four straight routes back to the origin. In order to ensure that the simulation of AGV in the simulation environment can be seen intuitively in the interface.

The motion control problem of AGV requires the motion process as an evaluation factor. Experiments need to use the information during the movement. The settings of the reward function may also use this information. The topic communication mechanism of ROS can realize the communication of AGV in the gazebo physical simulation environment and algorithm [20]. The coordinate (x, y) information, angle information, speed information, angle error and displacement error information of the AGV during movement can be obtained in real time.

The experiments in this paper are carried out in the above simulation environment. DQN is adopted as the network model of reinforcement learning algorithm. We set up multiple sets of comparative experiments to determine the appropriate reward function. In each set of experiments, the basic parameters of the network model remain the same, including learning rate, greediness, number of samples, experience pool capacity, and target network update frequency. Modify the reward function and the number of training sessions in the Markov decision process. It is assumed that the condition for the end of each episode is that the position deviation sum of the AGV running on a 10m straight line is less than or equal to 5 or the number of state changes reaches 50 times. dy is the position deviation sum. dth is the angular deviation sum. Ep_r is the cumulative sum of the reward value of each episode.

5.2. Experiment 1

In Experiment 1, the reward function adopts a sparse reward setting which is similar to the CartPole-v0 and MountainCar-v0. Our goal is to make the AGV adjust the optimal PID faster and make the AGV run smoothly. So we set a negative reward value for each step and give a positive reward value when reaching the final state. The reward function is set as follows.

$$r = \begin{cases} 1, & \text{not reach final state,} \quad dy \leq 0.5 \\ -1, & \text{reach final state,} \quad dy > 0.5 \end{cases} \tag{8}$$

In this experiment, we train with 500 episodes and 5000 episodes. Figure 1 and Figure 2 show the sum of reward values and model losses for 500 episodes training. Figure 3 and Figure 4 show the sum of reward values and model losses for 5000 episodes training. It can be seen from the comparison that after training the model 500 times, the sum of reward values did not converge and the loss of the model did not decrease. When the number of training times reaches 5000, the reward value accumulates and gradually increases and converges. The loss of the model also has a downward trend. The experimental results are as follows.

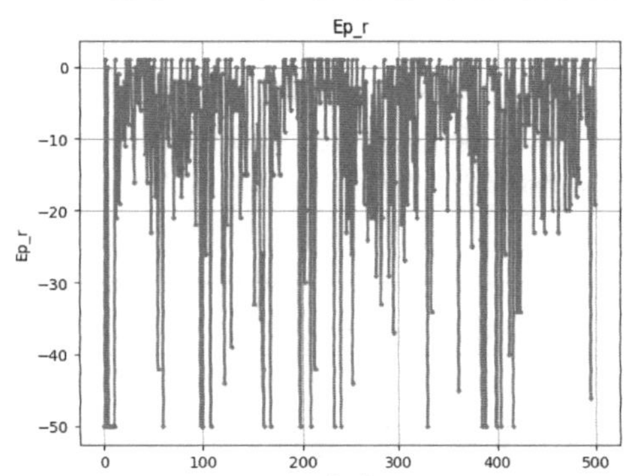

Figure 1. Sum of reward values in Experiment 1 (a)

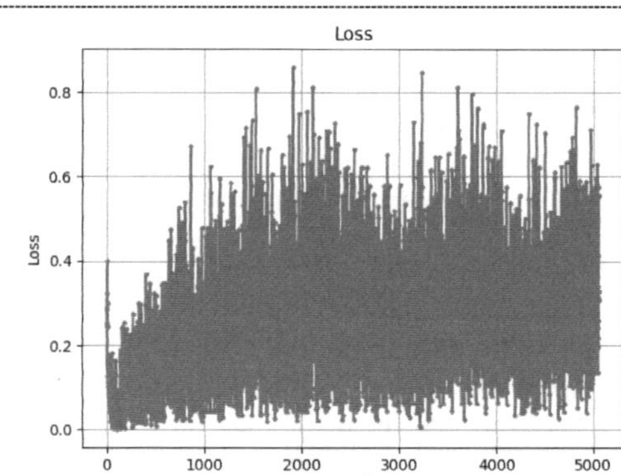

Figure 2. Loss of model in Experiment 1 (a)

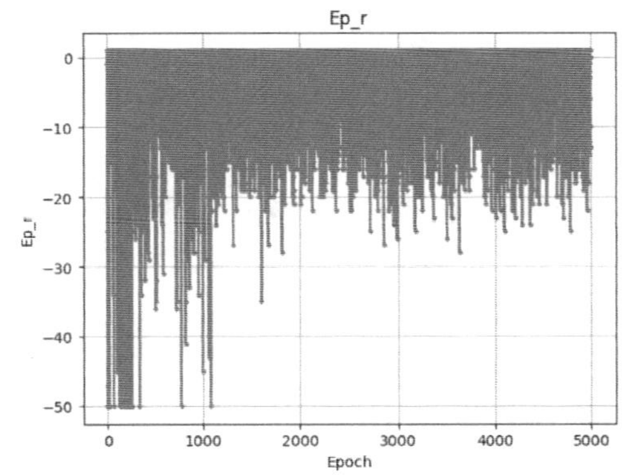

Figure 3. Sum of reward values in Experiment 1 (b)

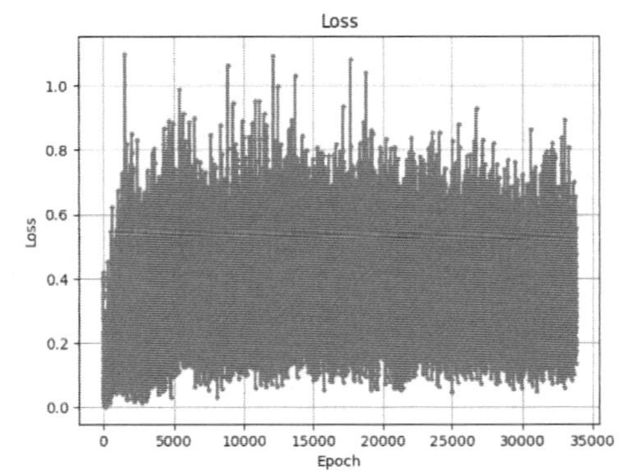

Figure 4. Loss of model in Experiment 1 (b)

5.3. Experiment 2

In Experiment 2, the reward function adopts the setting method of formal reward. The AGV runs in a straight line. The smaller the deviation from the straight line, the better it is. The smaller the change in the angle at which the trolley runs, the better it is. In the reward setting, the position offset is more influential than the angle offset. The reward function is set as follows.

$$r1 = \begin{cases} 10 \quad , & dy \leq 5 \\ 10 / dy , & 5 < dy \leq 10 \\ 5 / dy \quad , & 10 < dy \leq 40 \\ -1 \quad , & 40 < dy \end{cases} \tag{9}$$

$$r2 = \begin{cases} 200 / dth , & 200 \leq dth \leq 1000 \\ 100 / dth , & 1000 < dth \leq 2000 \\ -0.5 \quad , & 2000 < dth \end{cases} \tag{10}$$

$$r = r1 + r2 \tag{11}$$

It can be seen from Experiment 1 that too few training times cannot make the model converge faster. Therefore, we directly train 5000 episodes in Experiment 2. The experimental results are shown in Figure 5 and Figure 6. It can be seen from Figure 5 that after training 2000 times, the cumulative sum of reward values is basically greater than 10. It explains that you can reach the final state every time and find the best PID. Compare the model loss diagram of experiment one in Figure 4 and the model loss diagram of experiment two in Figure 6. Figure 6 shows the downward trend more clearly. The reward function setting of Experiment 2 can make the model reduce the loss faster.

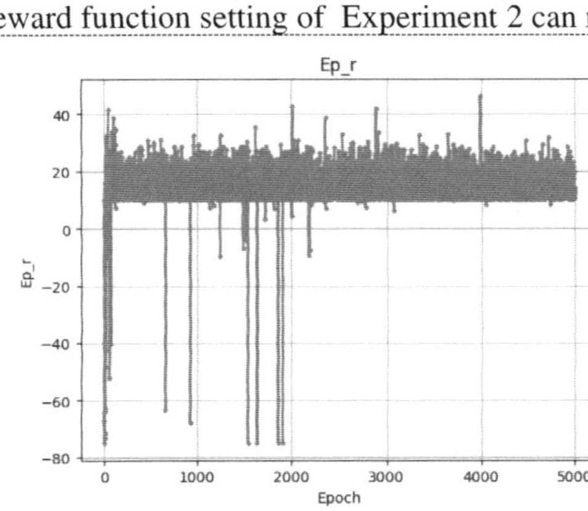

Figure 5. Sum of reward values in Experiment 2

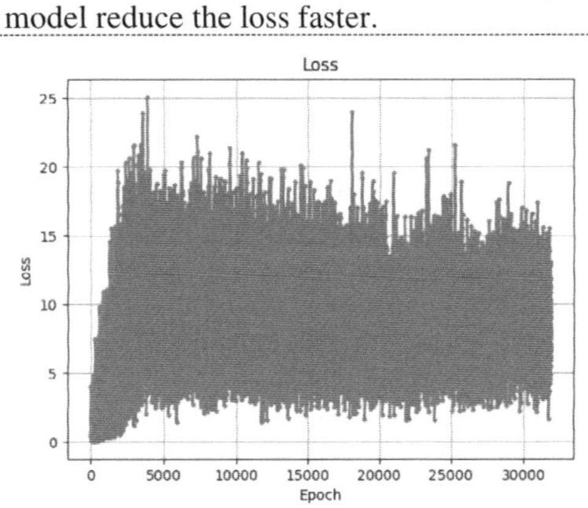

Figure 6. Loss of model in Experiment 2

5.4. Experiment 3

In Experiment 3, the reward function adopts the setting method of reward coefficient change reward which is similar to the Pendulum-v1. The larger the position deviation sum, the smaller the reward value obtained. The larger the angle deviation sum, the smaller the reward value obtained. So make the coefficients a negative number. In order to increase the degree of reward for reaching the final state, a negative reward value is given when the cumulative sum of the positions is less than or equal to 5. And a positive reward value is finally obtained by multiplying by a factor of -0.1. The range of dy and dth is large. Extreme values are rare. Two reward functions are designed in this experiment. And we add testing during training. The model is tested every 10 training runs. The testing process uses the current model for 8 episodes. And record the cumulative sum of the average reward value during the test. The reward function set is shown in Table 1.

Table 1. Reward function settings

	Reward function 1	Reward function 2
r1	$\begin{cases} -100, & dy \le 5 \\ dy, & dy > 5 \end{cases}$	$\begin{cases} -100, & dy \le 5 \\ dy, & 5 < dy < 40 \\ 50, & 40 \le dy \end{cases}$
r2	dth	$\begin{cases} dth, & dth \le 10000 \\ 20000, & 10000 < dth \end{cases}$
r	-0.1 * r1 - 0.00001 * r2	-0.1 * r1 - 0.00001 * r2

In the experimental results of reward function 1, Figure 7 is the cumulative sum of reward values for 5000 episodes. Figure 8 is a graph of model loss. Figure 9 is the average reward value cumulative sum plot for the test. As you can see from the three graphs, the way the reward function is set up makes the algorithm less stable. There will be periods of accumulation of rewards and particularly poor situations. The key point of model convergence cannot be found.

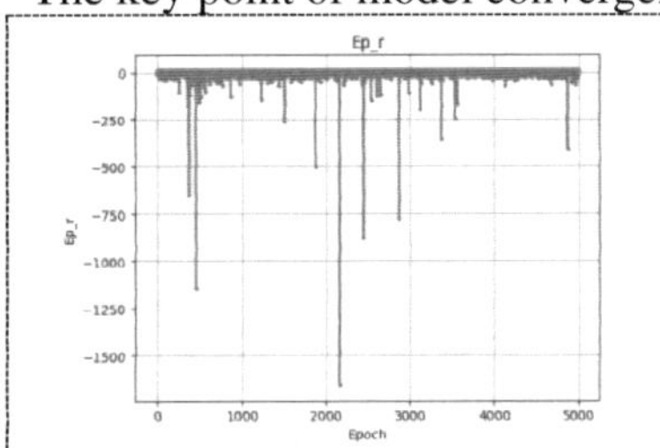

Figure 7. Sum of reward values in Experiment 3 (a)

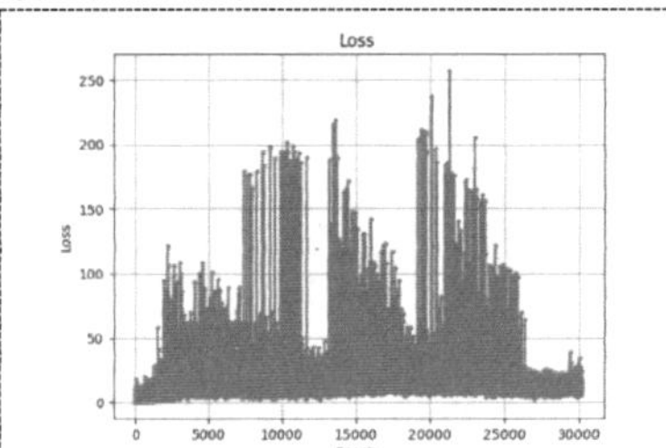

Figure 8. Loss of model in Experiment 3 (a)

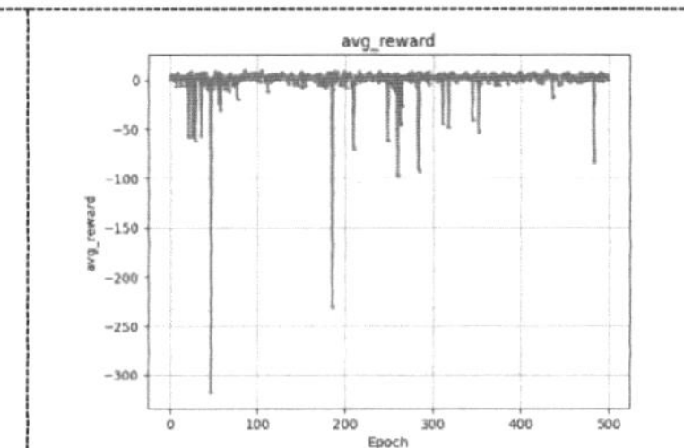

Figure 9. Sum of average reward values in Experiment 3 (a)

The experimental results of reward function 2 are shown in Figure 10, Figure 11, and Figure 12. It can be seen from the figure that the performance of the algorithm is improved after setting fixed reward values for some extreme values of dy and dth. Although there are still cases where the rewards are accumulated and extremely low in the first 2500 training sessions. But after that, it began to slowly converge. It can also be seen from Figure 11 that the loss of the model is gradually decreasing. It can be found in the average cumulative sum chart of the test that the whole is in a state of convergence.

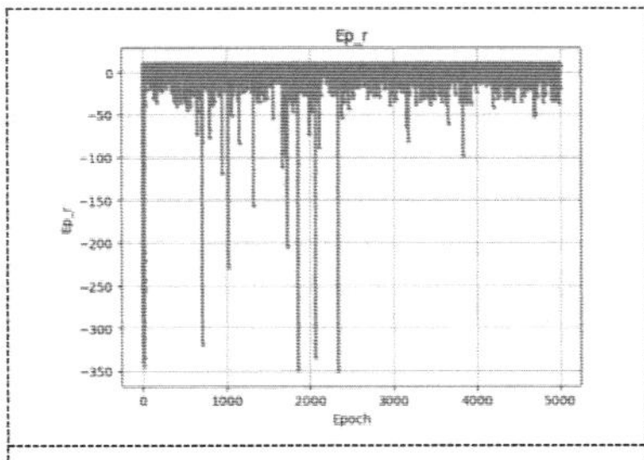

Figure 10. Sum of reward values in Experiment 3 (b)

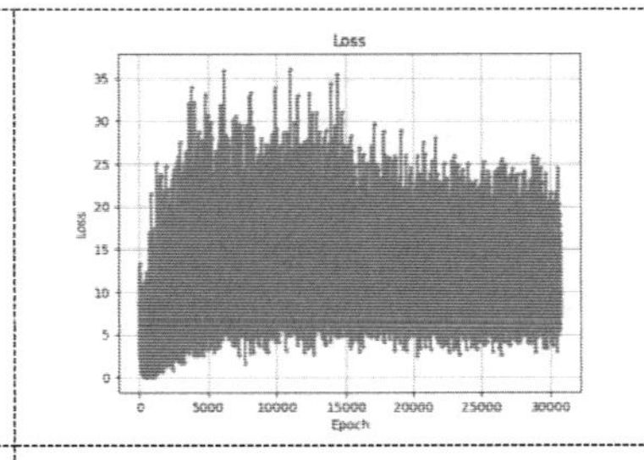

Figure 11. Loss of model in Experiment 3 (b)

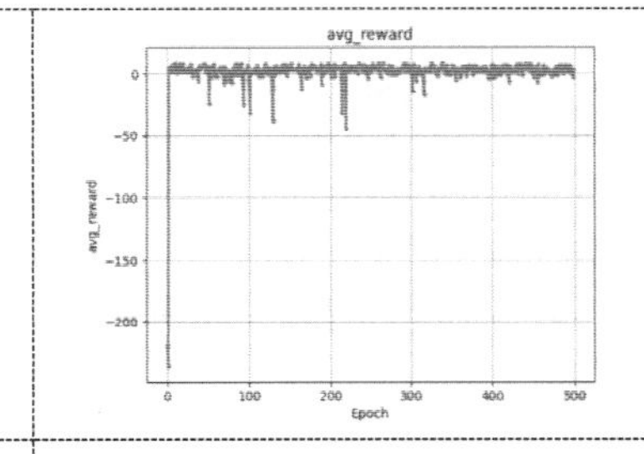

Figure 12. Sum of average reward values in Experiment 3 (b)

5.5. *Experiment 4*

In Experiment 4, the reward function is adopted in the same way as the CartPole-v0 formalizes the reward function. Combine the position deviation and angle deviation during the operation of the AGV to calculate the reward value obtained for each state change. The reward function is set as follows.

$$r1 = (10 - dy) / 10 \tag{12}$$

$$r2 = (2000 - dth) / 2000 \tag{13}$$

$$r = r1 + r2 \tag{14}$$

The results of this experiment are shown in Figure 13, Figure 14 and Figure 15. From Figure 15, we can find that after 1500 episodes, the cumulative reward sum has shown a convergent state. Due to the reinforcement learning DQN model, the characteristics of experience replay will cause the model to randomly sample not-so-good data during model training. Occasionally, the cumulative sum of rewards is small. A clear downward and converging trend can also be seen in the loss plot. The average reward cumulative sum plot formed by the test results shows that the reward function setting of this experiment is suitable.

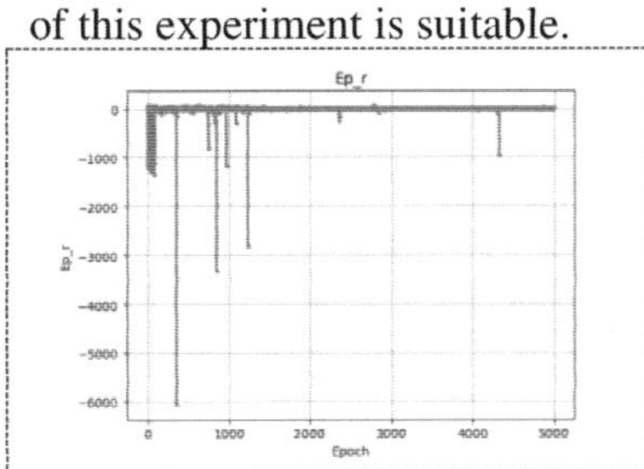

Figure 13. Sum of reward values in Experiment 4

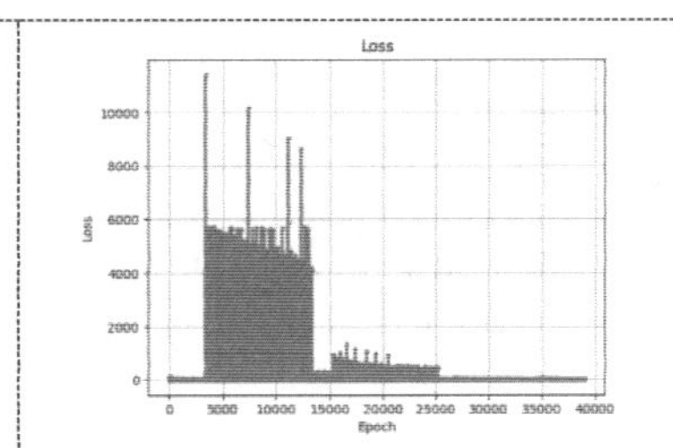

Figure 14. Loss of model in Experiment 4

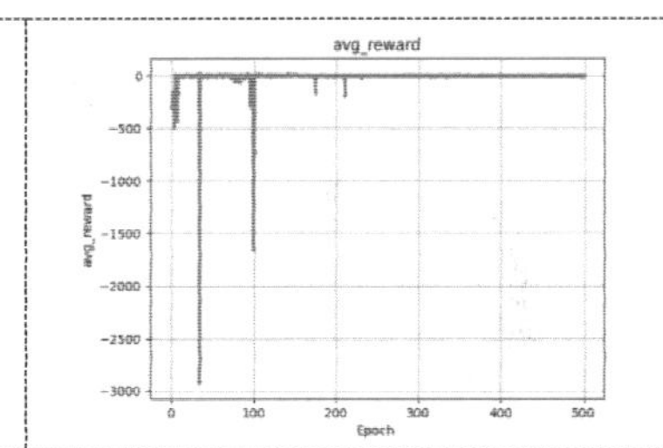

Figure 15. Sum of average reward values in Experiment 4

6. Conclusion

This paper studies and analyzes the reward function settings of several classical reinforcement learning models. We design several reward functions and set up a comparative experiment. Validation is performed in a given algorithm and a simulated simulation environment. Experiments show that sparse rewards are set in a slower way to converge. It is not conducive to improving the efficiency of the algorithm. Experiment 2 and Experiment 3 show results that are not as good as those of Experiment 4 and the algorithm is unstable. Only the reward function which including equation 12-14 used in Experiment 4 can make the algorithm converge faster and reduce the loss. It can make the AGV run smoothly faster. In the next research, We will consider whether it is possible to improve on the current best reward function. Adjust the coefficients to see if the algorithm will converge faster. In addition, consider other reward function setting methods, including intermediate difficulty starting reward and reward shaping.

7. References

[1] Ganesharajah T, Hall N G, Sriskandarajah C. Design and operational issues in AGV-served manufacturing systems[J]. *Annals of Operations Research*, 1998, **76**:109-154.

[2] Lei T, Ming L. A robot exploration strategy based on Q-learning network[C]// *2016 IEEE International Conference on Real-time Computing and Robotics (RCAR)*. IEEE, 2016:57-62.

[3] Sutton R, Barto A. *Reinforcement Learning:An Introduction*[M]. MIT Press, 1998.

[4] Ng A Y, Harada D, Russell S. Policy invariance under reward transformations: Theory and application to reward shaping[J]. Morgan Kaufmann Publishers Inc. 1999.

[5] Guo Siyu et al. Path Planning of Coastal Ships Based on Optimized DQN Reward Function[J]. *Journal of Marine Science and Engineering*, 2021, **9**(2):210-210.

[6] Mnih V, Kavukcuoglu K, Silver D, et al. Playing atari with deep reinforcement learning[J]. arXiv preprintarXiv:1312.5602, 2013.

[7] Watkins, C. J., Dayan, P. Technical Note: Q-Learning[J]. *Machine Learning*, 1992, **8**(3-4):279–292.

[8] Csaba Szepesvari. *Algorithms for reinforcement learning*[M]. Synthesis lectures on artifificial intelligence and machine learning, 2010, **4**(1):1-103.

[9] Mnih V, Kavukcuoglu K,Silver D,et al. Human level control through deep reinforcement learning[J]. *Nature*, 2015, **518**(7540):529-533.

[10] Jiayong Fang, Zhongliang Zhou, Huanbin Wang, et al. Modeling and Simulation of UAV Autonomous Obstacle Avoidance Based on DQN[C]//*Conference proceedings of 2021 4th International Conference on Algorithms, Computing and Artificial Intelligence (ACAI 2021)*, 2021:480-485.

[11] Szi Hui Tan, Khairi Ishak Mohamad, Fauzi Packeer Mohamed Mohamed, Mohd Fadzil Lokman, Afif Ahmarofi Ahmad. Balancing Excitation and Inhibition of Spike Neuron Using Deep Q Network (DQN)[J]. *Journal of Physics: Conference Series*, 2021, **1755**(1).

[12] Tingfei Huang, Guangquan Cheng, Kuihua Huang, et al. Task allocation method for multi-type interception equipment composite anti-UAV based on DQN[J]. *Control and Decision*, 2022, **37**(1):9.

[13] Greg Brockman, Vicki Cheung, Ludwig Pettersson, et al. OpenAI Gym[J]. arXiv:1606.01540, 2016.

[14] Barto Andrew G, Sutton Richard S, Anderson Charles W. Neuronlike adaptive elements that can solve difficult learning control problems[J]. *IEEE Transactions on Systems, Man, and Cybernetics*, 1983, SMC-**13**(5):834-846.

[15] Wiki. ROS Tutorials[EB/OL]. http://wiki.ros.org/ROS/Tutorials.

[16] Quigley M, Conley K, Gerkey B P, et al. ROS:an open-source Robot Operating System[C]// *ICRA Workshop on Open Source Software*, 2009.

[17] Denis Chikurtev. Mobile Robot Simulation and Navigation in ROS and Gazebo[J]. *International Journal of Mathematics and Computers in Simulation*, 2020, **14**:38-43.

[18] Pingping Hao, Lifeng Yao. Simulation realization of mobile robot path planning based on Player/Gazebo[J]. *Education*, 2010(10):138.

[19] Zhang Bowei,Liu Pengcheng. Control and benchmarking of a 7-DOF robotic arm using Gazebo and ROS[J]. *PeerJ. Computer science*, 2021, 7.

[20] Sokolov M, Gabdullin A, Afanasyev I, et al. 3D modelling and simulation of a crawler robot in ROS/Gazebo[C]// *International Conference on Control, Mechatronics and Automation*, 2016.

Acknowledgments

This work was financially supported by the Project of China's Ministry of Education and supported by Alibaba in 2020 (Grant No. 202002001063). Heartfelt thanks also to the reviewers for their constructive comments to the paper.

Journal of Physics: Conference Series **2320** (2022) 012003 doi:10.1088/1742-6596/2320/1/012003

Research and Analysis of Small Current Grounding Line Selection System

Jiang Xinzheng[1] and Jiang Nan[2]

[1]Guangzhou Institute of Science and Technology, No. 638, Xingtai Third Road, Baiyun District, Guangzhou,China

[2]China University of Political Science and Law, 27 Fuxue Road, Changping District, Beijing,China

[1]jxz8008@126.com, [2]1776799399@qq.com

Abstract. In the small current grounding system, because the short-circuit fault of single-phase grounding accounts for more than 80% of the total short-circuit fault, the selection of single-phase grounding fault line is still the main problem in electrical engineering. This paper studies the principles, advantages and disadvantages of various line selection technical methods in case of single-phase grounding fault, and puts forward the improved method of comprehensive line selection for the problem of misjudgment of fault line selection. On this basis, it puts forward the improved method of comprehensive line selection, verifies the line selection device of 10kV transmission line through the relevant data obtained from design calculation and simulation experiment, and puts forward a more reasonable design scheme for the hardware circuit and software circuit of the line selection device, The fault tolerance and accuracy of small current grounding line selection system are improved.

1. Introduction

Due to the rapid development of society, people have higher and higher requirements for the stability and security of power system. In small current grounding system, single-phase grounding fault is a common temporary fault. According to statistics, single-phase grounding fault accounts for more than 80% of the total faults of distribution lines [1]. When a fault occurs, because it cannot form a loop, the current at the fault point is very small, and the line voltage between the three phases is still symmetrical, which has no impact on the power supply of the load equipment, so the equipment in the system is allowed to operate for a short time. Generally, it can run for 1-2 hours, which improves the reliability of power supply, which is also the advantage of small current grounding system. However, when the power grid operates with single-phase grounding fault for a long time, the fault phase voltage drops to zero, while the non fault phase voltage rises to line voltage, which will pose a serious threat to the insulation of electrical equipment, and then lead to insulation breakdown or phase-to-phase short circuit and arc discharge.

Therefore, in case of single-phase grounding fault, rapid selection of fault line plays a great role in the safe operation of power grid. In the small current grounding system, how to quickly find the grounding fault line and eliminate the fault in time is worth studying. The development of reliable and accurate line selection method and automatic line selection device is very important for the development of special line selection technology of power system.

2. Basic characteristics of single-phase grounding fault in small current grounding system

At present, technicians have obtained many line selection methods through experimental research. The line selection of single-phase grounding fault in small current grounding system is mainly divided into three categories: Based on steady-state characteristic quantity, based on transient characteristic quantity and injected signal method [2]. Each method has limitations, and many line selection devices have been developed according to these line selection methods. At present, the line selection device is still based on a single line selection method, and the accuracy of line selection is not high.

By analyzing the fault characteristics of single-phase grounding in small current grounding system, it is very helpful for us to develop line selection method. Small current grounding system is a common grounding system in medium and low voltage power system. It is also divided into the following grounding modes: neutral point grounded through resistance, neutral point ungrounded and neutral point grounded through arc suppression coil system [3]. The comprehensive analysis of various grounding methods is shown in Table 1.

Table 1. Comparison of neutral grounding modes

Comparison items	Neutral grounding mode				
	Ungrounded	Resistance grounding	Arc suppression coil grounding	Arc suppression coil parallel (Series) resistance grounding	Direct grounding
Risk of personal electric shock	Highest	Low	High	Low	Minimum
Electromagnetic resonance overvoltage	Large	Large	Reduce	Reduce	Most dangerous
Single phase grounding protection	High	Low	High	Low	Low
Inductive hazards to communications	More difficult	Easy	Difficult	Easier	Easiest
Single phase grounding current	less	Large (related to grounding current)	Small	Small	Maximum
Operating overvoltage	Large	Large	Small (related to detuning)	Small (related to detuning)	Maximum
Overvoltage caused by high voltage in series with low voltage	Highest	Minimum	Higher	Lower	Low
Protective grounding safety	Danger	Danger	Security	Security	Danger

3. Line selection method of single-phase grounding fault in small current grounding system

Small current grounding line selection technology has always been a recognized technical problem. Many technicians and scholars have also proposed many line selection methods, but the accuracy of line selection has not been very high, and there are many misjudgments. The main problems are: when single-phase grounding fault occurs in different grounding modes, the fault characteristics are different, and the line selection method based on single fault characteristics can not achieve effective line selection; There are also some limitations in using steady-state and transient line selection methods. Under certain fault characteristics, these two line selection methods may also be unable to select lines; Whether there are various criteria for line selection, such as whether the fusion technology is reasonable or not, should also be considered. The comparative analysis of various fault line selection methods of small current grounding system is shown in Table 2.

Table 2. Comparative analysis of small current grounding line selection system technology

Classification	Line selection technology	Main features
Line selection method using steady-state characteristics	Zero sequence current amplitude comparison method Zero sequence current phase method Group amplitude and phase comparison method [3] Active component method [4] Fifth harmonic component method [4] Negative sequence current method [5] Zero sequence admittance method [6] Energy function method [7]	The principle of steady-state line selection is to select the line according to the size or phase of some steady-state characteristic quantities after fault, such as zero sequence current, active component, zero sequence admittance, fifth harmonic and negative sequence component. Both have certain limitations, and some of these steady-state quantities are relatively small and difficult to detect, which has a certain impact on the accuracy of line selection
Line selection method using transient characteristics	First half wave method [3] Wavelet analysis [8] Transient energy method [9] Transient zero sequence current comparison method [2]	Transient line selection uses the transient characteristics of fault, such as the amplitude and phase of transient current. The fault signal is large, which can improve the accuracy of line selection. However, when the fault occurs near the zero crossing point of line phase voltage, it is easy to misjudge
Line selection method of injection signal	S injection method [10] Injection frequency conversion signal method [11]	The injection signal method must add signal generation and detection devices on the basis of the original system, which has high cost and is greatly affected by the grounding of transition resistance
Intelligent line selection method	D-S evidence theory line selection method [12] Line selection method based on fuzzy theory [3] Neural network line selection method [13]	Information line selection methods can adapt to multi fault conditions and fuse different line selection results. These line selection methods have not been used in practice and need to be verified

The above studies the limitations of single line selection method. In fact, as far as the current fault line selection method of small current grounding system is concerned, there is still a problem of misjudgment in single line selection, which also leads to the low accuracy of single line selection technology.

Because the current single line selection method has its own limitations, only a single line selection method can not achieve accurate line selection in the face of a variety of fault conditions. Therefore, this paper proposes to adopt the comprehensive line selection method, which combines a variety of line selection methods to form a comprehensive criterion for line selection, and makes use of the advantages of various line selection to complement each other, so as to effectively select the line according to the comprehensive criterion in the face of multiple faults. The application scope of each line selection is different. Different line selection criteria are integrated to expand the application scope of line selection and make it suitable for a variety of fault conditions. When the fault meets one of the comprehensive line selection criteria, the line can also be selected correctly after comprehensive decision-making.

4. Criterion construction principle of comprehensive route selection method

Because the first half wave line selection method is based on the characteristics of transient fault, which occurs in the early stage of system fault, the first half wave criterion is used for line selection. After the transient period, the fault line with high possibility is selected by using the criteria of improved zero sequence current amplitude comparison method and zero sequence current phase comparison method.

Record the line selection result of the first half wave as "S"; The line selection result of the improved zero sequence current amplitude comparison method is recorded as "C"; The line selection result of the improved zero sequence current phase comparison method is recorded as "U"; The route selection result of comprehensive decision is recorded as "Z"; Then there are:

- Z=S||C||U

Where, the three line selection results are logically or, and finally the comprehensive line selection result is obtained. This not only forms the complementarity of various line selection, but also greatly improves the accuracy of line selection. When the three criteria do not effectively select the line, the final decision can be determined as bus fault.

Compared with the single line selection method, the comprehensive line selection method has significantly improved its applicability and can adapt to the line selection of various grounding modes; In terms of accuracy, the accuracy of using multiple criteria is higher than that of single criterion of single line selection method. Therefore, a variety of comprehensive line selection methods have high practical value in the fault line selection method of small current grounding system.

5. Simulation verification of comprehensive line selection

This paper puts forward a new idea of comprehensive line selection for small current grounding system by using three line selection methods: improved zero sequence current amplitude comparison method, improved zero sequence current phase comparison method and improved first half wave method. At the same time, it also puts forward the corresponding criterion method.

5.1. Construction of improved zero sequence current amplitude comparison criterion

The criterion of improving the amplitude comparison of zero sequence current is to judge whether the difference is close to the zero sequence current of the line by adding the zero sequence current of each line and subtracting the zero sequence current of the line when compared with the line. If it is close to or equal, it is a fault line, and if the difference is large, it is a non fault line. The logic criterion is constructed according to the above principles, as shown in Figure 1.

After it is encapsulated, the three inputs are the zero sequence current of three lines, and the three outputs correspond to the zero sequence current value of each line after criterion calculation. If it is the same as the zero sequence current of the line, it is a fault line; if it is different, it is a non fault line.

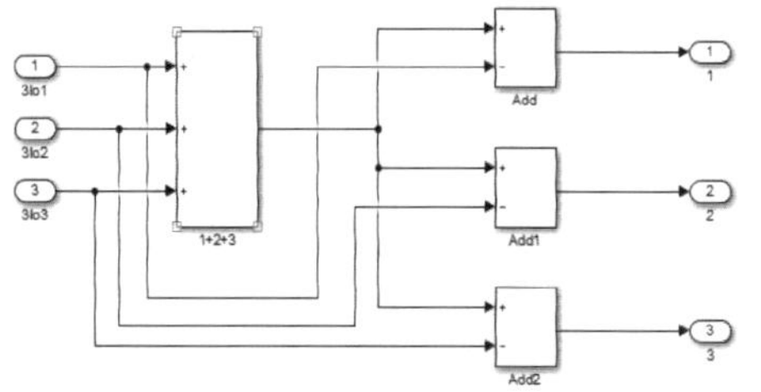

Figure 1. Constructed zero sequence amplitude comparison criterion

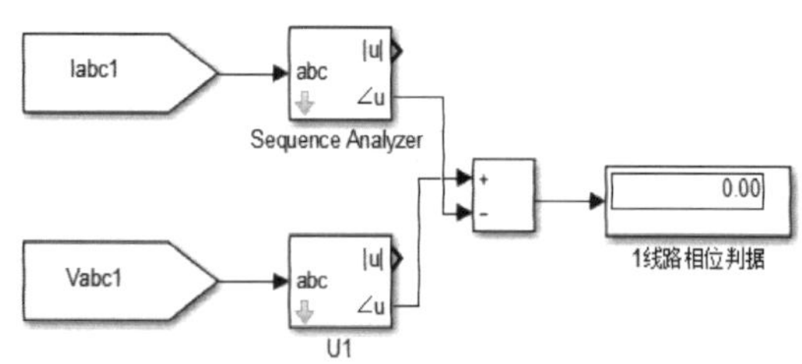

Figure 2. Constructed zero sequence phase comparison criterion

5.2. Construction of improved zero sequence current phase comparison criterion

The criterion of improving the phase comparison method of zero sequence current is to distinguish the phase of zero sequence voltage and zero sequence current. When the zero sequence current lags behind the phase of zero sequence voltage by 90 °, the line is a non fault line; on the contrary, the line is a faulty line. The phase comparison criterion of zero sequence current constructed according to this principle is shown in Figure 2.

5.3. Construction of improved first half wave criterion

The criterion of improving the first half wave line selection method is to sample the value of 0.5 ~ 1 cycle after the fault, and select the line by comparing the direction of the positive and negative components of the transient current when the fault occurs. If the transient current first appears negative value and then positive value, it indicates that the transient current first appears negative wave and then positive wave, indicating that the line is a non fault line; if the transient current first appears positive and then negative, it indicates that the transient current first appears positive wave and then negative wave, indicating that the line is a fault line. Based on the above characteristics, the criterion of the first half wave method is improved. First, the waveform value of each line is exported to the working area, and the oscilloscope setting is shown in Figure 3.

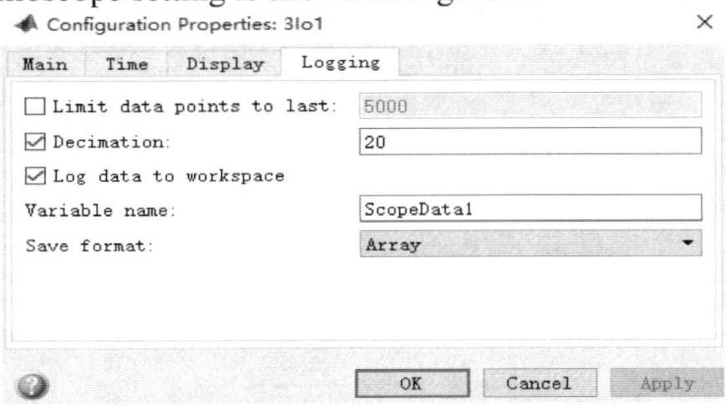

Figure 3. Oscilloscope value export setting

By checking "log data to workspace", the oscilloscope waveform value can be exported to the work area. Check "decision" again, set it to 20, set the export value to 1000, each data is separated by 200us, set the exported data type to matrix type and the data name to "scopedata1".

After the above operation, the waveform value has been exported to the working area, and then the program of the first half wave criterion is compiled according to the above criterion.

5.4. Simulation verification of single-phase fault in small current grounding system

According to the fault characteristics of single-phase grounding in small current grounding system, the effectiveness and accuracy of the proposed comprehensive line selection method can be simulated and verified by building the corresponding power system simulation model and parameter setting, simulating, observing and analyzing various waveforms when faults occur under various grounding conditions.

The zero sequence current waveform of the fault line and the grounding current waveform of the fault point are shown in Figs. 4 and Figs. 5. The current waveform flowing through the transition resistance and the current waveform flowing through the arc suppression coil are shown in Figs. 6 and Figs. 7

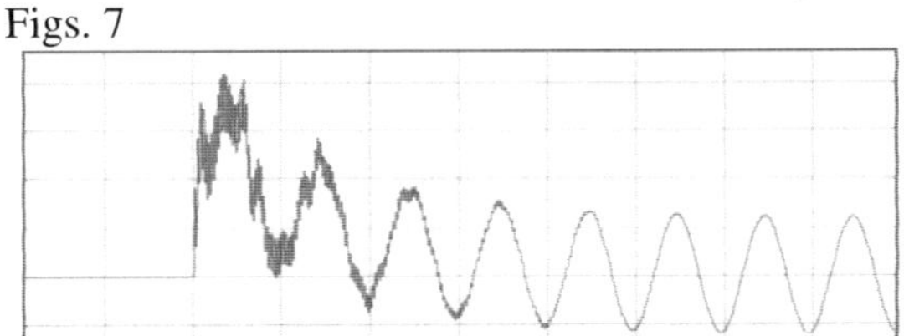

Figure 4. Zero sequence current waveform of fault line

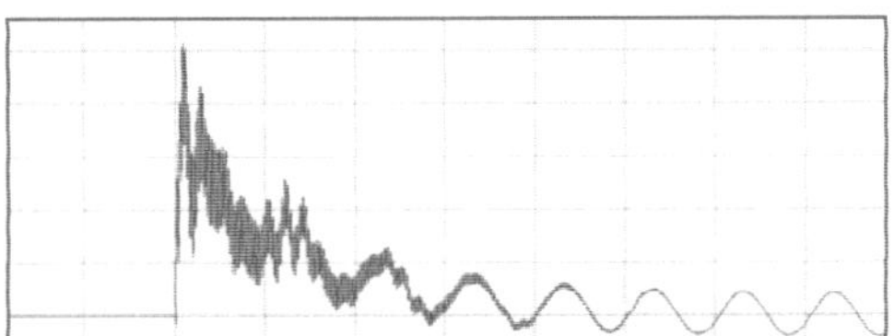

Figure 5. Grounding current waveform of fault point

Journal of Physics: Conference Series **2320** (2022) 012003 doi:10.1088/1742-6596/2320/1/012003

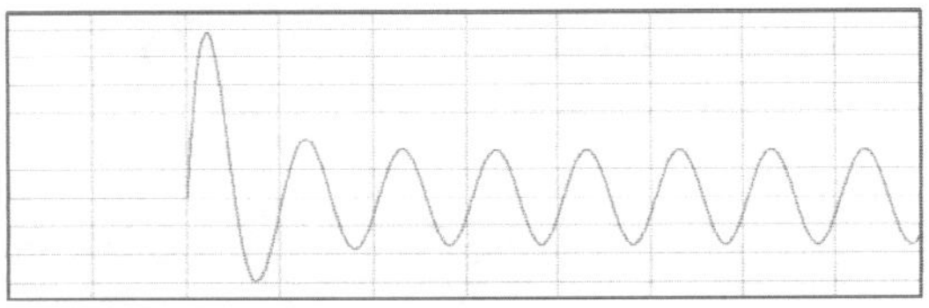

Figure 6. Current waveform through transition resistance

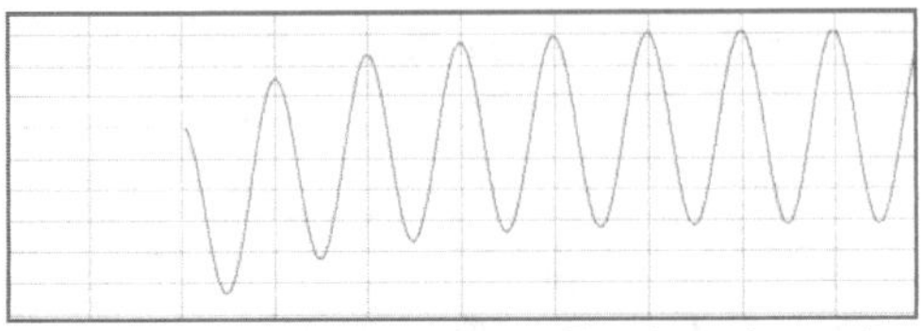

Figure 7. Current waveform flowing through arc suppression coil

6. Route selection results of comprehensive route selection method

Through the criteria of comprehensive route selection method constructed above, the three route selection results are integrated in a logical or way to form the final result of comprehensive route selection in this paper.

In the neutral ungrounded system, whether it is single-phase metallic grounding or grounding through 500 Ω transition resistance, the improved zero sequence current phase comparison method and the improved first half wave method can correctly select the line. The improved zero sequence current amplitude comparison method does not perform well in case of multiple line faults. The line selection results are represented by numbers, and 0 means that the line cannot be selected. The line selection results are shown in table 3:

Table 3. Single phase metallic grounding line selection results of neutral ungrounded system

Fault line selection method	Line 1 fault	Line 2 fault	Line 3 fault	Line 1 and 2 fault
Improved zero sequence current amplitude comparison method	1	2	3	0
Improved zero sequence current phase comparison method	1	2	3	1，2
Improved first half wave method	1	2	3	1
Comprehensive route selection results	1	2	3	1，2

In the neutral point wiring system through arc suppression coil, when metallic grounding occurs, because the improved zero sequence amplitude comparison method and zero sequence current phase comparison method are not applicable, the improved first half wave method is mainly used for line selection, and the line selection results can be obtained more accurately in case of multi line fault. When passing through the transition resistance grounding, only one result can be selected in case of multi line fault, and the line can be selected accurately in other cases. The line selection results are shown in table 4:

Table 4. Single phase metallic grounding line selection results of neutral point grounding system through arc suppression coil

Fault line selection method	Line 1 fault	Line 2 fault	Line 3 fault	Line 1 and 2 fault
Improved zero sequence current amplitude comparison method	0	0	0	0
Improved zero sequence current phase comparison method	0	0	0	2
Improved first half wave method	1	2	3	1
Comprehensive route selection results	1	2	3	1，2

7. Conclusion

In the small current grounding line selection system, through the above research and analysis, it can be found that there are still some limitations in constructing the comprehensive line selection method of improved zero sequence current amplitude comparison method, improved zero sequence current phase comparison method and improved first half wave method. However, from the above simulation results, the accuracy and misjudgment rate of the comprehensive line selection method have been improved, which has certain advantages over the single line selection method, and also verifies the correctness and effectiveness of the comprehensive line selection method.

The comprehensive line selection method also needs to improve the accuracy of multiple lines. Due to the complexity and variability of fault types and the limited number of criteria of the comprehensive line selection method synthesized by only three line selection methods, the improved comprehensive line selection method proposed in this paper still has some limitations, and the comprehensive line selection method of small current grounding line selection system needs to be further studied.

8. References

[1] Guo Jingwen Research on line selection method of single-phase grounding fault in small current grounding system Xi'an University of technology, 2019
[2] Li Donghui, Shi Lintong Summary of single-phase grounding fault line selection methods in indirect grounding system Relay, 2004, 32 (18) pp74-78,84
[3] Hao Yushan, Yang Yihan Group amplitude and phase comparison principle of Microcomputer Line Selection for small current grounding Power information, 1994, (2) pp15-19
[4] Zhang Fuli, Zhang Hongqi Grounding fault line selection method for small current grounding system Shanxi electric power, 2019,251 (2) pp1-5
[5] Yan Minmin, Chen Ping, Zhou Xinjun, Yan Yangang Small current grounding line selection based on instantaneous negative sequence component Guangdong electric power, 2017, 30 (3)pp 87-93
[6] Tang Yi, Chen Kui, Chen Qing, Dong Haibo Study on line selection of small current grounding by maximum method of absolute sum of mutual admittance difference Chinese Journal of electrical engineering, 2005, 25 (6)pp 49-54
[7] Maggie Development of line selection device for small current grounding system based on fuzzy theory Guangxi University, 2014
[8] Sun Zhe Analysis and Research on single-phase grounding fault line selection in small current grounding system Liaoning: Liaoning University of engineering and technology, 2018
[9] He Pentium, Hu Weijin Principle of small current grounding line selection by energy method Journal of Zhejiang University, 1998, 32 (4)pp 451-457
[10] Sang Zaizhong, Zhang Huifen Principle and application of "s injection method" line selection and positioning China electric power, 1997, 12 (6)pp 44-45,62
[11] Zeng Xiangjun, Yu Yongyuan A new method for control and protection of arc suppression coil grounding system based on injecting variable frequency signal Chinese Journal of electrical engineering, 2000, 20 (1)pp 29-32,36
[12] Wang Bin, Yang Jiang, Wang Tianhua Small current grounding line selection method based on D-S evidence theory Power grid technology, 2007, 31 (S1)pp 169-171
[13] Zhang Junfang, Liu Peng Research on fault line selection of distribution network based on improved fuzzy neural network Power system relay protection and control, 2010, 38 (22)pp 120-125

Acknowledgments

On the completion of this paper, there are too many people who have helped me. I sincerely thank all those who care about and help me, and thank them for giving me the strength to overcome myself, surpass myself and pursue constantly. Thank you!

Journal of Physics: Conference Series **2320** (2022) 012004 doi:10.1088/1742-6596/2320/1/012004

Negative Damping Mechanism Analysis of Power System Stabilizer Based on Damping Torque

Xueqin Zhu, Yiqiang Yang, Jiangtao Fu, Wenrui Bai and Xiaolin Su

Automation and Information Engineering, Sichuan University of Science and Engineering, Yibin 644000, China.

xueqin-zhu@qq.com; yiqiang_yang@outlook.com; 1593630684@qq.com; 1348032123@qq.com and xiaolin_su@qq.com

Abstract. Power system stabilizer plays an important role in suppressing the low frequency oscillations, but a PSS with poorly tuned parameters may produce negative damping and aggravate the system oscillation. The purpose of this paper is to study the mechanism of damping of a PSS in multi machine system and to analyze whether the PSS has produced negative damping. This paper adopts Park model for analysis. According to the concepts of complex frequency and generalized phasor, the correlation between the damping torque phasor and the real part of the complex eigenvalue is studied. With the auxiliary of Park equivalent circuits, the damping torque provided by the excitation system is separated from the electrical torque phasor. Then, the damping torque coefficient of the PSS is defined as an indicator of the damp effect of the stabilizer. Finally, the simulation results of single machine infinite bus system and 4-machine system demonstrate the validation of the proposed method for evaluating the damping effect of PSS on the system.

Nomenclature

ψ_{fd}, ψ_{kd} flux linkages of field winding and d-axis damper winding

ψ_{fq}, ψ_{kq} flux linkages of first and second q-axis damper windings

ψ_d, ψ_q flux linkages of d- and q-axis armature

ψ_{ad}, ψ_{aq} flux linkages of d- and q-axis air gap

u_d, u_q voltages of d- and q-axis armature

I_{fd}, I_{kd} currents of field winding and d-axis damper winding

I_{fq}, I_{kq} currents of first and second q-axis damper windings

I_d, I_q currents of d- and q-axis armature

V_t generator terminal voltage

$G_{ex}(s)$ transfer function of excitation system

V_s excitation auxiliary signal (or PSS output signal)

M inertia constant of generator

T_e generator electrical torque

T_m generator mechanical torque
ω_r rotor angular velocity
δ rotor angle of generator

1. Introduction

Power system low frequency oscillation[1] is the power swing on the tie line of power system after disturbance. It may directly lead to power system outage[2], or indirectly bring inconvenience to people's daily life and production[3]. In recent years, ultra-low frequency oscillation[4] has also been observed, which makes the low frequency oscillation more complex. Meanwhile, with the expansion of power grid scale and the emergence of wide-area complex power grid[5], the problem of low frequency oscillation has a great impact on the security and stability of power system.

The main reason for low frequency oscillation is the lack of sufficient damping[6]. In addition to the lack of damping, low frequency oscillation can also be explained by resonance mechanism[6], chaos mechanism[7] and bifurcation mechanism[7]. In addition, the energy function method[8,9] is also used for low frequency oscillation. In [8] and [9], a practical method for calculating energy flow was proposed. The damping torque coefficient is estimated by the energy consumption of the generator, and the energy flow is used to locate the oscillation source. In addition to analyzing the mechanism of oscillation, it is also necessary to obtain information related to oscillation according to the signal. So as to obtain information of oscillation, such as eigenvalue, damping ratio, etc., the method of signal analysis can be used. Common signal analysis methods include Prony, ESPRIT, HHT, etc.[10] . Reference [11] introduced that TLS-ESPRIT is a subspace based high-resolution signal analysis method, which can identify the low-frequency oscillation mode of power system with high accuracy. Reference [12] discussed a variety of equipment and tools that can be used to reduce damping under traditional grids and renewable energy integration ages.

Evaluating the generator damping is helpful to the study of power system stability. Reference [13] proposed a damping evaluation method based on oscillation energy and small signal analysis, and based on oscillation energy, it proposed a damping evaluation index. According to the eigenvector and eigenvalue, the expression of the index is given, and the effectiveness of the modified method is verified in the four-machine system. Reference [14] presented a calculation method for the overall damping characteristic design of excitation system during unit oscillation, and provides a reliable reference for the positioning analysis and control of negative damping low frequency oscillation source. Reference [15] studied the transient energy dissipation of resistors and its impact on power system damping. Reference [16] calculated the oscillation energy consumption based on eigenvalues and eigenvectors, and uses the energy consumption coefficient to evaluate the damping characteristics. Reference [17] used a new type of generator phantom measuring unit to provide an internal response during transient response of generator, and realize an online assessment of generator damping. Reference [18] analyzed the connection between damping torque analysis and energy flow analysis, which promotes the understanding of the mechanism of oscillating damping.

When low frequency oscillation occurs in power system, power system stabilizer (PSS)[19] , flexible AC transmission systems (FACTS)[20,21] , and high voltage direct current (HVDC)[22] are usually used to suppress oscillation. Compared with other measures to suppress low frequency oscillation, PSS is widely used due to its economy, robustness, and the simple structure. PSS was first proposed by F.P. deMello and C. Concodri[23,24] . It was verified in [25] that an optimized PSS has good inhibitory effect on low frequency oscillation with oscillation frequency of 0.1-2.0 Hz. Reference [26] applied collective decision optimization (CDO) algorithm to the optimization design of PSS, and the optimized PSS parameters improve the damping of weak damping oscillator, and improve the dynamic performance of system stability.

PSS, as the most widely used and mature generator excitation additional control technology in the world, is also facing great challenges. The eigenvalue analysis method[27] is the most common method. Improper adjustment of PSS parameters will bring negative damping to the power system and threaten the security and stability of the power system. Furthermore, understanding the mechanism of PSS in a multi-machine system is conducive to better PSS design and better parameter adjustment. Reference [28] used the non-dominated sorting genetic algorithm to obtain the optimal PSS parameters of specific work point. Since the energy does not satisfy the superposition theorem, and the energy provided by the stabilizer is difficult to separate, it is difficult to calculate the damping of PSS by energy method. Reference [29] used the phase relationship between the generator rotor angle oscillation component and its terminal bus voltage angle to assess the damping contribution of PSS. Based on the principle of PSS suppressing low frequency oscillation and using the damping torque method, this paper deduces a PSS damping evaluation method to judge the contribution of PSS to system damping.

The remainder is organized as follows. Section 2 introduces generator Park model and its equivalent circuits. Section 3 describes the phasor representation and defines the damping torque coefficient. In section 4, the PSS damping mechanism is studied with the aid of circuit diagram, and the damping torque coefficient of Park model is derived. In Section 5 the proposed damping evaluation method is verified by simulation in four-machine system. The numerical simulation results show that this method can assess whether the system has negative damping without removing PSS and doing simulation again. Conclusions are drawn in Section 6.

2. Park model of generator and its equivalent circuits

Consider Park generator model with one field winding and one damper winding in d-axis, identified with subscripts fd and kd respectively, and two damper windings in q-axis, identified with subscripts fq and kq. The equations are given as following.

Stator voltage equations

$$\begin{aligned} u_d &= \dot{\psi}_d - \psi_q \omega_r - r_a I_d \\ u_q &= \dot{\psi}_q + \psi_d \omega_r - r_a I_q \end{aligned} \tag{1}$$

Rotor voltage equations

$$\begin{aligned} u_{fd} &= \dot{\psi}_{fd} + r_{fd} I_{fd} \\ 0 &= \dot{\psi}_{kd} + r_{kd} I_{kd} \\ 0 &= \dot{\psi}_{fq} + r_{fq} I_{fq} \\ 0 &= \dot{\psi}_{kq} + r_{kq} I_{kq} \end{aligned} \tag{2}$$

Stator flux linkage equations

$$\begin{aligned} \psi_d &= x_{ad} I_{ad} - x_a I_d \\ \psi_q &= x_{aq} I_{aq} - x_a I_q \end{aligned} \tag{3}$$

Rotor flux linkage equations

$$\begin{aligned} \psi_{fd} &= x_{ad} I_{ad} + x_{fd} I_{fd} \\ \psi_{kd} &= x_{ad} I_{ad} + x_{kd} I_{kd} \\ \psi_{fq} &= x_{aq} I_{aq} + x_{fq} I_{fq} \\ \psi_{kq} &= x_{aq} I_{aq} + x_{kq} I_{kq} \end{aligned} \tag{4}$$

Where $I_{ad} = I_{fd} + I_{kd} - I_d$, $I_{aq} = I_{fq} + I_{kq} - I_q$, $\psi_{ad} = x_{ad} I_{ad}$, $\psi_{aq} = x_{aq} I_{aq}$.

And swing equations

$$\begin{aligned} \frac{d\delta}{dt} &= \Delta\omega_r \\ \frac{d\Delta\omega_r}{dt} &= \frac{1}{M}(T_m - T_e) \end{aligned} \tag{5}$$

Then the generator electrical torque is

$$T_e = \psi_d I_q - \psi_q I_d \tag{6}$$

To write Park model as state equations, currents, I_{fd}, I_{kd}, I_{fq} and I_{kq}, could be eliminated with (3). The voltage of the generator, $u_d + ju_q$, is usually selected as the state variable of the transmission network. The state variables of the generator are δ, $\Delta\omega_r$, ψ_{fd}, ψ_{kd}, ψ_{fq}, ψ_{kq}, ψ_d and ψ_q.The input variables are T_m, u_{fd}, I_d, I_q. Equations (1)-(4) could also be accurately represented with circuits[30], which are shown in Figure 1.

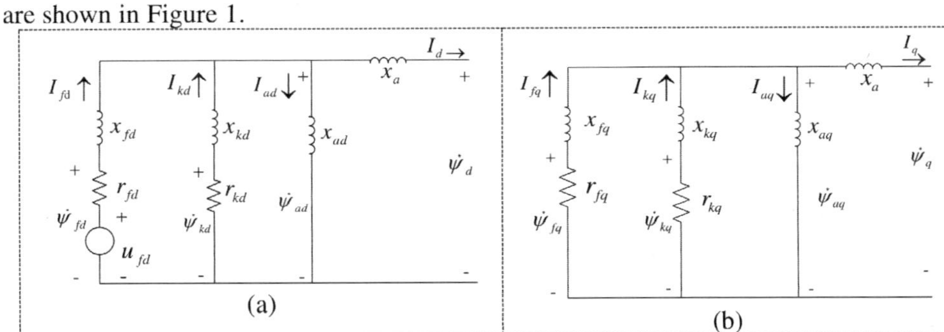

Figure 1. D- and q-axis equivalent circuits.

3. Phasor representation and damping torque

In oscillation analysis, it is a common practice to assume that any variable x to be expressed as the sum of its steady state value x_0, and the perturbation Δx.

$$x = x_0 + \Delta x \tag{7}$$

Where Δx is a linear combination of the time response of the oscillatory mode of the system and each component is a damp sinusoid.

$$\Delta x = \sum_i \Delta x_i = \sum_i \Delta x_{mi} e^{(\sigma_i t)} \cos(\omega_{di} t + \theta_i) \tag{8}$$

With the concept of complex frequency, for ith component, Δx_i could be represented with a phasor.

$$\Delta \tilde{x}_i = \Delta x_{mi} e^{j\theta_i} = \Delta x_{mi} \angle \theta_i \tag{9}$$

To study the damp effect of the PSS to the ith mode, only the mode of concern, $s_i = \sigma_i + j\omega_{di}$, is needed to be considered. Hence, the prefix Δ and subscript i of $\Delta\tilde{x}_i$ could be ignored, $\Delta\tilde{x}_i = \tilde{x}$.

With phasor representation, concept of damping coefficient could be used to evaluate the damp of multi-generator system. Usually, the change in electrical torque of a synchronous machine following a perturbation can be resolved into two components:

$$\Delta T_e = K_s \Delta\delta + K_D \Delta\omega_r \tag{10}$$

Considering the mode of concern only, variables in (10) are replaced with phasors.

$$\tilde{T}_e = K_s \tilde{\delta} + K_D \tilde{\omega}_r \tag{11}$$

Replace d/dt with $s = \sigma + j\omega_d \approx j\omega_d$ in swing equation $d\delta/dt = \Delta\omega_r$, and replace its variables with phasors, $s\tilde{\delta} = \tilde{\omega}_r$, $\tilde{\omega}_r$ is about 90° lead of $\tilde{\delta}$. Refer to the following Figure 2.

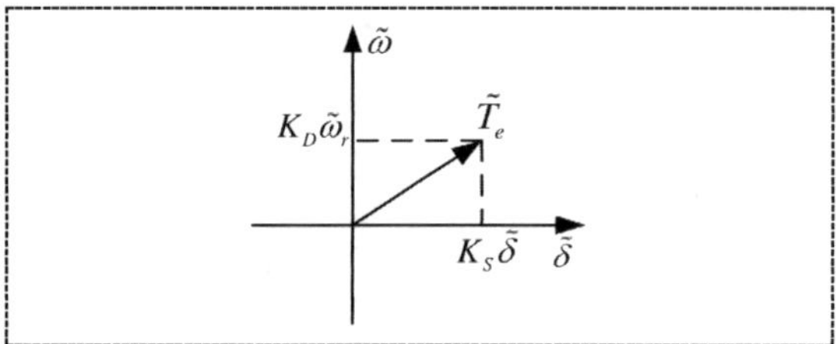

Figure 2. Electrical torque phasor projection

With phasor, the damping torque coefficient is defined as

$$K_D = \mathrm{Re}(\frac{\tilde{T}_e}{\tilde{\omega}_r}) \tag{12}$$

Replace d/dt with $s = \sigma + j\omega_d$ in swing equation $d\Delta\omega_r / dt = (T_m - T_e)/M$, and replace its variables with phasors, $\sigma + j\omega_d = (\tilde{T}_m - \tilde{T}_e)/(M\tilde{\omega}_r)$. Ignoring the damping effect of its mechanical torque, $\tilde{T}_m \approx 0$, the real part of the equation is

$$\sigma = -\frac{1}{M}\mathrm{Re}(\frac{\tilde{T}_e}{\tilde{\omega}_r}) = -\frac{K_D}{M} \tag{13}$$

The real part of the eigenvalue is proportional to the phasor based damping torque coefficient. Hence, it is an indication of the damping of the system. For a positively damped system, $\sigma < 0$. Then, $K_D > 0$ represents a positive damp effect of the power system.

4. Study of damping mechanism of PSS with circuit method

PSS takes one or more signals such as generator speed, frequency, and electrical real power as input signals, which are processed by a gain block, a washout filter and lead-lag filters to generate excitation auxiliary signals to act on excitation device, to ensure that PSS provides positive damping torque to the generator. Torque phasor could be used to study the mechanism of PSS in a multi-machine power system.

Linearizing (6) and replacing its variables with phasors, the following electrical torque phasor equation can be got.

$$\tilde{T}_e = I_{q0}\tilde{\psi}_d + \psi_{d0}\tilde{I}_q - I_{d0}\tilde{\psi}_q - \psi_{q0}\tilde{I}_d \tag{14}$$

There are four components in electrical torque equation. Two of them are input phasors terms, while the other two are state phasors terms. State phasors will be eliminated with input phasors of the generator, so that $\tilde{\psi}_d = \tilde{\psi}_d(\tilde{u}_{fd}, \tilde{I}_d, \tilde{I}_q)$ and $\tilde{\psi}_q = \tilde{\psi}_q(\tilde{u}_{fd}, \tilde{I}_d, \tilde{I}_q)$. To do this, Figure 3 could be used.

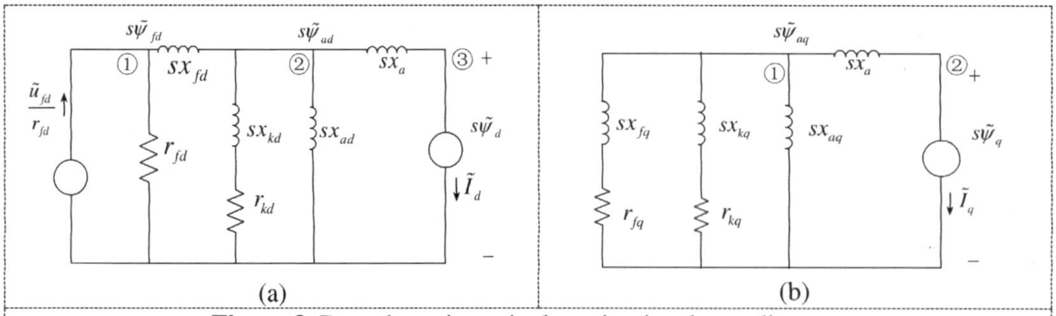

Figure 3. D- and q-axis equivalent circuits phasor diagram

Figure 3 is derived from Figure 1. To study the oscillatory course and to consider only the mode of concern, concepts of complex frequency could be applied to the circuits. The variables are replaced with their phasors, and d/dt for fluxes are replaced with $s=\sigma+j\omega_d$. In Figure 1 any reactance represents an inductance of DC circuits. Multiplying s, it becomes a generalized reactance of the AC circuits in Figure 3. With Norton's theorem, the voltage source is replaced with a current source. With substitution theorem, current sources are added to represent the input phasors, $\tilde{I}_d$ and $\tilde{I}_q$, of the generator.

According to the Figure 3, the d-axis node voltage equations can be written as follows:

$$\begin{bmatrix} \tilde{u}_{fd}/r_{fd} \\ 0 \\ \tilde{I}_d \end{bmatrix} = \begin{bmatrix} y_{d11} & y_{d12} & y_{d13} \\ y_{d21} & y_{d22} & y_{d23} \\ y_{d31} & y_{d32} & y_{d33} \end{bmatrix} \times \begin{bmatrix} s\tilde{\psi}_{fd} \\ s\tilde{\psi}_{ad} \\ s\tilde{\psi}_d \end{bmatrix} \tag{15}$$

Where $y_{d11}=1/r_{fd}+1/sx_{fd}$, $y_{d12}=-1/sx_{fd}$, $y_{d13}=0$, $y_{d21}=-1/sx_{fd}$, $y_{d22}=1/sx_{fd}+1/(r_{kd}+sx_{kd})+1/sx_{ad}+1/sx_a$, $y_{d23}=-1/sx_a$, $y_{d31}=0$, $y_{d32}=1/sx_a$, $y_{d33}=-1/sx_a$.

The q-axis nodal voltage equations are:

$$\begin{bmatrix} 0 \\ \tilde{I}_q \end{bmatrix} = \begin{bmatrix} y_{q11} & y_{q12} \\ y_{q21} & y_{q22} \end{bmatrix} \times \begin{bmatrix} s\tilde{\psi}_{aq} \\ s\tilde{\psi}_q \end{bmatrix} \tag{16}$$

Where $y_{q11}=1/(r_{fq}+sx_{fq})+1/(r_{kq}+sx_{kq})+1/sx_{aq}+1/sx_a$, $y_{q12}=-1/sx_a$, $y_{q21}=1/sx_a$, $y_{q22}=-1/sx_a$.

Solving (15) (16), $\tilde{\psi}_d$, $\tilde{\psi}_q$ could be represented by input phasors, the expressions are as follows.

$$\tilde{\psi}_d = Z_d(s)\tilde{I}_d + \frac{Z_{fd}(s)}{r_{fd}}\tilde{u}_{fd} \tag{17}$$

$$\tilde{\psi}_q = Z_q(s)\tilde{I}_q \tag{18}$$

Where $Z_d(s) = \dfrac{1}{sy_{d33} - \dfrac{sy_{d32}y_{d23}}{sy_{d22} - \dfrac{y_{d12}y_{d21}}{sy_{d11}}}}$, $Z_{fd}(s) = \dfrac{\dfrac{\dfrac{y_{d32}y_{d21}}{y_{d11}}}{y_{d22} - \dfrac{y_{d12}y_{d21}}{y_{d11}}}}{sy_{d33} - \dfrac{y_{d32}y_{d23}}{sy_{d22} - \dfrac{y_{d12}y_{d21}}{y_{d11}}}}$ and

$Z_q(s) = \dfrac{y_{q11}}{s(y_{q11}y_{q22} - y_{q21}y_{q12})}$.

The electrical torque can be expressed with input phasors:

$$\tilde{T}_e = \tilde{T}_{dis} + I_{q0}\frac{Z_{fd}(s)}{r_{fd}}\tilde{u}_{fd} \tag{19}$$

Where $\tilde{T}_{dis} = (I_{q0}Z_d(s)-\psi_{q0})\tilde{I}_d + (\psi_{d0}-I_{d0}Z_q(s))\tilde{I}_q$.

The electrical torque is divided into two parts. The first part, $\tilde{T}_{dis}$, is the disturbance torque component from the transmission network. The second part, $I_{q0}(Z_{fd}(s)/r_{fd})\tilde{u}_{fd}$, is the torque component generated by the excitation system. The second part could be further separated into two parts. Consider the block diagram of an excitation system with a PSS of Figure 4, in which $G_{ex}(s)$ represents the transfer function of the exciter and its regulator, and $G_{PSS}(s)$ is PSS transfer function, V_t is generator terminal voltage, and V_s is the output of the PSS.

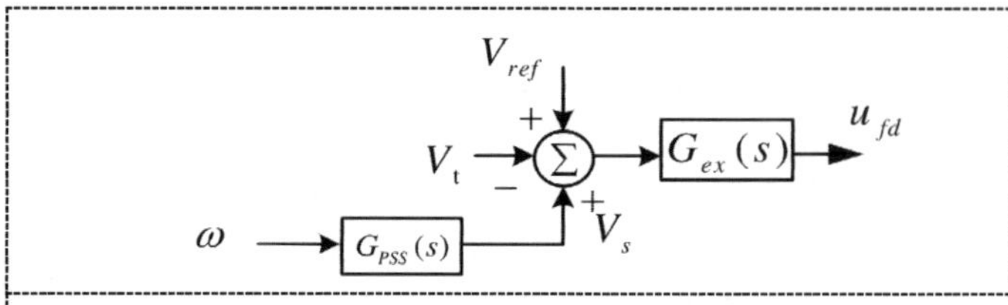

Figure 4. Block diagram of excitation system with PSS

With phasor representation, $\tilde{u}_{fd}$ is calculated by

$$\tilde{u}_{fd} = G_{ex}(\sigma + j\omega_d)(-\tilde{V}_t + \tilde{V}_s) \tag{20}$$

$-I_{q0}\dfrac{Z_{fd}(s)}{r_{fd}}G_{ex}\tilde{V}_t$ is the torque component from the fluctuation of the terminal voltage, and $I_{q0}\dfrac{Z_{fd}(s)}{r_{fd}}G_{ex}\tilde{V}_s$ is the torque component provided by the stabilizer. With (12) and (20), damping torque coefficient of PSS, K_{DP}, can be calculated:

$$K_{DP} = \mathrm{Re}(I_{q0}\frac{Z_{fd}(s)}{r_{fd}}\frac{G_{ex}\tilde{V}_s}{\tilde{\omega}_r}) \tag{21}$$

In (21), the damping torque coefficient of PSS is the projection of the electrical torque phasor generated by the PSS on the speed phasor of the generator. When K_{DP} is positive, it means that PSS provides positive damping torque to the system, indicating that PSS presents positive damping. Conversely, when K_{DP} is negative, it means that PSS provides negative damping torque to the system, indicating that PSS presents negative damping.

5. Case study

5.1. Single machine infinite bus system

Simulate in the single machine infinite bus system[30]. The applied excitation system is Alternator Supplied Rectifier Exciter #4 (AC4A) in PSCAD. The parameters of the excitation system are: $T_C = 1.0$, $T_B = 10.0$, $T_A = 0.015$, $V_{IMAX} = 10$, $V_{IMIN} = -10$, $V_{RMAX} = 5.64$, $V_{RMIN} = -4.53$, $K_A = 200$. AC4A model is shown in Figure5.

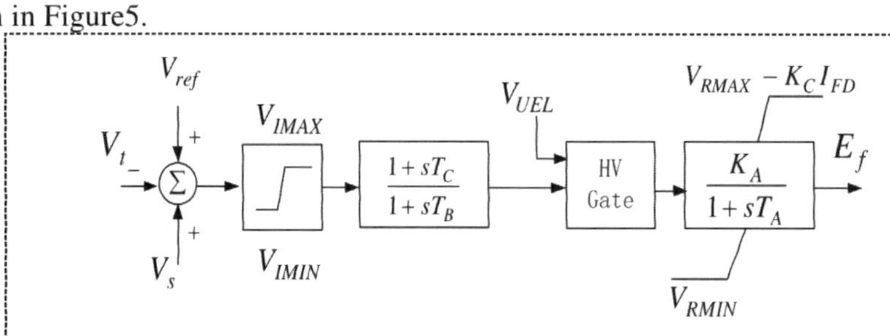

Figure 5. AC4A model of excitation model in PSCAD

The PSS model of the generator is Single Input Power System Stabilizer (PSS1A), the parameters of the PSS system are: $T_1 = 0.05$, $T_2 = 0.02$, $T_3 = 3$, $T_4 = 5.4$, $T_5 = 10$, $T_6 = 0$, $A_1 = 0$, $A_2 = 0$. PSS1A model is shown in Figure 6.

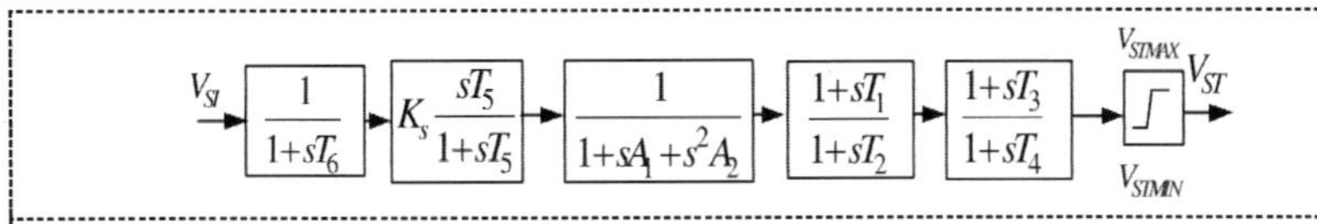

Figure 6. PSS1A model of PSS in PSCAD

Run the simulation. Select an observing window. Analyze u_{fd}, V_s, V_t and $\Delta\omega_r$ with TLS-ESPRIT program to get their phasors for the mode of concern. Repeat the process under different gain, K_s, of the PSS. Refer to the Table 1 for identification results.

Table 1. Identification results of single machine infinite bus system

Phasors	Amplitude/p.u.	Angle/°	K_s
$\tilde{u}_{fd}$	0.0024	-169.777	0
$\tilde{V}_t$	0.0740	14.358	
$\tilde{V}_s$	0	0	
$\tilde{\omega}_r$	1.6794	-26.260	
$\tilde{u}_{fd}$	0.0001	62.662	3
$\tilde{V}_t$	0.0358	39.438	
$\tilde{V}_s$	0.0386	47.149	
$\tilde{\omega}_r$	2.4047	-23.249	
$\tilde{u}_{fd}$	0.0012	77.795	8
$\tilde{V}_t$	0.0258	8.764	
$\tilde{V}_s$	0.0573	64.862	
$\tilde{\omega}_r$	1.1676	-5.811	
$\tilde{u}_{fd}$	0.0017	84.608	13
$\tilde{V}_t$	0.0288	-13.294	
$\tilde{V}_s$	0.0668	74.110	
$\tilde{\omega}_r$	0.8062	-3.471	
$\tilde{u}_{fd}$	0.0025	84.699	18
$\tilde{V}_t$	0.0381	-28.681	
$\tilde{V}_s$	0.0857	77.233	
$\tilde{\omega}_r$	0.7034	-7.864	

Using $s=\sigma+j\omega_d$ represents $G_{ex}(s)$, the values of K_{DP} under different parameters are calculated by Table 1 and (21), as shown in Table 2.

Table 2. PSS damping torque coefficient of single machine infinite bus system

K_s	ζ	K_{DP}
0	0.0216	0
3	0.0527	0.006
8	0.1586	0.016
13	0.2649	0.025
18	0.3683	0.030

Where ζ is damping ratio.

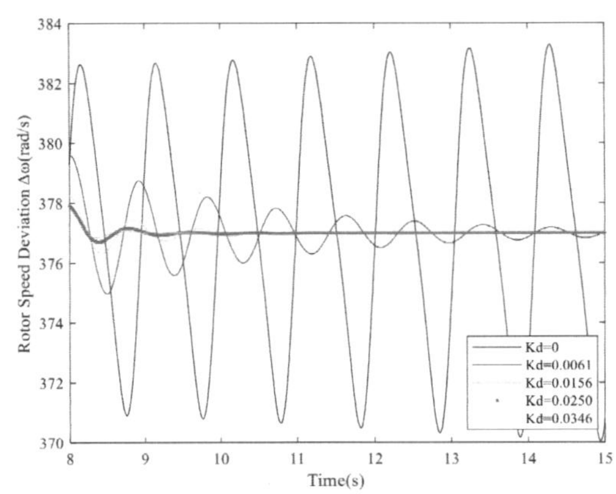

Figure 7. Rotor speed deviation of generator of single machine infinite bus system

The attenuation velocity of rotor speed deviation curve reflects the damping effect of PSS in Figure7. PSS damping torque coefficient K_{DP} is shown in Table 2. Without PSS, K_{DP} is equal to 0 and there is a constant amplitude oscillation in the system. With the increase of PSS gain K_s, the damping ratio ζ increases, the damping coefficient $K_{DP} > 0$ indicates that PSS provides a positive damping torque for the system. Meanwhile, the attenuation velocity becomes faster, and the system tends to be stable faster. Therefore, K_{DP} can be used to measure the positive or negative damping provided by PSS for the system. The sets of data in Table 2 can be obtained by identifying relevant data (such as u_{fd}) through TLS-ESPRIT program. These data sets show that the damping of PSS could be evaluated by the damping torque method proposed.

5.2. A simple four-machine two-area system

Simulate in the four-machine two-area system[30], the system is shown in Figure 8.

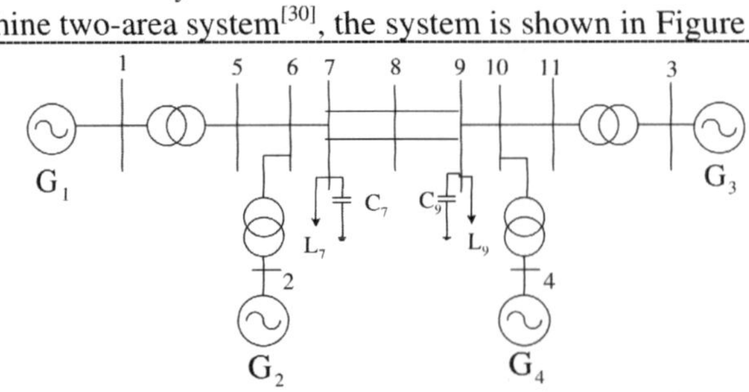

Figure 8. Four-machine two-area system

The generator G1 is analyzed. The excitation system of G1 is IEEE Static Excitation System #1 (ST1A) in PSCAD. The parameters of the excitation system are: $T_C = 0$, $T_B = 0$, $T_{C1} = 0$, $T_{B1} = 0$ $T_A = 0$, $T_F = 0$, $K_F = 0$, $K_A = 200$. ST1A model is shown in Figure 9.

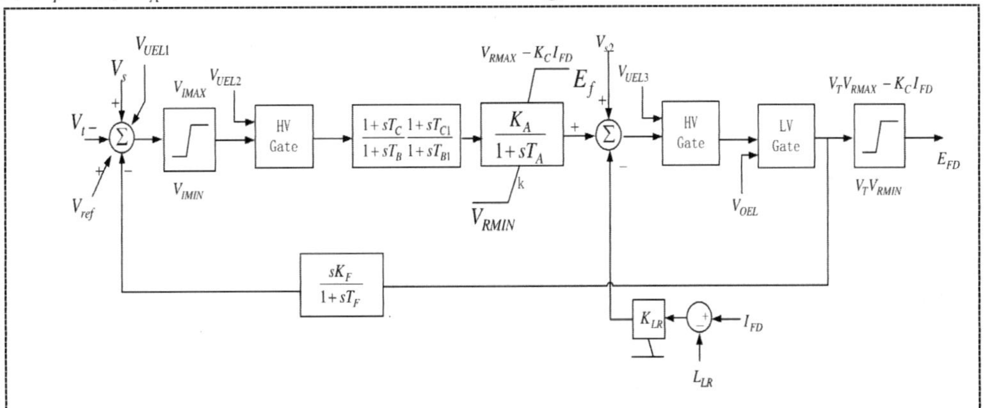

Figure 9. ST1A model of excitation model in PSCAD

The PSS model of G1 in four-machine two-area system is the same as that in the single machine infinite bus system. Run the simulation. Select an observing window. Analyze u_{fd}, V_s, V_t and $\Delta\omega_r$ with TLS-ESPRIT program to get their phasors for the mode of concern. Repeat the process under different gain, K_s, of the PSS. Refer to the results in Table 3.

Table 3. Identification results of four-machine two-area system

Phasors	Amplitude/p.u.	Angle/°	K_s
$\tilde{u}_{fd}$	6.49E-06	119.019	0
$\tilde{V}_t$	23.7E-06	-55.613	
$\tilde{V}_s$	0	0	
$\tilde{\omega}_r$	0.0076	-19.056	
$\tilde{u}_{fd}$	8.22E-06	109.425	3
$\tilde{V}_t$	54.0E-06	-29.300	
$\tilde{V}_s$	31.5E-06	-7.873	
$\tilde{\omega}_r$	0.0068	-19.151	
$\tilde{u}_{fd}$	10.5E-06	105.656	10
$\tilde{V}_t$	104E-06	-15.683	
$\tilde{V}_s$	82.1E-06	-1.727	
$\tilde{\omega}_r$	0.0053	-13.764	
$\tilde{u}_{fd}$	11E-06	104.202	13
$\tilde{V}_t$	119E-06	-13.683	
$\tilde{V}_s$	96.9E-06	-1.784	
$\tilde{\omega}_r$	0.0048	-13.157	
$\tilde{u}_{fd}$	12E-06	103.881	20

$\tilde{V}_t$	142E-06	-9.128
$\tilde{V}_s$	123E-06	0.627
$\tilde{\omega}_r$	0.0039	-10.166

The above data is used to perform PSS damping torque coefficient analysis. Then the results calculated under different parameters are shown in Table 4.

Table 4. PSS damping torque coefficient of four-machine two-area system

K_s	ζ	K_{DP}
0	0.0085	0
3	0.0119	0.008
10	0.0191	0.029
13	0.0226	0.034
20	0.0291	0.047

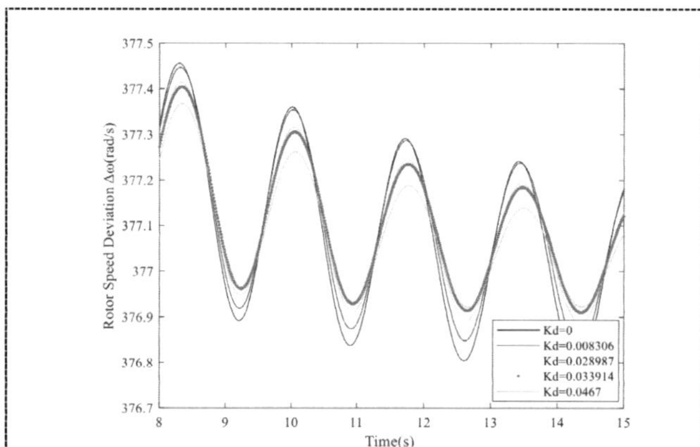

Figure 10. Rotor speed deviation of G1

Similar to the damping analysis of the single machine infinite bus system, it can be known from the damping ratio in Table 4 that the damping ratio of the four sets of data is greater than the damping ratio without PSS 0.0085, indicating that the PSS presents positive damping. The results of KDP which are greater than 0 can be analyzed to show that PSS also presents positive damping. The damping ratio analysis method and the damping torque method are consistent. These sets of data also show the correctness of the proposed theory.

6. Conclusion

Improper parameter adjustment of PSS may bring negative damping to power system, aggravating system oscillation and threatening the safety and stability of the system. Therefore, it is necessary to judge the positive or negative damping provided by PSS. This paper studies the mechanism of PSS suppressing low frequency oscillation with the damping torque analysis method. The main conclusions are as follows.

1) The analysis is carried out for 8th order Park generator model. The equivalent circuit of model is helpful for deriving formula.

2) The concepts of complex frequency and generalized phasor are used for analysis and calculation. The damping torque coefficient is defined with the phasor representation, and the relationship between the coefficient and system damping is analyzed. The variables in Park equations are expressed as phasors,

and the equivalent circuits are transformed from DC circuits to AC circuits, so as to prepare for deriving the equations.

3) The mechanism of PSS suppressing low frequency oscillation is studied with circuit method. The state variables in electrical torque are eliminated with input variables. Then, the electrical torque is decomposed into the component provided by the excitation system and the component injected from the transmission network. The expression of PSS damping torque coefficient is deduced.

The calculated damping torque coefficient can be used to judge the contribution of PSS to system damping. Simulation results verify the effectiveness of the method.

7. References

[1] Prasertwong Komkrit, Mithulananthan N, Thakur Devbratta. Understanding low-frequency oscillation in power systems[J]. *International Journal of Electrical Engineering Education*, 2010, 47(3): 248-262.

[2] Zuo Jian, Tang Jihong, Guo Hu, Hu Dijun, Zhang Keren, Xiang Meng. Low-Frequency Oscillation Mode Identification with OpenPDC Platform[C]. 2019 6th *International Conference on Systems and Informatics (ICSAI),* 2019: 223-227.

[3] Khattak Muhammad Adil, Yahya Tuan Mohamad Hakimi Tuan, Sallehhudin Muhammad Wazif Mohd, et al. Energy Security Policy Shift of North America and Ontario, Canada Following 2003 Power Blackout: A Review[J]. *Progress in Energy and Environment*, 2018, 4: 14-24.

[4] Zheng Chao, Ding Guoqiang, Liu Baisi, et al. Analysis and control to the ultra-low frequency oscillation in southwest power grid of China: A case study[C]. 2018 *Chinese Control And Decision Conference (CCDC)*, 2018: 5721-5724.

[5] Chang Shujun, Peng Chao, Hu Yuyang, Mou Yaxue, Cazzulani Gabriele. An Improved Prony Prediction Compensation-Based Wide-Area Damping Control Approach for Power System Low-Frequency Oscillation Suppression[J]. *Journal of Sensors,* 2021, 2021: 1-10.

[6] Wei Yunbing, He Ping, Li Shande, Zhang Wenzhong, Zhu Xiangqian. Study on the Mechanism and Control Policy of Low Frequency Oscillation for Power Systems [J]. *Journal of Shandong University of Science and Technology (Natural Science)*, 2008, 2.

[7] Qiang Wang, Baohua Yang, Chengwu Zhang. Summary of bifurcation and chaos research in electric power system [J]. *Transactions of China Electrotechnical Society*, 2005, 7.

[8] Chen Lei, Min Yong, Hu Wei. Low frequency oscillation analysis and oscillation source location based on oscillation energy part one mathematical foundation and energy flow computation[J]. *Automation of Electric Power Systems*, 2012, 36(3): 22-27.

[9] Chen Lei, Chen Yiping, Min Yong, Hu Wei, Zhang Kun. Low frequency oscillation analysis and oscillation source location based on oscillation energy part two method for oscillation source location and case studies[J]. *Automation of Electric Power Systems,* 2012, 36(4): 1-5.

[10] Ray Papia. Power system low frequency oscillation mode estimation using wide area measurement systems[J]. *Engineering science and technology, an international journal*, 2017, 20(2): 598-615.

[11] Zhang Jing, Xu Zheng, Wang Feng, Chang Yong. TLS-ESPRIT based method for low frequency oscillation analysis in power system[J]. *Automation of Electric Power Systems,* 2007, 20: 84-88.

[12] Rafique Zimran, Khalid Haris M, Muyeen SM, Kamwa Innocent. Bibliographic review on power system oscillations damping: An era of conventional grids and renewable energy integration[J]. *International Journal of Electrical Power & Energy Systems*, 2022, 136: 107556.

[13] Zhu Hanxing, Ye Hhua, Gao Lei. Evaluation of Damping of Generator by Using Oscillation Energy[C]. 2020 *IEEE/IAS Industrial and Commercial Power System Asia (I&CPS Asia)*, 2020: 727-731.

[14] Wang Maohai, Guo Dengfeng, Jiang Changming. An analysis method to evaluate damping characteristics of excitation systems in low frequency oscillation process[J]. *Automation of Electric Power Systems,* 2013, 37(4): 47-50.

[15] Chen Lei, Xu Fei, Min Yong, Wang Maohai, Hu Wei. Transient energy dissipation of resistances and its effect on power system damping[J]. *International Journal of Electrical Power & Energy Systems*, 2017, 91: 201-208.

[16] Lei Chen, Yong Min, Yiping Chen, Mingfu Lu, Yutong Zhang. Relationship between oscillation energy analysis and eigenvalue analysis and assessment of generator damping [J]. *Automation of Electric Power Systems*, 2013, 37(19): 33-40.

[17] Sun Zhenglong, Cai Guowei, Yang Deyou, Liu Cheng, Wang Bo, Wang Lixin. A method for the evaluation of generator damping during low-frequency oscillations[J]. *IEEE Transactions on Power Systems*, 2018, 34(1): 109-119.

[18] Hu Yong, Bu Siqi, Zhang Xin, Chung Chiyung, Cai Hui. Connection Between Damping Torque Analysis and Energy Flow Analysis in Damping Performance Evaluation for Electromechanical Oscillations in Power Systems[J]. *Journal of Modern Power Systems and Clean Energy*, 2021.

[19] Shahgholian Gh. Review of power system stabilizer: Application modeling analysis and control strategy[J]. *International Journal on Technical and Physical Problems of Engineering*, 2013, 5(3): 41-52.

[20] Zhang Li, Liu Y. Bulk power system low frequency oscillation suppression by FACTS/ESS[C]. *IEEE PES Power Systems Conference and Exposition*, 2004., 2004: 219-226.

[21] Mithulananthan Nadarajah, Canizares Claudio A, Reeve John, Rogers Graham J. Comparison of PSS, SVC, and STATCOM controllers for damping power system oscillations[J]. *IEEE transactions on power systems,* 2003, 18(2): 786-792.

[22] Chen Feng, Yang Xiaonan, Liu Liang, Ma Xiaochen, Zhang Lei, Yu Yiping. Study on the Influence of HVDC to Low Frequency Oscillation in Interconnected Power System[C]. 2018 2nd *IEEE Conference on Energy Internet and Energy System Integration* (EI2), 2018: 1-5.

[23] Demello Francisco P, Concordia Charles. Concepts of synchronous machine stability as affected by excitation control[J]. *IEEE Transactions on power apparatus and systems,* 1969, 88(4): 316-329.

[24] De Mello FP, Nolan PJ, Laskowski TF, Undrill JM. Coordinated application of stabilizers in multimachine power systems[J]. *IEEE Transactions on Power Apparatus and Systems*, 1980(3): 892-901.

[25] Lu Shengyang, Zhang Wuyang, Wang Tong, et al. Parameter Tuning and Simulation Analysis of PSS Function in Excitation System with Suppression of Low Frequency Oscillation[C]. 2019 IEEE 8th *International Conference on Advanced Power System Automation and Protection (APAP)*, 2019: 474-479.

[26] Dey Prasenjit, Bhattacharya Aniruddha, Das Priyanath. Tuning of power system stabilizer for small signal stability improvement of interconnected power system[J]. *Applied Computing and Informatics*, 2017.

[27] Abdel-Magid YL, Abido MA, Mantaway AH. Robust tuning of power system stabilizers in multimachine power systems[J]. *IEEE transactions on Power Systems*, 2000, 15(2): 735-740.

[28] Kahouli Omar, Jebali Mariam, Alshammari Badr, Hadj Abdallah Hsan. PSS design for damping low-frequency oscillations in a multi-machine power system with penetration of renewable power generations[J]. *IET Renewable Power Generation*, 2019, 13(1): 116-127.

[29] Jiang Xinyu Tony, Chow Joe H, Wilches-Bernal Felipe. A synchrophasor measurement based method for assessing damping torque contributions from power system stabilizers[C]. 2015 *IEEE Eindhoven PowerTech*, 2015: 1-6.

[30] P. Kundur, Power System Stability and Control. New. York, NY, USA: McGraw-Hill In, 1994.pp.813-814.

Journal of Physics: Conference Series **2320** (2022) 012005 doi:10.1088/1742-6596/2320/1/012005

Lithium-ion Batteries RUL Prediction Based on Temporal Pattern Attention

Haochen Qin, Xuexin Fan, Yaxiang Fan*, Ruitian Wang and Fengyuan Tian

National Key Laboratory of Science and Technology on Vessel Integrated Power System, Naval University of Engineering, 717 Jiefang Avenue, 430033, China

*E-mail:fanyaxiang_1989@163.com

Abstract. Accurate prediction of battery remaining useful life (RUL) under various operating conditions is essential for battery management systems to evaluate battery reliability, reduce the risk of battery usage and provide a rationale for battery maintenance. However, RUL prediction is a challenging problem since battery degradation is a nonlinear process and is influenced by external factors. In order to improve the prediction speed and accuracy, the research proposes a new Li-ion batteries RUL prediction method based on temporal pattern attention-based, which can take into account the influence of different variables for prediction. To model time-invariant patterns across multiple time steps, it combines a gated recurrent unit (GRU), a convolutional neural network, and an attention mechanism. Battery capacity, impedance and temperature are taken as input to train the model. Experiments are validated on public datasets and the results are compared with state of art methods. The experimental results show that the proposed method achieves the lowest MAE with 8.99, which proves the effectiveness of the method.

1. Introduction

In last few years, with the improvement and optimization of materials, lithium-ion batteries use high-safety battery materials and high-stability electrolyte systems, and the improvements in manufacturing technology have enabled lithium-ion batteries as the primary energy supply or storage element. Compared with other types of batteries, lithium-ion batteries have the advantages of high specific energy, low self-discharge rate, long life, fast charging, and high safety. They are widely used in electronic products, electric vehicles, military equipment such as missiles and torpedoes, aerospace applications such as aircraft and space stations, power stations and major energy storage systems. However, due to the dissolution of the electrode material, the decomposition of the electrolyte solution, the environmental temperature, the discharge rate or the harsh external environment, the internal micro-short circuit and the activity of the battery are reduced, resulting in a gradual deterioration of the battery. This directly and severely affects the life and safety of the battery, and then indirectly leads to the performance degradation or failure of the electrical system and equipment. Therefore, it is necessary to improve the Remaining Useful Life (RUL) prediction precision of lithium-ion batteries, which could obtain relevant information about the battery life in advance, so as to replace the batteries that reach the threshold in time, to avoid the failure of the battery performance caused by the power distribution that is far from the expectation. Furthermore, predicting the RUL of lithium-ion batteries has practical application value for extending battery life, reducing equipment maintenance costs and improving the economics of energy storage equipment management.

The RUL of a Li-ion battery refers to the number of cycles required for the battery's performance state (such as capacity or internal resistance) to decay and degrade to a specified threshold (such as 80% of capacity) under existing operating conditions. At present, there are many researches on RUL prediction, which are roughly composed of two categories: model-based methods and data-driven methods. The model-based approach seeks a complete mathematical model to follow the law of battery degradation. Therefore, this kind of method realizes RUL prediction by analyzing the failure principle and electrochemical reaction of the battery. These methods does not need a lot of data in the prediction process, and has strong adaptability, but it is easy to damage the battery and has poor real-time performance. These degradation models include electrochemical models, equivalent circuit models and empirical statistical models. Xu et al. [1] proposes to model the empirical mechanism between battery capacity and number of charge-discharge cycles for RUL prediction after evaluating the life loss of the battery operating curve. Ref.[2] proposed a method to iteratively predict battery RUL by combining electrochemical model and particle filter to establish a prediction model. This method uses SOH as the state variable, and the capacity decay model and the fitting result of the state variable are used as the measurement equation and process equation, respectively.

The data-driven approach eliminates the need to build a specific predictive model based on the complex laws of battery degradation, but instead uses the charge and discharge data to fully train a machine learning or deep learning model. It has strong adaptability, high real-time performance, and strong generalization ability. Data-driven methods [3][6] include auto regressive integrated moving average (ARIMA), Gaussian process regression (GPR), relevance vector machine (RVM), support vector regression (SVR), artificial neural network (ANN), etc. Liu et al. [4] proposed to optimize the GPR model using a combined kernel function to map the relationship between HI and RUL prediction. To suppress the measurement the prediction results noises, [5] proposed combining particle filtering (PF) to use the SVR method to represent battery aging dynamics, after building a state space model. Similarly, Zhou [7] proposed to build a capacity and mean voltage falloff regression model through relevance vector machine (RVM). Recently, deep learning methods have realized fruitful marks in related fields such as speech and image recognition, natural speech systems, and recommender systems. Inspired by this, various studies have used deep learning technology to explore their broad application in RUL prediction due to its advantages in modeling complex non-linear mapping and achieves better accuracy. These works employ a recurrent neural network (RNN) [8], one dimension convolution neural network (1D-CNN) [9], dual-stage attention gate recurrent unit neural network [10], and other hybrid neural networks [11] [12]

However, one disadvantage of these methods is their weakness in managing long-term dependencies. As to this issue, this research proposes a new battery RUL prediction method by utilizing temporal pattern attention [13], which could model time-invariant pattern over long periods of time. Attention mechanism simulates human thought, and is often used in deep learning to extract key information from a large amount of data. Compared with other attention mechanisms, the temporal pattern attention can take into account the influence of different variables to prediction. First, the proposed method learns new temporal latent representations from the three internal states including battery capacity, impedance and temperature wich the Gated Recurrent Unit (GRU). Second, in order to abstract contextual features, multiple 1-D convolutional neural networks is applied to extract the temporal patterns of various features in the latent state. Finally, an attention mechanism is incorporated, enabling the model to reinforce effective features through attention distribution. This model differs from existing methods mainly in that it uses an attention mechanism to fuse temporal information with feature information, thereby achieving RUL prediction with higher accuracy and better interpretability.

2. Method

2.1. Problem formulation

For RUL prediction problem, given the temporal intermediate state $x=\left(x_{t-t_0},\cdots,x_{t-1},x_t\right)$, where x_i represents the temporal internal state, the mission is to predict the time t RUL y_t with the window size t_0 Through training the forecasting model $f\left(\bullet\right)$, the relationship function between RUL y_t and temporal intermediate state $x=\left(x_{t-t_0},\cdots,x_{t-1},x_t\right)$ is:

$$y_t = f\left(x_{t-t_0},\cdots,x_{t-1},x_t\right) \tag{1}$$

2.2. Lithium-ion battery RUL prediction based on temporal pattern attention

This section will introduce the detail of RUL prediction based on temporal pattern attention. Inspired by related research of time series forecasting, the prediction model could achieve a better forecasting effect by capturing the time-series patterns of the three internal state characteristics such as battery capacity, impedance and temperature and drilling down important characteristics. As shown in figure as below, the model can be divided into four parts: Gated Recurrent Unit (GRU) neural network, convolutional neural network (CNN), Temporal Pattern Attention (TPA), and final prediction. The models will be described in detail below in this order.

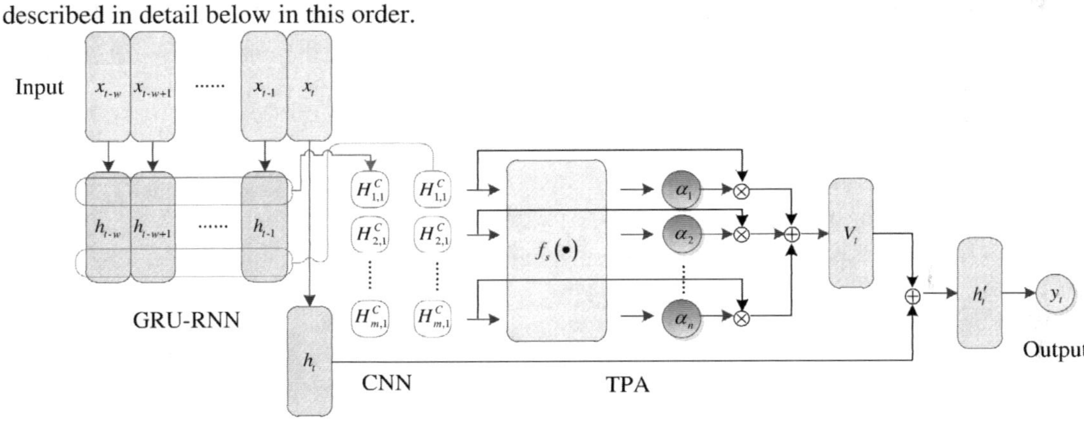

Figure 1. The structure of the Temporal Pattern Attention for RUL prediction

2.2.1. Gated Recurrent Unit (GRU) neural network. The GRU is a transformation form of recurrent neural network, which could capture the long-term temporal dependencies. Actually, to make predictions at the current step, it is able to control input, memory, etc., through dependency gating mechanisms The input x_t, reset gate r_t, update gate z_t, candidate hidden state $\tilde{h}_t$, hidden state h_t constitutes the GRU unit. The details can be described as follows:

$$\tilde{h}_t = \tanh(W_h x_t + U_h(r_t * h_{t-1}) + b_h) \tag{2}$$

$$r_t = \sigma(W_r x_t + U_t h_{t-1} + b_r) \tag{3}$$

$$h_t = z_t * h_{t-1} + (1 - z_t) * \tilde{h}_t \tag{4}$$

$$z_t = \sigma(W_z x_t + U_z h_{t-1} + b_z) \tag{5}$$

Where U_h is the weight matrix; σ is the gate activation function, which is set to the sigmoid function, and the output activation function is the tanh function.

2.2.2. Temporal Pattern Detection using CNN. CNN has the ability to obtain a variety of important signal patterns, which can be used to strengthen the model's generalization ability. After obtaining the latent state representation at each time stage, CNN is used to further extract the temporal patterns of various features over the entire time stage. Specifically, k filters $C_i \in \boldsymbol{R}^{1\times T}$, T is the maximum length. In detail, the convolution operation is defined as

$$H_{i,j}^{C} = \sum_{l=1}^{H} H_{i,(t-\omega-1+l)} \times C_{j.T-\omega+l} \tag{6}$$

Where $H_{i,j}^{C} \in \boldsymbol{R}^{n\times k}$ is the convolutional value of the i-th row vector and the j-th filter. The result of this transformation in the convolutional layer will be used in the next stage, and the attention mechanism will act on these latent vectors that contain various feature information.

2.2.3. Temporal Pattern Attention. Relying on deep learning, the attention mechanism mimics the human mind, extracting key information from large amounts of data. Typical attention mechanisms usually focus on information related to the current time step, while ignoring the attention to important features. Furthermore, typical attention mechanisms average feature information over multiple temporal stages, making it difficult to mine temporal patterns for forecasting.

For the temporal pattern attention, it can take into account the influence of different variables for prediction. With the previous states $H^{C} = \left\{H_1^C, H_2^C, \ldots, H_{t-1}^C\right\}$, a weighted sum v_t of each column h_i is extracted to represent information about the current time step. Firstly the relevance is defined by a scoring function as below:

$$f\left(H_i^C, h_t\right) = \left(H_i^C\right)^T W_a h^t \tag{7}$$

Where h^t represents the query vector, H_i^C is the key vector, W_a is the parameter that could learn. Then the sigmoid function is applied to compute the attention weight α_i as below:

$$\alpha_i = sigmoid\left(f\left(H_i^C, h_t\right)\right) \tag{8}$$

Finally, use the attention weight α_i to make a weighted summation of each row H^C to obtain the output

$$v_t = \sum_{i=1}^{n} \alpha_i H_i^C \tag{9}$$

2.2.4. Prediction. To obtain the final prediction, the v_t and output of the GRU h_t are combined through linear transform as below:

$$h_t' = W_h h_t + W_v v_t \tag{10}$$

$$y_{t-1+\Delta} = W_{h'} h_t' \tag{11}$$

where $h_t, h_t' \in \boldsymbol{R}^m$, $W_h \in \boldsymbol{R}^{m\times m}$, $W_v \in \boldsymbol{R}^{m\times k}$, and $W_{h'} \in \boldsymbol{R}^{m\times m}$ and $y_{t-1+\Delta} \in \boldsymbol{R}^n$.

3. Experiment

In this section, the test battery parameters is introduced at first. Next, the detail of the experiment is provided. And the experimental results and experimental analysis are provided in the end.

3.1. Datasets
To validate the TPA based RUL prediction method, the method is evaluated on public datasets from MIT and Stanford University [15]. This dataset was collected from cycle life testing of 124 commercial lithium iron phosphate (LFP)/graphite batteries manufactured by A123 System (APR18650M1A). The nominal capacity and nominal voltage of the test battery were 1.1Ah and 3.3V respectively. More detail is shown in Table 1. The cells were rapidly charge-discharged to failure at a constant temperature of 30°C, and experiments were performed on an Arbin battery test cycler with 48 channels.

Table 1. Test battery parameters.

Type	Matiral	Specification	nominal capacity	nominal voltage	Voltage operating range	Working Temperature
APR18650M 1A	phosphate (LFP)/graphite	18mmx65mm	1.1Ah	3.3V	2.0-3.6V	-30°C-60°C

Under the fast charge strategy, it is expressed as a two-step fast charge strategy C1(Q1)-C2" until 80% SOC. C1 and C2 denote the first and second constant current steps, respectively, and Q1 is the state of charge (SOC) for current switching. After charging reaches 80% SOC, all cells are charged in 1C CC mode until the voltage reaches the upper limit voltage, and then switch to CV mode to charge until the charging current decreases to C/50. Then in the discharge stage, all the cells were discharged by CC-CV at 4C to 2.0V with a current cutoff of C/50. 12 cells were chosen for the experiment and divided into two groups, 8 cells are used to train the network and 4 cells are used for testing. The details of these cells are shown in Table 2 as below.

Table 2. Summary of battery charging strategies used in experiments.

Type	Cell	Strategies	Cycles
Group 1 (Training set)	1	5.4C(40%)-3.6C	1054
	2	5.4C(50%)-3.6C	897
	3	5.4C(50%)-3C	788
	4	5.4C(60%)-3.6C	862
	5	5.4C(60%)-3C	880
	6	5.4C(70%)-3C	691
	7	5.4C(70%)-5.4C	788
	8	5.4C(50%)-3.6C	906
Group 2 (Testing set)	9	5.4C(60%)-3.6C	840
	10	5.4C(80%)-5.4C	559
	11	5.4C(80%)-5.4C	534
	12	5.4C(60%)-3C	719

3.2. Implementation Details
The battery capacity, impedance and temperature data normalization is applied to reduce redundancy and improve data integrity first by the Min-Max method as beblow

$$x^{'} = \frac{x - x_{\min}}{x_{\max} - x_{\min}} \tag{12}$$

Then the sliding time window technique is used to generate fixed-length training samples, where the length of the sliding time window ω is 20 and the step size is 1, and the normalized battery data are input into the model for training. Inspired by [16], the CNN filters number is fixed at 16. The number of hidden units m is set to 20. The method select Adam optimizer during training for adaptive optimization and the learning rate, weight decay, hyperparameter β_1, hyperparameter β_2, and total iterations number are setting as 0.001, 0.9 per 100 epochs, 0.9, 0.999 and 1000, respectively. The experiments is performed on a desktop computer equipped with an Intel Core i7-8700 3.7GHz CPU, NVIDIA 3080TI GPU, and 32GB RAM.

The total mean absolute error (MAE) is used to evaluate the error between the estimated result and the reference value as below:

$$MAE = \frac{\sum_{i=1}^{n} |\hat{y}_i - y_i|}{n} \quad (13)$$

Where y_i and $\hat{y}_i$ are the predicted value and the actual value, respectively.

3.3. Experiment result

This subsection presents the performance of our approach on public datasets. Effects of the proposed model is compared with the standard autoregressive (AR) model, Gaussian process (GP) model, support vector regression (SVR) model, one-dimensional convolutional neural network (CNN) model, gated recurrent unit (GRU) model. Among these methods, the first three methods are traditional baseline methods and the last method that uses deep learning technology. The experiment result is presented in Fig. 2 and Table 3.

In Figure 2(a)(b), the red solid lines are used to show the true values of the relationship between RUL and cycles, and the blue solid lines show the predicted values. From Fig. 2(a)-(d), it is obvious that the proposed method achieves accurate RUL prediction results. For cell9, the predicted curve is less volatile compared to the ground truth, while for Cell 11 the predicted curve and the ground truth curve overlap over most of the loops. However, from Fig. 2(c)(d), it can be found that the proposed method achieves a smaller prediction error. From the quantitative results in Table 3, it can be concluded that the MAEs of this method are 11.98, 10.34, 7.13 and 6.53, respectively. Obviously, the proposed method achieves the best results compared to the state-of-the-art methods with a smaller average MAE of four cells. It is worth noting that this method fuses CNN and GRU-RNN architectures, and the prediction performance of this method is significantly improved compared to simple CNN and GRU-RNN methods. This is attributed to Temporal Pattern Attention architecture. Experimental results on open-source datasets to verify the validity of the proposed method.

Table 3. Results of RUL prediction compared with the state of art methods.

Methods	Cell 9	Cell 10	Cell 11	Cell 12	Average
AR	20.25	19.57	12.41	10.29	15.63
GP	13.24	13.68	7.97	7.64	10.63
SVR	16.42	15.38	11.04	7.32	12.54
1D-CNN	13.61	14.35	8.46	8.07	11.12
GRU-RNN	14.27	14.69	9.23	8.94	11.78
Our method	11.98	10.34	7.13	6.53	8.99

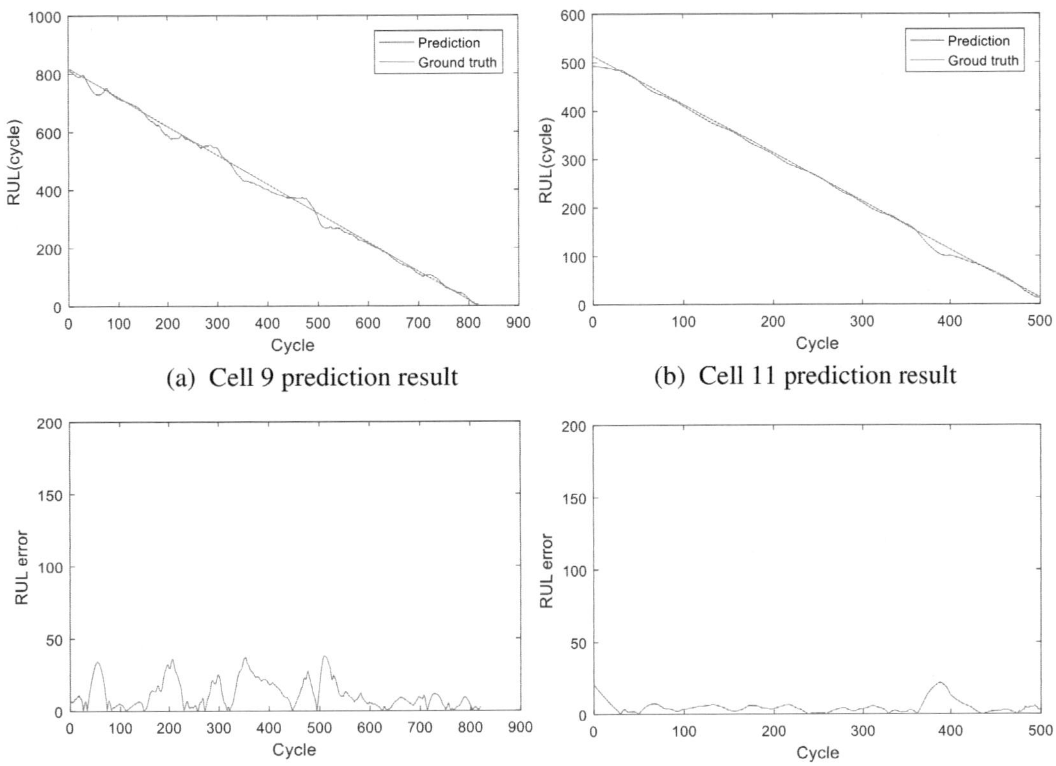

(a) Cell 9 prediction result (b) Cell 11 prediction result

(c) Cell 9 prediction error (d) Cell 11 prediction error

Figure 2. RUL Prediction of the two cells

4. Conclusions

Accurate and fast prediction of RUL of lithium-ion batteries can ensure safe and reliable operation of batteries, extend battery life, reduce equipment maintenance costs, and improve the economy of energy storage equipment management, which has practical application value. This work proposes an approach for Li-ion battery RUL prediction with an advanced intelligence method called temporal pattern attention, which can model any time-invariant pattern across multiple time. Using battery capacity, impedance, and temperature as input to train the model, the open-source dataset predicts an average MAE of 8.99, which is significantly better than the other five latest approach. In future, further work plans to study predicted results on actual test data of batteries used in energy storage systems in our microgrid energy management projects.

5. References

[1] B Xu, A Oudalov, A Ulbig, G Andersson, D Kirschen. Modeling of lithium-ion battery degradation for cell life assessment. IEEE Transaction on Smart Grid, 2018, 9: 1131–40.

[2] C. Lyu, Q. Lai, T. Ge, H. Yu, L. Wang, N. Ma. A lead-acid battery's remaining useful life prediction by using electrochemical model in the Particle Filtering framework. Energy, 2017, 120(1): 975-984.

[3] B. Zraibi, C. Okar, H. Chaoui, and M. Mansouri. Remaining Useful Life Assessment for Lithium-Ion Batteries Using CNN-LSTM-DNN Hybrid Method. IEEE Transaction on Vehicular Technology, 2021, 70(5): 4252-4261.

[4] J Liu, Z Chen. Remaining Useful Life Prediction of Lithium-Ion Batteries Based on Health Indicator and Gaussian Process Regression Model. IEEE Access, 7: 39474 – 39484.

[5] J. Wei, G. Dong, Member, and Z. Chen. Remaining Useful Life Prediction and State of Health Diagnosis for Lithium-ion Batteries Using Particle filter and Support Vector Regression. IEEE Transactions on Industrial Electronics, 2018, 65(7): 5634 – 5643.

[6] H. Rauf, M. Khalid, N. Arshad. Machine learning in state of health and remaining useful life estimation: Theoretical and technological development in battery degradation modelling. Renewable and Sustainable Energy Reviews, 2022(156): 111903.

[7] Y. Zhou, H. Gu, T. Su, X. Han, L. Lu, Y. Zheng. Remaining useful life prediction with probability distribution for lithium-ion batteries based on edge and cloud collaborative computation. Journal of Energy Storage, 2021, 44(A-1), 103342.

[8] C Lin, J Cabrera, F Yang, M Ling, K Tsui, S Bae. Battery state of health modeling and remaining useful life prediction through time series model. Appl Energy 2020;275(June):115338.

[9] J. Hong, D. Lee, E. Jeong, Y. Yi. Towards the swift prediction of the remaining useful life of lithium-ion batteries with end-to-end deep learning. Applied Energy, 2020, 278: 115646.

[10] K. Yang, Y. Tang, S. Zhang, Z. Zhang. A deep learning approach to state of charge estimation of lithium-ion batteries based on dual-stage attention mechanism. Energy, 2022, 244: 123233.

[11] X. Li, L. Zhang, Z. Wang, P. Dong. Remaining useful life prediction for lithium-ion batteries based on a hybrid model combining the long short-term memory and Elman neural networks. Journal of Energy Storage, 2019(21): 510–518.

[12] B. Zraibi, C. Okar, H. Chaoui, and M. Mansouri. Remaining Useful Life Assessment for Lithium-Ion Batteries Using CNN-LSTM-DNN Hybrid Method. IEEE Transactions on Vehicular Technology. 2021, 7(5): 4252-4261.

[13] S. Shih, F. Sun and H. Lee. Temporal Pattern Attention for Multivariate Time Series Forecasting. Machine Learning, 2019(108): 1421–1441.

[14] K Cho, B V Merrienboer, C Gulcehre, D Bahdanau, F Bougares, H Schwenk, et al. Learning phrase representations using RNN encoder-decoder for statistical machine translation. In: EMNLP 2014 - 2014 conf empir methods Natang process proconf; 2014, 1724-1734.

[15] A K Severson, M P Attia, N Jin, et al. Data-driven prediction of battery cycle life before capacity degradation. Nature Energy, 2019, 4(5): 383–391

[16] S. Shih, F. Sun,·H. Lee. Temporal Pattern Attention for Multivariate Time Series Forecasting. Machine learning. 2019, 108(8/9): 1421-1441.

Acknowledgements
This research was funded by The National Natural Science Foundation of China (Grant No. 52007196).

VSC-HVDC Incorporated Corrective Security-Constrained Optimal Power Flow

Cong Fu, Yashan Zhong, Pingping Zhang, Xiangzhen He, Yun Yang and Jian Zuo
Power Dispatching Control Center of Guangdong Power Grid Co., Ltd., Guangzhou, Guangdong, China

411011167@qq.com

Abstract. With the rapid development and diverse engineering applications of VSC-HVDC technology, development in power industry has put forward higher economic and security requirements for the security-constrained optimal power flow. Under these circumstances, this paper proposes a security-constrained optimal power flow model for hybrid AC/DC power grids that incorporates the VSC-HVDC corrective control. The linear relationship between the pre-contingency state power flow and post-contingency state power flow is established based on the shift factor and the reliability constraints are described as the linear inequality constraints of the post-contingency state active power flow which is expressed by the pre-contingency state branch power flow. With the inherent ability of the VSC-HVDC converter station (short as VSC station) to rapidly regulate active power taken into account, this paper utilizes the linear relationship between the active power adjustment of the VSC station and the power flows of transmission lines and electrical equipment and corrects the reliability constraints with the sensitivity coefficient. Hence the correction of the post-contingency state power flow is reflected in the constraints. By solving the proposed model with the primal-dual interior-point method, the pre-contingency state optimal security operating point that considers the corrective capability of the VSC stations is obtained. Finally, the correctness and validity of the proposed model is verified based on a VSC-HVDC modified IEEE 14-bus system.

1. Introduction

With a focus on the safe operating conditions both in the normal state and in scenarios with anticipated faults, the security-constrained optimal power flow (SCOPF) ensures the security both in the normal state and with anticipated faults, and on this basis, incorporates economic benefits as well. SCOPF problems are divided into two categories: preventive security-constrained optimal power flow (PSCOPF) [1] and corrective security-constrained optimal power flow (CSCOPF) [2]. PSCOPF requires that the calculated operating parameters meet the safe and stable operating requirements both in the normal state and in scenarios with anticipated faults. That is, in the event of a contingency, the operator does not need to adjust the operating parameters of the system and the control variables remain unchanged. To guarantee the secure operation of the power grid, optimization results are slightly conservative. On the contrary, CSCOPF allows overflow in the electrical equipment in the event of a contingency and the operator adjusts the operating parameters of the system so that the system returns to a static safe and stable state. That is, control variables are adjusted within an allowable range to make the required operating parameters satisfy the security operating conditions of the anticipated fault scenarios. In practical use, both categories face the problem of upgraded computational scale and increased difficulty in finding a solution. Therefore, reduction in the problem-solving scale [3, 4] and optimization of computational method [5-7] have become the research hotspots for SCOPF problems. Although CSCOPF is more economical, it is more complicated to implement. Apart from the calculation aspect, the traditional thermal power unit is slow in responding to failures and is weak in rapid adjustment of line power flows, which also brings more risks to the rapid post-contingency adjustment [8].

With the wide application of the Voltage Source Converter based High Voltage Direct Current (VSC-HVDC) technology in a variety of areas such as new energy integration, long-distance power transmission, and large-scale hybrid AC/DC power grid construction, etc., more and more traditional SCOPFs originally intended for AC power grids are considering VSC-HVDC converter stations as well [9, 10]. Compared with thermal units, the active and reactive power settings of VSC stations can be adjusted flexibly and quickly, hence providing new possibilities for quickly changing the distribution of line power flow after a contingency and thus preventing overflows. Reference [11] established an optimal power flow (OPF) model of the hybrid AC/DC grid and corrected the constraints of the optimization model with the linear correction coefficients. It sacrificed a certain degree of economy in exchange for security and was essentially preventive. For the CSCOPF problems, [12, 13] adopted the DC power flow model to deal with all the power flow constraints of the hybrid AC/DC grids. Although the DC power flow model is beneficial to improve the calculation speed, it brings huge errors to large-scale power systems since network losses are ignored. For HVDC-VSC incorporated hybrid AC/DC power grids, [14] put forward the nonlinear equality and inequality constraints of the AC power flow model, using the reactive power capacity of the VSC stations and the transformer turns ratio as the main adjustment methods. Through the fault filtering technology, it also identified to which extend the anticipated faults influenced the optimization results and emphasized on the most severe faults, hence improving the calculation efficiency. With the continuous progress and improvement of the control and optimization of the VSC-HVDC technology, more in-depth research will be conducted on VSC-HVDC incorporated CSCOPF problems.

In this paper, an OPF model of hybrid AC/DC power grids that incorporates VSC-HVDC is put forward and based on the shift factor, the linear relationship between the pre-contingency state power flow and post-contingency state power flow is established. Linear reliability constraints are added to the OPF to simplify the optimization problem. With the rapid regulating ability of the VSC stations, the linear relationship between the active power adjustment of the VSC stations and the power flow of transmission lines and network equipment is utilized and the sensitivity coefficient is introduced herein to correct the reliability constraints. Therefore, the correction of the post-contingency power flow is reflected in the constraints. In the end, the primal-dual interior-point method is adopted to solve the CSCOPF model.

2. OPF with VSC-HVDC incorporated

2.1. Equivalent circuit of the VSC

In this paper, the VSC equivalent model in the polar coordinate system is used, and the equivalent circuit of voltage source converter k (VSC_k) is depicted in Figure 1.

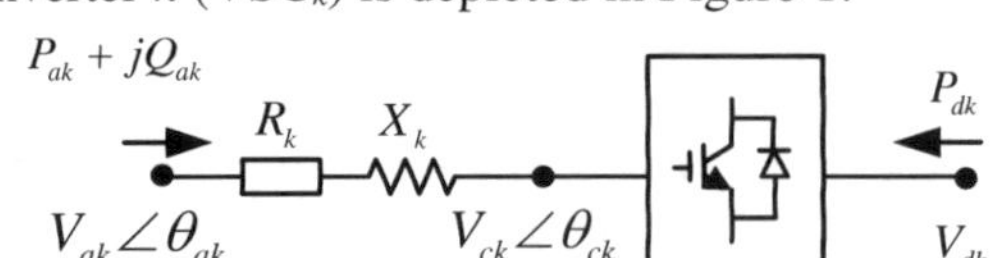

Figure 1. Equivalent circuit of voltage source converter k .

In Figure 1, R_k and X_k are the equivalent internal resistance and equivalent internal reactance, respectively; P_{ak} , Q_{ak} and P_{dk} are the AC real power, the AC reactive power and the DC real power, respectively; V_{ak} and θ_{ak} are the voltage amplitude and voltage angle on the AC side, respectively; V_{ck}

and θ_{ck} are the equivalent internal voltage amplitude and angle, respectively; V_{dk} is the voltage amplitude on the DC side. The power flow equations are listed below:

$$P_{ak} = V_{ak}^2 g_k - V_{ak} V_{ck} \left(g_k \cos\theta_{ck,ak} + b_k \sin\theta_{ck,ak} \right) \tag{1}$$

$$Q_{ak} = -V_{ak}^2 b_k + V_{ak} V_{ck} \left(g_k \sin\theta_{ck,ak} + b_k \cos\theta_{ck,ak} \right) \tag{2}$$

$$P_{dk} = V_{ak}^2 g_k - V_{ak} V_{ck} \left(g_k \cos\theta_{ck,ak} - b_k \sin\theta_{ck,ak} \right) + P_{vscloss}^k \tag{3}$$

where $g_k = \frac{R_k}{R_k^2 + X_k^2}$ and $b_k = -\frac{X_k}{R_k^2 + X_k^2}$. $P_{vscloss}^k$ is the internal loss of the VSC station, which can be calculated by [17]:

$$P_{vscloss}^k = \mathrm{A}_{lk} + \mathrm{B}_{lk} \mathrm{I}_{vk} + \mathrm{C}_{lk} \mathrm{I}_{vk}^2 \tag{4}$$

where A_{lk}, B_{lk} and C_{lk} are the constant coefficient, the linear coefficient and the quadratic coefficient of the active power loss of VSC$_k$, respectively; I_{vk} is the phase current of the converter valve.

2.2. Objective function

In the optimal power flow calculation of the VSC-HVDC integrated hybrid AC/DC power grids, reduction in the network losses or the generation cost is often the optimization goal. The minimum of the total generation cost is used herein as the objective function:

$$\min \sum_{n=i}^{NG} (c_0 + c_{li} P_{Gi} + c_{qi} P_{Gi}^2) \tag{5}$$

where c_0, c_{li} and c_{qi} are the constant coefficient, the linear coefficient and the quadratic coefficient of the generation cost of unit i, respectively; P_{Gi} is the active power output of unit i; NG is the number of generator units.

2.3. Power balance constraints on AC nodes

According to the Kirchhoff's current law, the active and reactive power flows of node i are balanced:

$$\Delta P_{ai}^{inj} = P_{Gi} - P_{Li} - \sum_{j \in N_a^i} P_{ij} \left(V_{ij}, \theta_{ij} \right) = 0 \tag{6}$$

$$\Delta Q_{ai}^{inj} = Q_{Gi} + Q_{ci} - Q_{Li} - \sum_{j \in N_a^i} Q_{ij} \left(V_{ij}, \theta_{ij} \right) = 0 \tag{7}$$

where N_a^i is the set of AC nodes adjacent to node i; V_{ij} and θ_{ij} are the differences in voltage magnitude and voltage angle between node i and adjacent node j, respectively.

2.4. Power balance constraints on DC nodes

The active power flow balance constraints of the DC nodes are as follows:

$$\Delta P_{di}^{inj} = V_{di}^2 G_{ii} + V_{di} \sum_{j \in N_d^i} V_{dj} G_{ij} + \sum_{k \in N_c^i} P_{dk} = 0 \tag{8}$$

where N_d^i is the set of DC nodes adjacent to node i; N_c^i is the set of nodes on the DC side of VSC stations which are adjacent to node i.

2.5. Constraints on node voltage

The voltage amplitude of each node should satisfy the upper/lower limit constraints:

$$\underline{V}_i \le V_i \le \overline{V}_i \tag{9}$$

where $\underline{V}_i$ and $\overline{V}_i$ are the lower and upper limits of the voltage amplitude at node i, respectively.

For DC nodes, the following constraints apply:

$$\underline{V}_{di} \le U_{di} \le \overline{V}_{di} \tag{10}$$

where $\underline{V}_{di}$ and $\overline{V}_{di}$ are the lower and upper limits of the voltage amplitude at DC node i, respectively.

2.6. Constraints on unit output

The active and reactive power output of the generator units satisfy the upper/lower limit constraints:

$$\begin{cases} \underline{P}_{gi} \le P_{gi} \le \overline{P}_{gi} \\ \underline{Q}_{gi} \le Q_{gi} \le \overline{Q}_{gi} \end{cases} \tag{11}$$

where P_{gi} and Q_{gi} are the active and reactive power output of the generator units at node i, respectively; $\underline{P}_{gi}$ and $\overline{P}_{gi}$ are the maximum leading and lagging active power output, respectively; $\underline{Q}_{gi}$ and $\overline{Q}_{gi}$ are the maximum leading and lagging reactive power output, respectively.

2.7. Constraints on branch power flow

There are two implementations of branch power flow constraints in practice: 1) the branch current does not exceed the upper limit; 2) the branch active power flow does not exceed the upper limit. The second implementation is adopted here, that is:

$$-\overline{P}_{ij} \le P_{ij} \le \overline{P}_{ij} \tag{12}$$

where $\overline{P}_{ij}$ is the long-term allowable current carrying capacity of the branch connecting node i and node j.

3. The CSCOPF model with VSC correction

When a device in the power system fails and goes out of operation, the power flow of the power grid is redistributed, which may cause a dramatic increase in the power flow of some devices and threaten the secure operation of the system. To ensure that the power grid can resist the power flow impact after a contingency, apart from enhancing the quality of the devices, post-contingency adjustment methods can be adopted to maintain the line power flow within a reasonable range.

Traditional CSCOPF model is illustrated as follows:

$$\begin{aligned} \min_{\boldsymbol{x}_c,\boldsymbol{u}_c,\forall c\in\{0\}\cup C} \quad & f(\boldsymbol{x}_0,\boldsymbol{u}_0) \\ \text{s.t.} \quad & g_c(\boldsymbol{x}_c,\boldsymbol{u}_c)=0, \quad c\in\{0\}\cup C \\ & h_c(\boldsymbol{x}_c,\boldsymbol{u}_c)\le 0, \quad c\in\{0\}\cup C \\ & \left|\boldsymbol{u}_c-\boldsymbol{u}_0\right| \le \overline{\boldsymbol{u}}_c^{\max}, \; c\in C \end{aligned} \tag{13}$$

where 0 and c denote the pre-contingency state and the post-contingency state, respectively; C is the set of anticipated faults; $\boldsymbol{x}$ and $\boldsymbol{u}$ are the vector of state variables and the vector of control variables, respectively; $\overline{\boldsymbol{u}}_c^{\max}$ is the upper bound of the adjustable range of the control variables after a contingency c.

The traditional CSCOPF model includes all the post-contingency constraints, so the scale of the constraints is C+1 times that of the pre-contingency state OPF, which brings about computational issues for large-scale systems. For actual systems, local compensation of reactive power is available to quickly sustain the node voltage after a fault occurs, so this paper only considers post-contingency power flow constraints—only linear reliability constraints are added to the above OPF model—which simplifies the traditional CSCOPF model and makes it easier to solve the optimization problem. The following section introduces both the reliability constraints based on shift factors and the correction of the reliability constraints by corrective control.

3.1. Reliability constraints

In high-voltage power grids, the active power flow of transmission lines has a relatively good linear relationship, which makes it convenient to establish the connection between the active power flows before and after an anticipated fault. The devices are grouped so that the parallel transmission lines in operation or the main transformers running in parallel are divided into a group. For the devices in the same group, only one device is selected for monitoring purpose, which further reduces the number of constraints. The device is selected according to the ratio between the current carrying capacity and the branch coefficient.

The selection of the anticipated fault set usually uses N-1 security criterion in practice, mainly considering the N-1 fault of the AC lines, the main transformers, the busbars, as well as the N-1 fault of the DC lines, the converter transformers and the converter valves when VSC-HVDC stations are integrated.

According to [15], after the occurrence of an anticipated fault k, the post-contingency state active power flow of branch j can be described as a linear function of the pre-contingency state active power flow of that branch and the pre-contingency state active power of the branches switched to open state, that is:

$$P_{Bj}^{k} = P_{Bj}^{0} + \sum_{i \in S_F^k} R_{ij}^{k} P_{Bi}^{0} \quad \forall k \in C, \forall j \in S_B^k \tag{14}$$

where C is the set of anticipated faults; P_{Bj}^{k} is the active power flow of branch j after anticipated fault k occurs; P_{Bi}^{0} and P_{Bj}^{0} are the pre-contingency state active power flow of branch i and branch j, respectively; R_{ij}^{k} is the power flow shift factor from the open branch i to the closed branch j after anticipated fault k occurs; S_B^k is the set of branches in operation (i.e., the closed branches) after anticipated fault k occurs.

The shift of power flow can be calculated via AC power flow [15] or DC power flow [16].

Reliability constraints considering the N-1 security criterion are added to the OPF model:

$$-\bar{E}_{Bj} \le P_{Bj}^{k} \le \bar{E}_{Bj} \quad \forall k \in C, \forall j \in S_B^k \tag{15}$$

where $\bar{E}_{Bj}$ is the short-term allowable current carrying capacity of branch j.

3.2. Corrective Control Modification of Reliability Constraints

Based on the OPF model, the SCOPF model adds the reliability constraints of equation (14) and equation (15). The reduction in the feasible region of the optimization model will lead to an increase in the generation cost. The purpose is to ensure that the active power flow of each power device will not exceed its short-term overload capacity after an anticipated fault occurs, thus sparing enough time to handle the contingency. However, with the development of power electronics, VSC stations can quickly adjust the setting values of active power with load so that the power flow changes stably and rapidly, correcting the overloaded devices.

As mentioned earlier, the VSC active power setting also has a good linear relationship with the branch power flow. After a fault occurs, the amount of change in the VSC active power setting is proportional to the branch power flow. The sensitivity coefficient ε is used herein to represent branch power flow increment to the per unit increase in the VSC active power setting value. The sensitivity coefficient can be calculated in the same way as the shift factor. Therefore, equation (14) is modified as:

$$P_{Bj}^{k} = P_{Bj}^{0} + \sum_{i \in S_F^k} R_{ij}^{k} P_{Bi}^{0} + \sum_{l \in T^k} \varepsilon_{jl} (P_{Tl}^{k} - P_{Tl}^{0}) \quad \forall k \in C, \forall j \in S_B^k \tag{16}$$

where T^k is the set of regulatable VSC stations after fault k occurs; ε_{jl} is the sensitivity coefficient between VSC station l and branch j; P_{Tl}^{0} and P_{Tl}^{k} are the active power settings before and after the fault, respectively. Substitute equation (16) into equation (15) and we obtain a new reliability constraint considering VSC corrective control.

The post-contingency state active power setting is restricted by the adjustable range of active power by the VSC station, and the relationship with the pre-contingency setting value can be established:

$$\Delta \underline{P}_{Tl} \le P_{Tl}^{k} - P_{Tl}^{0} \le \Delta \overline{P}_{Tl} \tag{17}$$

$$\Delta \underline{P}_{Tl} + P_{Tl}^{0} \le P_{Tl}^{k} \le \Delta \overline{P}_{Tl} + P_{Tl}^{0} \tag{18}$$

where $\Delta \overline{P}_{Tl}$ and $\Delta \underline{P}_{Tl}$ are the upper and lower limits of the adjustable range of active power by VSC station l. The correction of the power flow of branch j after the fault is realized with equation (16) and equation (18). The essence is to control the power flow distribution by adjusting the active power setting of the VSC station, so that the power flow of non-fault lines can be adjusted after the fault occurs and stay within the short-term allowable current carrying capacity.

Based on the shift factor and the sensitivity coefficient between the VSC station and the branches, the post-contingency state branch power flow is expressed as a linear relationship with the relevant pre-contingency state parameters, which realizes the unification of the three stages: pre-contingency state, post-contingency state and rapid regulation of VSC stations. The solution of the model above ensures that the pre-contingency optimal operating state is the optimal solution that fully considers the regulation capability of the VSC stations. The calculation results are more economical compared with PSCOPF.

4. Case study

Based on the proposed hybrid AC/DC power grid security-constrained power flow model in which corrective control of the VSC-HVDC converter station is integrated, this paper developed a security-constrained optimal power flow software with C++ and applied it to a test case. The simulation environment is Microsoft Visual Studio 2018, and the computer hardware configuration is Intel(R) Core(TM) i7-4710MQ CPU@2.5GHz, 8GB RAM. In the simulation, the DC power flow method is adopted to calculate the branch coefficient and the power flow shift. The central path parameter of the primal-dual interior-point method is 0.2 and the tolerance is 10^{-5}.

The test case selected herein is the IEEE 14-bus system and VSC stations are implemented to form a hybrid AC/DC system, as illustrated in Figure 2 [12]. The load remains unchanged during the optimization process. Bus 1 to 14 are AC buses; bus 15 to 18 are DC buses, which are connected to corresponding AC buses through VSC stations; bus 19 is a pure DC bus, which is not directly connected to the AC buses. Both bus 15 and 19 are directly connected to generators, the generation cost coefficients of which are the same as those of the generators on bus 2 and 1, respectively, and the upper limits of the active power output are both 20MW. For the specific parameters of the VSC stations and the calculation of the losses in the VSC stations, we refer to [17]. The capacity of the VSC stations is taken as 100MW, the upper limit of the DC line capacity is taken as 50MW, and the resistance of the DC lines is taken as 0.0149p.u. (estimated with the DC cable cross-section area as 2000mm^2 and the rated voltage as ± 100kV). According to the default values of actual grids, the short-term allowable current carrying capacity is set to 1.3 times the long-term allowable current carrying capacity. For bus 1 to 7, the line capacity between those buses is set to 100MVA (shown as the dotted lines in Figure 2). The line capacity of the rest of the lines, including AC lines with transformers, is set to 50MVA. Since the active power flow is the focus of the calculation, 0.9 times the line capacity is taken as the active power flow limit. The generation output limits and generation cost coefficients of the five generators connected to the AC system are summarized in Table 1.

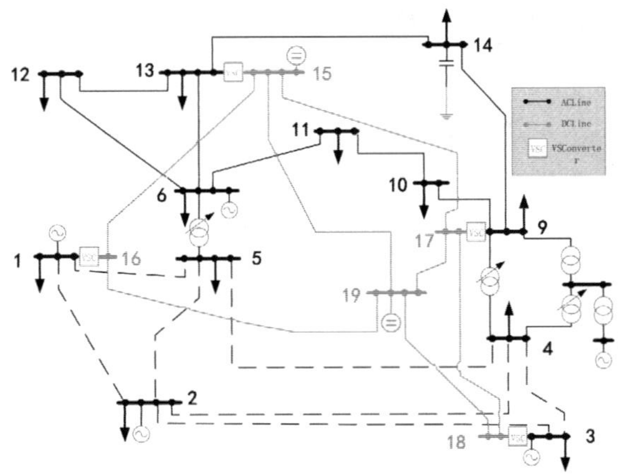

Figure 2. IEEE 14-bus system with VSC stations.

Table 1. Modified generation output limits and generation cost coefficients.

Bus no.	Pmin (MW)	Pmax (MW)	Qmin (MVar)	Qmax (MVar)	Generation cost coefficients		
					Constant ($)	Linear ($/MWh)	Quadratic ($/MWh^2)
1	0	300	0	10	0	10	0.043
2	0	150	-40	50	0	20	0.25
3	0	100	0	40	0	40	0.01
6	0	100	-6	24	0	40	0.01
8	0	100	-6	24	0	40	0.01

Five cases are selected to represent different OPF problems, as outlined in Table 2, and normalized generation costs for these cases are depicted in Figure 3.

Table 2. Details of the five cases.

Description \ Case no.	1	2	3	4	5
Does the system contain VSC stations?	No	No	Yes	Yes	Yes
Does the system contain security constraints?	No	Yes	No	Yes	Yes
Are VSC stations involved in the correction?	No	No	No	No	Yes

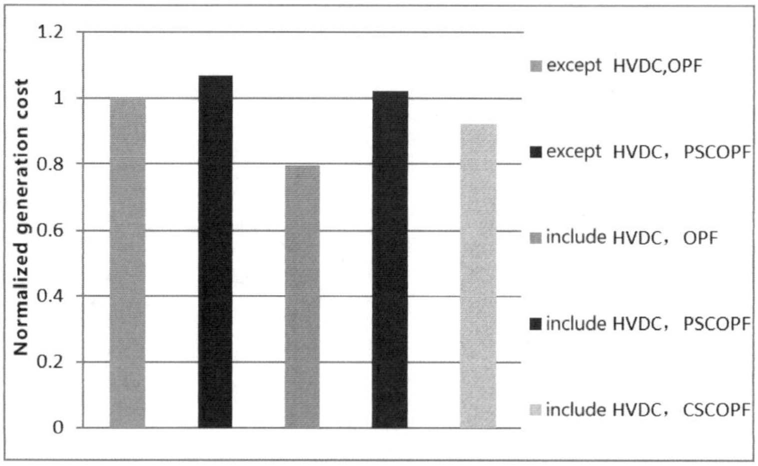

Figure 3. Normalized generation cost for different cases.

As observed in Figure 3, there are significant differences in generation costs among the five cases. In case 1 and 2, VSC stations are not included. Case 1 is the most basic OPF that ignores security constraints. Generation cost of case 1 is taken as the base value. Based on case 1, case 2 takes one step further by adding security constraints after anticipated faults occur. Compared with case 1, the generation cost increases by nearly 7%, as shown in Figure 3. This is mainly because after considering the possible impact by the line faults, part of the active power originally provided by generators at bus 1 and 2 shifts to generators at bus 6 and 8, as indicated in Table 3, while the generation costs of the latter are significantly higher than those of the former. After incorporating VSC stations, three cases are scrutinized—representing OPF, PSCOPF and CSCOPF, respectively—according to whether the security constraints are considered and whether the VSC stations participate in the corrective control. After adding the VSC stations, the generation cost of case 3 is reduced by 21%. Although VSC stations have internal losses, the losses at DC cables are relatively low and the power flow channels provided fully exploit the generation cost advantage of generators at bus 1 and 19.

Table 3. Generator active power output (MW).

Bus no. \ Case no.	1	2	3	4	5
1 (AC)	149.89	123.50	181.62	91.37	134.09
2 (AC)	44.28	22.92	38.99	39.63	38.62
3 (AC)	70.94	53.12	6.16	53.88	31.85
6 (AC)	0.00	22.54	0.00	10.84	10.20
8 (AC)	0.07	40.79	0.00	26.80	22.51
15 (DC)	\	\	19.99	19.99	6.40
19 (DC)	\	\	20.00	19.99	19.99

Comparing case 2 and 4, with the influence of the security constraints, the generation cost of the PSCOPF with VSC stations is relatively low, but it is close to that of the PSCOPF of the pure AC system. This is because the AC lines adjacent to the VSC stations need to retain a degree of transmission redundancy and the low-cost generators at bus 1 reduce the output (as indicated in Table 3), since the set of anticipated faults includes the faults at the VSC stations and the DC cables. Comparing case 4 and 5, it can be observed that the CSCOPF utilizing the post-contingency correction capability of the VSC stations does have better economic benefits under the premise of ensuring security, with generation cost reduced by 15%.

Figure 4 and Figure 5 depict the load factor curves of the DC lines and the VSC stations in different cases. Compared with case 3, due to the limitation of security constraints, the DC line load factors of case 4 and 5 are more balanced. For case 4, the addition of security constraints leads to the redistribution of generator output. When the active power settings of the VSC stations are not adjustable after a fault, it is necessary to ensure that the grid can maintain normal operation in the event of an anticipated fault. Therefore, the DC lines should retain a relatively large degree of redundancy, and hence the line load factors tend to be more conservative, as indicated in Figure 4. However, for the CSCOPF in case 5, the VSC stations are more flexible in adjustment and the DC line load factors can be higher than those of the PSCOPF, as depicted at line 18 and 19 in Figure 4, so that low-cost units are encouraged to deliver more so as to achieve better economic benefits. In the event of AC line failures, to bear the shifted active power flow, the VSC stations in case 5 have the lowest load factors to ensure the adjustment capability after the failure, as shown in Figure 5.

Based on the analysis of the modified IEEE 14-bus system that incorporates the VSC-HVDC model, it can be observed that the integration of the VSC-HVDC model to form a hybrid AC/DC power grid is rather effective in the reduction of generation cost, compared with the original AC grid. Under the premise of satisfying the security constraints, the CSCOPF that considers the correcting ability of the VSC stations has more economic benefits.

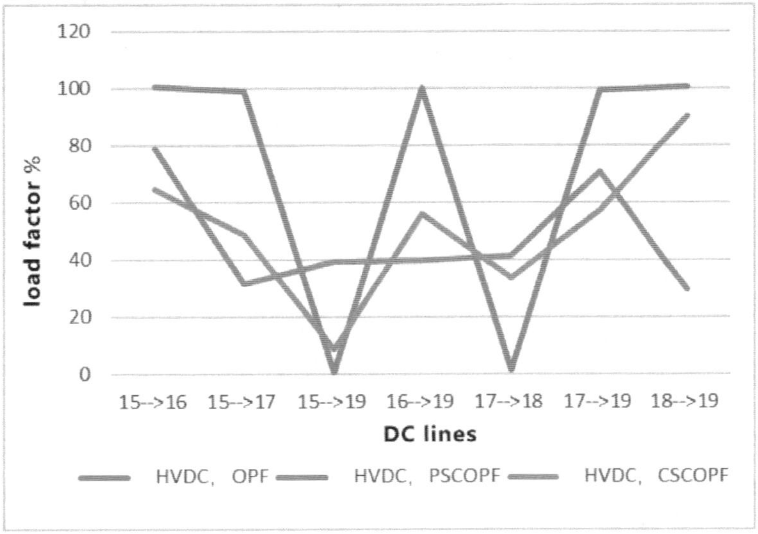

Figure 4. The load factor curves of DC lines for different cases.

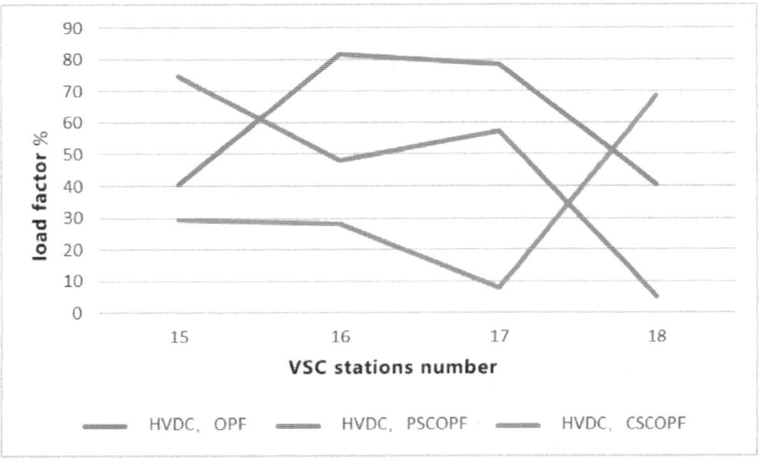

Figure 5. The load factor curves of VSC stations for different cases.

5. Conclusion

In this paper, a security-constrained optimal power flow model for hybrid AC/DC power grids which considers the VSC-HVDC corrective control is proposed, and the correctness and effectiveness of the proposed model are verified by a modified IEEE 14-bus system that integrates the VSC-HVDC model. Through the analysis of the results, the following conclusions are drawn:

- The VSC equivalent model is established in this paper, and on this basis, the OPF model is modified to incorporate the VSC model. By adding linear reliability constraints, the proposed model is simplified compared with the traditional model and the optimization problem is easier to solve.
- In the context of the reliability constraints established based on the shift factors, the linear relationship between the active power adjustment of the VSC stations and the power flows of transmission lines and electrical equipment is utilized, and the reliability constraints are corrected with the sensitivity coefficient so that the correction of the post-contingency state power flow is reflected in the constraints, fully exploiting the rapid adjustment ability of the VSC stations.
- Compared with the traditional PSCOPF, the model proposed herein takes into account the rapid correction and adjustment ability of the VSC stations after the equipment failure and has more economic benefits while satisfying security constraints.

References

[1] Scott B and Hobson E, 1978 Power system security control calculations using linear programming *IEEE Transactions on Power Apparatus and Systems* **PAS-97** 1713-20
[2] Burchett R C and Happ H H, 1983 Large scale security dispatching: an exact model *IEEE Power Engineering Review* **PER-3** 32–3
[3] Zhong S, Han X, Liu D and Wang M, 2011 Benders decomposition algorithm for corrective security-constrained optimal power flow *Chinese Journal of Electrical Engineering* **31** 65–71
[4] Capitanescu F, Glavic M, Ernst D and Wehenkel L, 2007 Contingency filtering techniques for preventive security-constrained optimal power flow *IEEE Transactions on Power Systems* **22** 1690–7
[5] Li J, Wang P and Yang L, 2014 Parallel multiple centrality correctors interior point method for ESC-OPF *Microelectronics & Computer* **31** 137–41
[6] Li Y, Zhang B, Sun H and Wu W, 2007 Security constrained optimal power flow based on nonlinear interior point method part one theory analysis *Automation of Electric Power Systems* **31** 7–13
[7] Li Y, Zhang B, Sun H and Wu W, 2007 Security Constrained optimal power flow based on nonlinear interior point method part two implementation *Automation of Electric Power Systems* **31** 6–11
[8] Wen Y, Jiang Y, Shen C, Liao J, Guo C, Wang Y and Wu Q, 2014 Research on the model and algorithm of multistage coordinated risk-based dispatch *Chinese Journal of Electrical Engineering* **34** 153–60
[9] Xu Z, Tu Q and Qiu P, 2010 New trends in HVDC technology viewed through CIGRE 2010 *High Voltage Engineering* **36** 3070–7
[10] Tang G 2010 *Voltage Source Converter based HVDC Transmission Technology* (Beijing: China Electric Power Press) pp 46–9
[11] Lin Y, Lin Z, Ju Y and Zhang M, 2017 A novel approach on security-constrained optimal power flow for hybrid AC/DC grids based on MMC-HVDC *Power System Technology* **41** 2854–61
[12] Roger W, Emil I and Göran A, 2014 Security constrained optimal power flow for HVAC and HVDC grids *2014 Power Systems Computation Conf.* Wroclaw Poland

[13] Spyros C and Göran A, 2014 Security constrained OPF incorporating corrective control of HVDC *2014 Power Systems Computation Conf.* Wroclaw Poland
[14] Vasileios S, Roger Wiget and Göran A, 2015 Security constrained optimal power flow for mixed AC and multi-terminal HVDC grids *2015 IEEE Eindhoven PowerTech* Eindhoven Netherlands
[15] Chen W, Guo R, Ye L and Sun W, 2012 Calculation model and method for thermal stability control limit to transmission interfaces in a power grid *Automation of Electric Power Systems* **36** 20–4
[16] Zhao J, Shen X, Wu L, Fu H and Xie Y, 2015 Fast evaluation method on thermal stability limit of power grid cross-section with DC power flow model *Power System Protection and Control* **43** 97–101
[17] Feng W, Tuan L A, Tjernberg L B, Mannikoff A and Bergman A, 2014 A new approach for benefit evaluation of multiterminal VSC-HVDC using a proposed mixed AC/DC optimal power flow *IEEE Transactions on Power Delivery* **29** 432–43

Acknowledgments
This work was funded by Science and Technology Project of China Southern Power Grid Corporation (GDKJXM20190139).

Electric Vehicle Charging Station Load Prediction Based on IWOA-LSSVM Combined Model with Improved Variational Mode Decomposition

Feng Ru, Xiaoqian Yang, Haoquan Zou, Lina Zhang and Xianfeng Xu *

School of Electronics and Control Engineering, Changan University, Xian, China

* xxf_chd@163.com

Abstract. Accurately predicting the electric vehicle charging station load is the basis for the optimal design of power supply planning. Aiming at the problem that the load data of electric vehicle charging station has strong randomness and volatility due to the interaction of various factors. The electric vehicle charging load prediction model is constructed, which considers the highest temperature, day type and weather type factors to select similar days. Firstly, an algorithm for optimizing variational modal decomposition parameters with relative entropy is proposed to decompose the extracted similar daily charging load data into multiple sub-sequences. Then, the decomposed data are respectively input into the combined prediction model of the least squares support vector machine optimized by the whale algorithm improved by the adaptive probability threshold and the random difference mutation strategy, and finally the predicted value of each subsequence is superimposed and summed to obtain the final predicted value. Based on the measured charging load data and weather forecast data of an electric vehicle charging station in Shanghai, simulation experiments are carried out to compare and verify the accuracy and effectiveness of the model in this paper.

1. Introduction

With the continuous expansion of the scale of electric vehicles, numerous charging loads of electric vehicle charging stations will have an important impact on the load scheduling and reliable operation of the regional power grid. Therefore, accurate prediction of the charging demand of electric vehicle charging stations can provide important information for the operation and scheduling of the power system. At the same time, it provides effective data support for the construction planning, site selection and capacity determination of charging stations [1-3].

Electric vehicle charging load has strong randomness and volatility, and is affected by many factors, which increases the difficulty of charging demand forecasting [4]. At present, there are two main categories of electric vehicle load forecasting: physical model-driven [5-7] and data-driven [8-10] charging demand forecasting methods. The prediction method based on physical model has strong generality and generalization, but it is difficult to comprehensively consider complex factors in modeling, resulting in difficult data management, poor flexibility, and it is difficult to reflect the reality [11]. Compared with prediction methods based on physical model-driven, the prediction method driven by actual data and respective characteristic data is closer to the actual load operation of the charging station. This method makes full use of deep learning and artificial intelligence technology according to the actual data, thereby improving the prediction accuracy [12].

Scholars have conducted related research on artificial intelligence algorithms in the field of electric vehicle charging station load forecasting. BP neural network is a supervised learning method suitable for multi-layer neural networks, which can obtain better prediction results, but it converges slowly and is easy to optimize locally. Least Squares Support Vector Machine [13] (LSSVM) is an extension of standard SVM, which can effectively reduce the upper limit of model generalization error and achieve better load forecasting effect. However, the kernel function parameters σ and punishment parameters γ in the LSSVM algorithm rely on empirical selection, resulting in low prediction accuracy and slow convergence speed. In order to solve this problem, the reference [14] applied the whale optimization algorithm (WOA) to optimize the parameters of LSSVM. Compared with other optimization algorithms, WOA is simple and easy to implement, with few parameters to adjust and the ability to jump out of the local optimum is strong. However, the generation of the initial population in the standard WOA algorithm is random, and the initial population is not uniformly distributed in the space, resulting in low convergence accuracy and slow convergence speed of the algorithm.

The strong volatility and randomness have a great impact on the load prediction of electric vehicle charging stations. The corresponding pre-processing of the load data with large fluctuation can greatly improve the prediction accuracy. The load is decomposed into natural mode functions with physical meaning by the time-frequency domain analysis method, in order to achieve the purpose of reducing the randomness and fluctuation of the electric vehicle charging load. Variational modal decomposition [15] (VMD) adopts a quasi-orthogonal and completely non-recursive method, applies the variational problem solving method to signal decomposition, and iteratively obtains the center frequency and bandwidth of each modal function, the local characteristics of the signal are displayed, which makes the signal more stable when stripped, and the VMD has strong anti-interference ability. Iteratively obtains the center frequency and bandwidth of each modal function, and displays the local characteristics of the signal, which makes the signal more stable when stripped, and the VMD has strong anti-interference ability. However, the parameters in the decomposition basically need to be set by experience, which makes it difficult to control the error effectively.

According to the above analysis, first of all, this paper considers the influence factors of day type, temperature and weather type, and uses the grey relational analysis method to select similar days to improve the prediction accuracy; Secondly, the improved whale optimization algorithm (IWOA) improves the ability of WOA to coordinate global search and local mining by introducing adaptive probability threshold and random difference mutation strategy, and optimizes the LSSVM kernel parameters σ and punishment parameters γ ; In order to further improve the prediction accuracy, relative entropy (also known as KL divergence) is proposed to optimize VMD decomposition to obtain the optimal K and penalty factor α . Then, the subsequences obtained by KL-VMD decomposition are respectively input into the improved whale optimization algorithm optimized least squares support vector machine (IWOA-LSSVM) prediction model, and finally the predicted values of each component are superimposed to obtain the final predicted value. The experimental results show that the IWOA-LSSVM combined prediction model based on the improved variational modal decomposition has high prediction accuracy for the actual load prediction of electric vehicle charging stations.

2. Analysis of influencing factors of electric vehicle charging load and selection of similar days

Considering that the charging load of electric vehicles will be affected by various factors during actual charging. In this paper, the day type factor, temperature factor and weather type factor are regarded as the main influencing factors affecting the charging load prediction of electric vehicles.

- Day type factor

Due to the different types of days, people's travel situations are also different, so the changes in the charging load of electric vehicles are also different. Taking the weekly load data of a charging station in Shanghai in the second week (June 12-18) and the third week (June 19-25) in June 2017 as an example, it can be seen from Figure 1 that the load changes closely related to the day type, the Nth day load curves of each week are similar.

- Temperature factor

The load fluctuation of the electric vehicle charging station is greatly affected by the temperature. For example, when the temperature is too hot or too cold in summer and winter, people will use the on-board air conditioner when using electric vehicles, which will increase the energy consumption of electric vehicles, which in turn leads to an increase in the charging load of electric vehicles. Figure 2 shows the daily load curve of a charging station in Shanghai for three consecutive Mondays on June 12, June 19, and June 26, 2017. The highest temperatures in these three days were 24°C, 29°C, and 26°C, respectively. It can be seen from Figure 2 that with the increase of temperature, the daily load curve shows an upward trend, so temperature is an important factor affecting the load characteristics of electric vehicle charging stations.

- Weather type factor

The type of weather will also have a greater impact on the charging load. It is mainly reflected in the reduction of residents' going out in extreme weather such as rain and snow, and the low speed of traffic congestion, which will lead to a reduction in the charging load of the charging station. Figure 3 shows the daily load curves on June 17 and June 27, 2017. The weather types were sunny and thunderstorms, respectively, and the average temperature was 25 °C. It can be seen from Figure 3 that the charging load in sunny days is significantly higher than that in rainy days. Obviously, this is because bad weather reduces people's travel, and the charging load of charging stations naturally decreases.

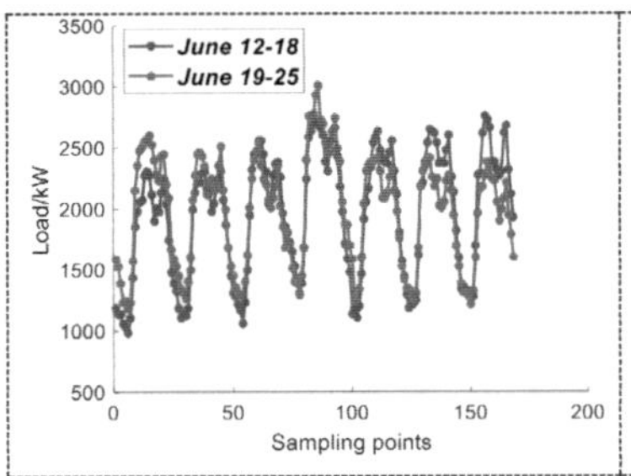

Figure 1. Weekly load comparison.

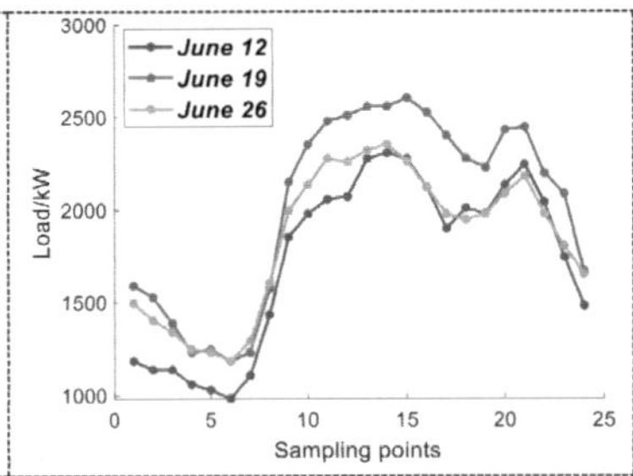

Figure 2. Daily charging load curve at different temperatures.

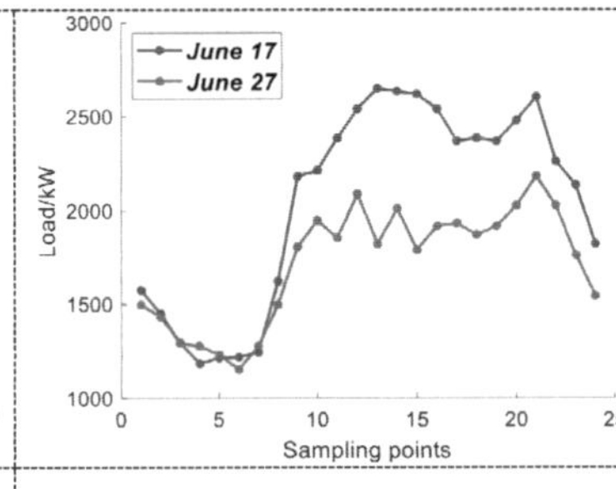

Figure 3. Daily charging load curve under different weather conditions.

Reasonable selection of similar days can improve the prediction accuracy. In this paper, the method of calculating the grey correlation degree is used to select similar days. Similar days will be selected based on the maximum temperature, day type and weather type factors, so that there is a high similarity in these features between the training samples and the predicted days.

3. Predictive model of electric vehicle charging load based on KL-VMD and IOWA-LSSVM

The flow chart of the IOWA-LSSVM combined prediction model based on improved variational modal decomposition is shown in Figure 4, The specific implementation steps are as follows:

- Step1: Pre-processing the load data of the electric vehicle charging station, and select similar days according to the grey relational analysis method.
- Step2: Use the KL-VMD algorithm to modal decomposition of the load data to obtain a series of sub-sequence IMFs.
- Step3: The IWOA-LSSVM prediction model is established for each subsequence obtained by decomposition, the optimal punishment parameters γ and kernel function parameters σ in LSSVM are selected by the improved whale optimization algorithm, and each subsequence is trained and predicted to get the predicted value of each subsequence.
- Step4: The predicted value of each subsequence is superimposed and summed to obtain the final predicted value Y.

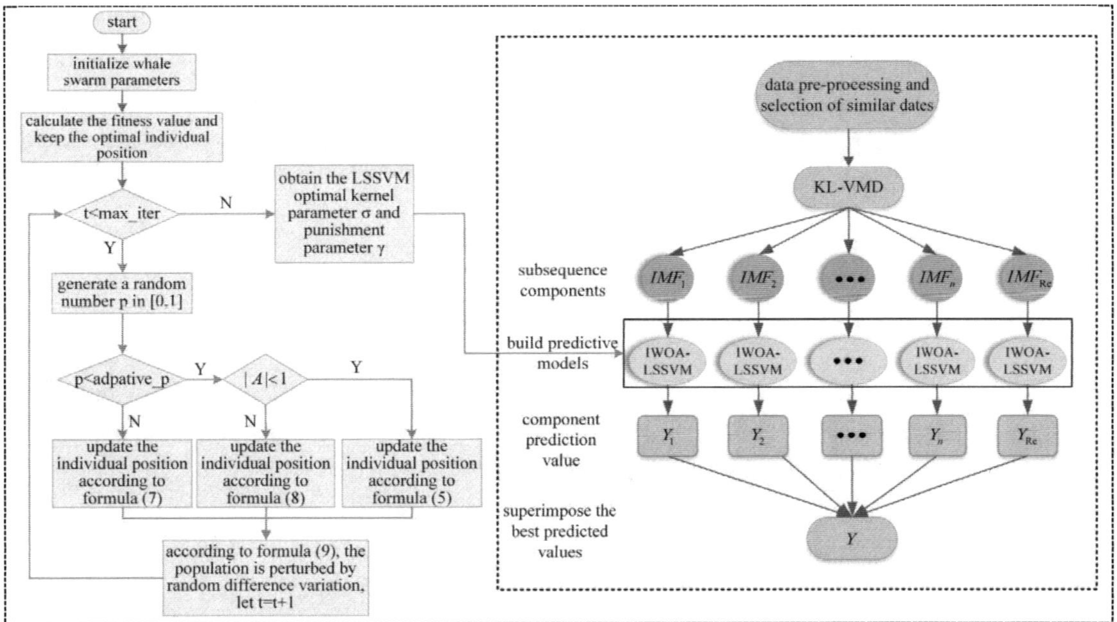

Figure 4. KL-VMD and IWOA-LSSVM combined predictive model.

3.1. Optimizing VMD parameters with relative entropy

The VMD can adaptively match the optimal center frequency and limited bandwidth of each mode to achieve effective separation of intrinsic mode components, which can decompose the original signal f into K intrinsic mode function (IMF) components. Its core idea is to construct and solve variational problems. However, before decomposing the signal, it is necessary to set the appropriate number of decomposition layers K and the penalty factor α. If the value of K is too large, it will lead to over-decomposition. Otherwise, it will be under-decomposed. If the value of α is too large, the frequency band information will be lost. Otherwise, the information will be redundant. Therefore, it is necessary to determine the optimal parameter combination [K, α].

In this paper, the relative entropy is used as the iterative stop condition, and the parameters of VMD are optimized by fixing the number of decomposition layers, and the K value and penalty factor of the VMD decomposition signal are determined. Relative entropy can be expressed as the difference between the information entropy of two probability distributions, the formula is as follows:

$$D_{KL}\left(P \| Q\right)=\sum_{i=1}^{N} P\left(x_i\right)\log\frac{P\left(x_i\right)}{Q\left(x_i\right)} \tag{1}$$

In the above formula: the $P\left(x_i\right)$ is the probability distribution of the real data, the $Q\left(x_i\right)$ and is the probability distribution of theoretical data. The difference in the distribution of the data between the two groups is proportional to the relative entropy. Therefore, by calculating the relative entropy, the degree of correlation between the reconstructed signal and the original signal can be obtained, and the optimal parameters can be selected. The specific operation method of parameter optimization is mainly divided into two steps: first, determine the penalty factor α, set the value range of modal K to [2,10], the initial value is 2, Perform variational modal decomposition on the original signal, then calculate and determine the relative entropy of the modal components, and finally obtain the optimal solution of K corresponding to the minimum relative entropy. Then according to the optimal K value, determine the optimal solution of the penalty factor α. The set range of α is [100, 2000], the step size is set to 50, and the value corresponding to the minimum relative entropy is found to be the optimal value α. The values of K and α obtained at this time are all optimal.

3.2. Least squares support vector machine

Given a set of samples $D=\{(\mathbf{x}_i, y_i)|i=1,2,3\cdots N\}$, where $\mathbf{x}_i \in R^n$ is the input vector, including historical load, temperature, day type, weather type and other factors closely related to prediction; $y_i \in R$ is the corresponding prediction output; N is the total number of samples; n is the dimension of the space vector. Map the sample dataset to the corresponding high-dimensional feature space, and construct a regression prediction model:

$$f(\mathbf{x})=\boldsymbol{\omega}^T\varphi(\mathbf{x})+b \tag{2}$$

In the above formula: the $\boldsymbol{\omega}^T$ is the weight vector in the feature space; the $\varphi(\mathbf{x})$ is the spatial mapping; b is the paranoia. According to the principle of risk minimization, the optimization model can be expressed as:

$$\begin{cases} \min \quad J(\boldsymbol{\omega},\xi)=\frac{1}{2}\|\boldsymbol{\omega}\|^2+\frac{1}{2}\gamma\sum_{i=1}^{N}\xi_i^2 \\ s.t. \quad \boldsymbol{\omega}^T \ \varphi(\mathbf{x}_i)+\mathrm{b}+\xi_i=y_i \end{cases} \tag{3}$$

In the above formula: the ξ_i is a slack variable; the γ is a punishment parameter. By introducing the Lagrange multiplication operator λ_i, using the radial basis kernel function, the optimized formula can be obtained as:

$$f(\mathbf{x})=\sum_{i=1}^{N}\lambda_i\left[\exp\left(-\frac{\|\mathbf{x}-\mathbf{x}_i\|^2}{2\sigma^2}\right)\right]+b \tag{4}$$

In the above formula: the σ is the kernel function parameter. The prediction performance of LSSVM is closely related to the kernel function parameters σ and punishment parameters γ. Improper selection of parameters will reduce the prediction accuracy and convergence speed of the algorithm. In order to improve the prediction performance, this paper will use the IWOA algorithm to optimize the σ, γ.

3.3. Improve whale optimization algorithm

Stage 1: Surrounding the prey. Its mathematical model is:

$$\mathbf{X}(t+1)=\mathbf{X}^*(t)-A\cdot|C\cdot\mathbf{X}^*(t)-\mathbf{X}(t)| \tag{5}$$

In the above formula: the t is the current number of iterations; the D is the distance between the currently obtained optimal solution and the search body; the $\mathbf{X}$ is the whale individual position vector; the $\mathbf{X}^*$ is the prey position vector; the A and C represent coefficient variables.

Stage 2: Round-up hunting. Whales wrap their prey in a spiral shape while shrinking the circle. The traditional whale optimization algorithm sets the probability of selecting a strategy to 50%. By comparing the randomly generated p in the interval [0,1] with the probability threshold, the corresponding hunting strategy is selected according to the comparison result. In the iterative process, the selection scheme of the equal probability optimization strategy may not be able to select the predation strategy suitable for the current whale population, which makes the algorithm suffer from problems such as slow convergence speed and low accuracy. Therefore, this paper introduces adaptive parameters to replace the original probability threshold. As the number of iterations of the algorithm changes, the adaptive parameters change correspondingly in [0, 1]. Through this method, the whales can choose the predation strategy suitable for the current group, thereby improving the convergence speed of the algorithm. The expression is:

$$adapative\_p=1-\frac{e^{\frac{t}{\max\_iter}}-1}{e-1} \tag{6}$$

where: the $\max\_iter$ is the maximum number of iterations.

The mathematical model of the spiral update position is:

$$\mathbf{X}(t+1)=\begin{cases}\mathbf{X}^*(t)-A\cdot D & \text{p} < \text{adapative\_p} \\ \left|\mathbf{X}^*(t)-\mathbf{X}(t)\right|\cdot e^{bl}\cdot\cos(2\pi l)+\mathbf{X}^*(t) & \text{p}\geq \text{adapative\_p}\end{cases} \tag{7}$$

In the above formula: the b is a constant coefficient that defines the shape of a logarithmic spiral; the l represents a random number in the interval [0,1].

Stage 3: Search for prey. The algorithm is set, when $|A|<1$, the whale group surrounds and attacks the prey according to formula (5); when $|A|\geq 1$, a new whale position is randomly selected, forcing the whale group to stay away from the current prey, and then looking for new prey. The random search model is:

$$\mathbf{X}(t+1)=\mathbf{X}_{rand}(t)-A\cdot\left|C\cdot\mathbf{X}_{rand}(t)-\mathbf{X}(t)\right| \tag{8}$$

In the above formula: the $\mathbf{X}_{rand}$ is the position of a randomly selected individual whale in the whale swarm.

Random differential mutation strategy. In the traditional whale optimization algorithm, the position update method of other whale individuals in the group is to generate new individuals near the current individual and the current optimal individual. If the current optimal individual is the local optimal solution, as the iterative search continues, groups will gather around the local optimal solution, resulting in a loss of group diversity, which in turn leads to premature convergence of the algorithm and low convergence accuracy. In order to solve this problem, this paper adopts the random difference mutation strategy to enhance the diversity of the population and jump out of the local optimum. Inspired by the mutation strategy of the differential evolution algorithm, the current whale individual, the current optimal individual and the randomly selected whale individual in the group are used to perform random differences to generate new individuals, and the expression is:

$$\mathbf{X}(t+1)=r\cdot\left(\mathbf{X}^*(t)-\mathbf{X}(t)\right)+r\cdot\left(\mathbf{X}'(t)-\mathbf{X}(t)\right) \tag{9}$$

where: r is a random number in the [0,1] interval; $\mathbf{X}^*$ is the current optimal individual position vector; $\mathbf{X}'$ is the individual position vector randomly selected in the group.

4. Experimental simulation and result analysis

In order to effectively verify the feasibility of the above-established model, the load data [16] and meteorological data of a charging station in Shanghai at 24 o'clock every day in May and June in 2017 were used as experimental data and applied to the above prediction model to test the prediction effect. The algorithm is programmed and implemented on the Matlab2020a version, running on a PC with 4GB memory and 2.5GHz processor. This paper uses mean absolute error (MAE), root mean square error (RMSE), and mean relative percent error (MAPE) as error evaluation indicators to evaluate the prediction results.

Using the IWOA-LSSVM combined model of the relative entropy optimized VMD for load forecasting on June 30. Firstly, according to the grey relational analysis method selecting training samples that have a high similarity to the days to be predicted. Then, using KL-VMD to decompose the load data, the decomposed subsequence data curve is shown in Figure 5.

Finally, the three sub-sequences obtained after being decomposed by KL-VMD are input into the IWOA-LSSVM prediction model respectively. In order to verify the prediction effect of the improved model in this paper, the prediction results of the traditional LSSVM, WOA-LSSVM, IWOA-LSSVM, KL-VMD-LSSVM, and BP neural network model were compared with the prediction results of the IWOA-LSSVM model decomposed by KL-VMD. A comparison of the prediction results is shown in Figure 6. It can be seen from Figure 6. that the error between the BP prediction curve and the real curve is the largest, followed by the traditional LSSVM prediction model; the absolute error gap between the prediction value of the LSSVM model optimized by the standard whale optimization algorithm and the real value at most prediction points is larger; Using the improved whale optimization algorithm to optimize the LSSVM prediction curve and the LSSVM prediction curve decomposed by KL-VMD will have a relatively large vibration amplitude when the real data fluctuates violently, but the absolute error with the real curve is relatively small on the whole; However, after KL-VMD decomposition and using the IWOA to optimize the LSSVM, the absolute error between the predicted curve and the real curve is smaller. Except that the predicted value of individual predicted points has a certain deviation from the sampling point, the other predicted values are very close to the real values.

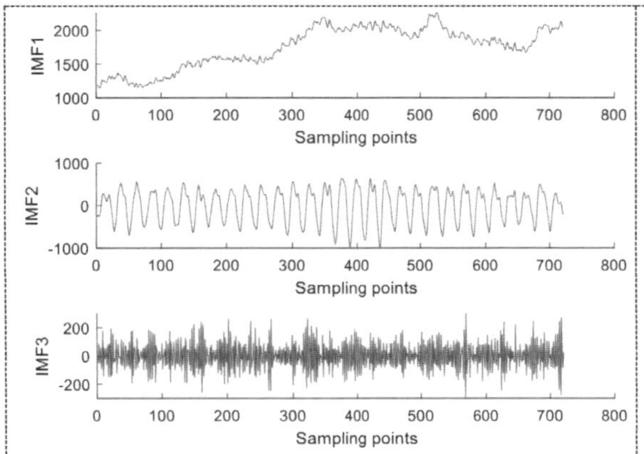

Figure 5. KL-VMD decomposition result.

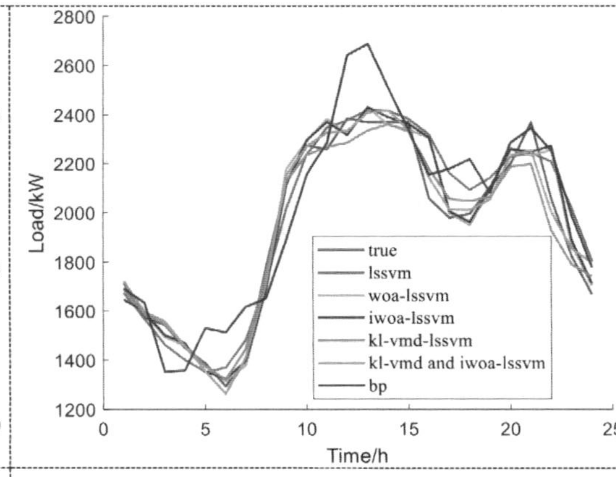

Figure 6. Comparison of prediction results of each algorithm.

In order to have a more intuitive display of the prediction performance of several prediction models, three error evaluation indicators are used to evaluate the prediction results. The specific evaluation values are shown in Table 1.

Table 1. Comparison of error evaluation indicators of six prediction models.

	MAE(kW)	RMSE(kW)	MAPE(%)
BP	405.2113	484.1768	20.7379
LSSVM	83.0275	104.1003	4.1347
WOA-LSSVM	60.8761	85.254	3.2
IWOA-LSSVM	58.9334	84.188	3.0843
KL-VMD-LSSVM	46.1447	63.3093	2.4343
KL-VMD and IWOA-LSSVM	**42.3426**	**52.7083**	**2.196**

It can be seen from Table 1 that the prediction effect of the BP model is the worst. Compared with the BP model, the LSSVM prediction effect is better, and the performance of the WOA-LSSVM is much improved, which once again proves the necessity of optimizing the LSSVM prediction model. The improved IWOA optimization ability is further improved, the average values of the three error evaluation indicators (MAE, RMSE, MAPE) of IWOA-LSSVM are 58.9334, 84.188, 3.08433, while the prediction model error evaluation indicators after combining KL-VMD and IWOA-LSSVM have decreased by 16.5908, 31.4797, 0.8883. This demonstrates the effectiveness and accuracy of the combined prediction model of KL-VMD and IWOA-LSSVM.

5. Conclusion

Aiming at the complexity of electric vehicle charging demand, a prediction model of electric vehicle charging demand based on improved variational modal decomposition and IWOA-LSSVM is proposed. Firstly, by analyzing the influencing factors of electric vehicle charging load, the gray correlation analysis method is used to select similar days. Then, the preprocessed load data is decomposed into a series of subsequences by KL-VMD. Finally, the decomposed data are input into the IWOA-LSSVM model respectively, and the predicted values of each subsequence are superimposed and summed to obtain the final predicted value. Comparative experiments based on real data show that the method in this paper has achieved good prediction results and is an effective method for forecasting electric vehicle charging demand.

6. References

[1] Zhang Y, Li W, and Du X. Outlook of Electric Vehicles and Grid Interaction in Energy Internet. *2019 IEEE 3rd International Conference on Green Energy and Applications (ICGEA)*. IEEE, 2019.

[2] Li F, Dou C, and Xu S. Optimal scheduling strategy of distribution network based on electric vehicle forecasting [J]. *Electronics*, 2019, 8(7): 816.

[3] Cao F, Li S, and Zhang Y. Simulation of EV charging load space-time distribution considering the attractiveness of charging stations [J]. *Power Grid Technology*, 2021, 45 (1): 75-87.

[4] Zhu J, Yang Z, Mourshed M, Guo Y, Zhou Y and Chang Y. Electric vehicle charging load forecasting: A comparative study of deep learning approaches [J]. *Energies*, 2019, 12 (14): 2692.

[5] Guo G, Xu T, Han Y H, and Zhao Q. A review of grid-transportation network collaborative optimization in the era of electric vehicles [J]. *Control and Decision*, 2021, 36(9): 2049-2062.

[6] Wu C S, Gao S, Liu Y, Song T E, and Han H. A model predictive control approach in microgrid considering multi-uncertainty of electric vehicles [J]. *Renewable Energy*, 2021, 163: 1385-1396.

[7] Wang H L, Zhou G B, Xue R, Lu Y J, and McCann J A. A driving-behavior-based SOC prediction method for light urban vehicles powered by supercapacitors [J]. *IEEE Transactions on Intelligent Transportation Systems*, 2020, 21(5): 2090-2099.

[8] Munshi A A. and Mohamed Y A R I. Unsupervised nonintrusive extraction of electrical vehicle charging load patterns [J]. *IEEE Transactions on Industrial Informatics*, 2018, 15(1): 266-279.

[9] Morlock F, Rolle B, Bauer M, and Sawodny O. Forecasts of electric vehicle energy consumption based on characteristic speed profiles and real-time traffic data[J]. *IEEE Transactions on Vehicular Technology*, 2019, 69(2): 1404-1418.

[10] Wang R, Gao X, Li J L, Xu J H, Ai G Q, and Jing X. Electric vehicle charging load prediction method based on cluster analysis [J]. *Power System Protection and Control*, 2020, 48(16): 37-44.

[11] Scheubner S, Thorgeirsson A T, Vaillant M, and Gauterin F. A stochastic range estimation algorithm for electric vehicles using traffic phase classification [J]. *IEEE Transactions on Vehicular Technology*, 2019, 68(7): 6414-6428.

[12] Tao Y, Huang M, Chen Y, and Yang L. Orderly charging strategy of battery electric vehicle driven by real-world driving data [J]. *Energy*, 2020, 193: 116806.

[13] Zhang H, Yang Y, Zhang Y, He Z, Yuan W, Yang Y, et al. A combined model based on SSA, neural networks, and LSSVM for short-term electric load and price forecasting [J]. *Neural Computing and Applications*, 2021, 33(2): 773-788.

[14] Chen J, Zhuo X, Xu F, Wang J, Zhang D, Zhang L. A novel multi-classifier based on a density-dependent quantized binary tree LSSVM and the logistic global whale optimization algorithm [J]. *Applied Intelligence*, 2020, 50(11): 3808-3821.

[15] Dragomiretskiy K, Zosso D. Variational mode decomposition[J]. *IEEE transactions on signal processing*, 2014, 62(3): 531-544.

[16] Huang J J. Research on short-term load forecasting and scheduling of new energy electric vehicle charging stations [D]. *Shanghai Jiaotong University*, 2019.

Journal of Physics: Conference Series **2320** (2022) 012008 doi:10.1088/1742-6596/2320/1/012008

The Method for Automatic Adjustment of AGV's PID Based on Deep Reinforcement Learning

Yubin Chen[1], Dancheng Li[1], Huagang Zhong[2] and Ruihang Zhao[1]

[1] Software College, Northeastern University, Shenyang, China
[2] Four Leaves Clover(Suzhou) Intelligent Technology Co.,Ltd, Suzhou, China

Ldc@mail.neu.edu.cn, 2071267@stu.neu.edu.cn

Abstract. This paper proposes a method for automatic adjustment of PID(Proportion Integration Differentiation) based on deep reinforcement learning in order to solve the problem of smooth movement of AGV(Automated Guided Vehicle). First, based on reinforcement learning, the problem of AGV smooth operation is transformed into the solution of PID adjustment operation sequence. The action value model is constructed using the Deep Q-Learning Network(DQN) algorithm. Then, take the AGV adjustment PID as the research object. The goal is to achieve smooth movement of AGV. The specific design of AGV automatic PID adjustment method based on deep reinforcement learning is introduced. Finally, the simulation data is used to train the AGV action value network model. The method is validated in the ROS(Robot Operating System) simulation environment. Then compare with the actual environment. The results show that the automatic adjustment method of AGV's PID based on deep reinforcement learning can adjust PID to the optimum without manual work. It can make AGV run smoothly. The feasibility and effectiveness of the method are illustrated and the problem of automatic PID determination in AGV movement control is solved.

1. Introduction

About AGV control algorithm, PID algorithm is a basic and important part. PID algorithm is simple to use. However, it is difficult to positive parameters. It is necessary to constantly try to find the best parameters. PID algorithm not only needs to consider the current state value of the control object, but also needs to consider the state value of the control object in the past period of time and the state value change in the most recent period. The output control signal is jointly determined by these three aspects. The result of PID control algorithm is a set of values which are used to control the state of the controlled object in various jobs, such as the control of running speed, turning speed, acceleration, etc. There are two types of parameter adjustment methods for PID control. One is the theoretical calculation setting method. It is mainly based on the mathematical model of the system to determine the controller parameters through theoretical calculation. The calculated data obtained by this method may not be used directly. They must be adjusted and modified through engineering practice. The second is the engineering setting method. It mainly relies on engineering experience and is carried out directly in the test of the control system. The method is simple and easy to master. It is widely used in engineering practice. The engineering setting method of PID controller parameters [1] mainly include critical proportional method, response curve method and decay method. Each of the three methods has its own characteristics. And the common point is to pass the test. Then the controller parameters are set according to the engineering experience formula. But no matter which method is adopted, the controller parameters obtained need to be adjusted and perfected in the actual operation. The critical ratio method is now generally used.

After AGV leaves the factory, it needs to be initialized. AGV movement is controlled using a series of PID parameters so that it can run smoothly. If the movement situation deviates, it needs to be re-adjusted in the subsequent use process [2]. In the production environment, the method of engineering tuning is generally used to determine PID parameters. This needs to rely on engineering experience. According to the historically set PID parameters, the best PID parameters are determined through continuous experiments and adjustments. In this process, a lot of labor costs need to be invested. To set the parameters once, you need to observe the movement of AGV. Then make adjustments and repeat the experiment many times to get the optimal parameters. And tiny deviations in movement are indistinguishable to the naked eye. Every AGV is unique. The environment in which AGV works has a variety of different influencing factors, such as ground material, friction, etc. If the existing method is adopted, PID parameters need to be modified many times for each AGV. Moreover, the discovery and summary of parameter adjustment rules, the accumulation of experience and the setting of parameters all require the participation of personnel, the consumption of resources and time. Therefore, it is of great significance to study an efficient method for AGV to automatically adjust PID for improving the efficiency of use and intelligent transformation.

2. Deep reinforcement learning

Reinforcement learning [3] has achieved many successes in games, robotics, smart homes, etc [4, 5, 6]. Some technicians use reinforcement learning to solve the navigation and path planning problems of AGV [7, 8, 9]. Reinforcement learning tasks can use Markov Decision Process to describe the task. It learns by trial and error through an agent. Reward obtained by interacting with the environment guide behavior so that the agent gets the maximum reward. In reinforcement learning, the reinforcement signal provided by the environment is an evaluation of the quality of the action instead of telling the reinforcement learning system how to generate the correct actions. It enables the algorithm to make the correct decision on the adjustment of PID.

Q-Learning is one of the classic algorithms of reinforcement learning. It uses Q table to store the Q value of the action value function. It can solve the problem that the state space and action space are discrete and the dimension is not high. If the state space and action space dimensions are high, Q table will be very large and take up a lot of space [10]. Google DeepMind proposed a combination of Q-Learning and deep neural network, namely Deep Q-Learning Network (DQN). It can solve problems with high dimensionality of state space and behavior space. DQN introduces experience pools [11] that support experience replay [12]. It makes up for the problem of insufficient data during training. And two network models with the same structure are constructed which make the process of network training more stable. In this paper, DQN is used to realize the automatic PID adjustment method of AGV.

3. The method for automatic PID adjustment of AGV based on deep reinforcement learning

Automatic adjustment method of AGV's PID is to find the appropriate PID through adjustment. And the ultimate goal is to make AGV run smoothly. The process of AGV adjusting PID is as follows. Randomly set PID to AGV. AGV runs on a given path. The running process is evaluated after running to the target point. And adjust PID. Set new PID for AGV. Repeat the above operations until AGV running process meets requirements and achieves the goal of smooth movement. The PID adjustment completion process of AGV is actually a combined sequence of multiple sequential operation actions. Solving the most suitable PID can be transformed into solving the sequential operation action sequence of each AGV. Reinforcement learning is an effective way to solve such decision-making problems. As shown in Figure 1, AGV is abstracted into a single agent. The route of AGV movement is abstracted into the environment. The action of the agent is the operation action of PID. The state observed by the agent is the set PID information. Reward function is designed through the objective function to guide AGV to quickly adjust PID.

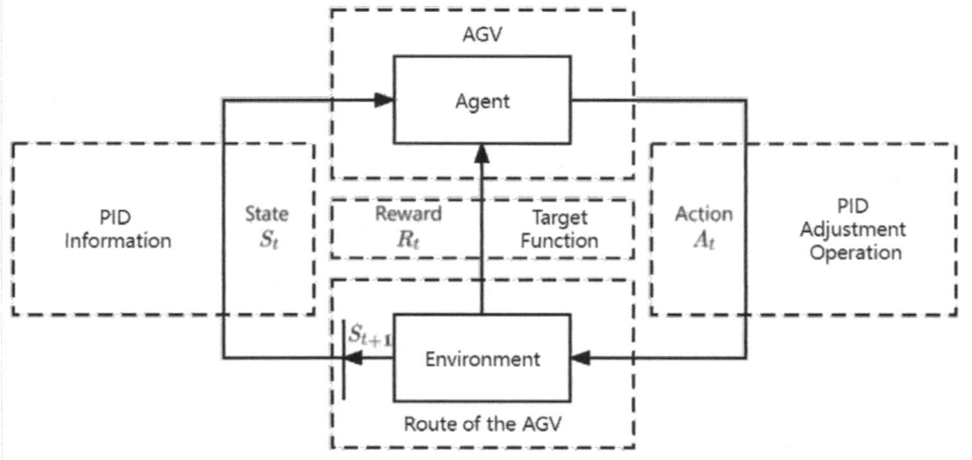

Figure 1. The mechanism of PID adjustment of AGV based on reinforcement learning

Under the framework of reinforcement learning, the adjustment operation of PID not only affects the current reward, but also affects the state of the environment at the next moment and the final accumulated reward value. In order to maximize the final cumulative reward value, the globally optimal action must be selected in each state. Therefore, the adjustment problem of PID can be transformed into the problem of solving the optimal strategy. DQN is used to solve the action value network to indirectly solve the optimal policy.

4. The specific design of automatic adjustment PID method

Taking AGV to adjust PID as the research object, the goal is to achieve smooth movement of AGV. This paper introduces the automatic adjustment method of AGV's PID based on deep reinforcement learning. A Markov Decision Process is a quintuple consisting of $<S, A, P, R, \gamma>$. S is an action set. A is an action set. P is the state transition probability matrix. R is a reward function. γ is a reward discount factor. Markov Decision Process introduces policy π. Policy π is a probability distribution based on a set of actions in a state. The core of automatic PID adjustment method is the trained Q network. Basic elements such as state space, action space, reward function, the structure of DQN and training method need to be designed.

4.1. State space

The state space contains all the states in the environment and is the input of the Q network model. For AGV automatic adjustment of PID problem, the required environmental information is a series of PID set. For example, Pd, Id, Dd for position control, Pv, Iv, Dv for speed control and Pa, Ia, Da for angle control. In order to find a suitable PID more quickly, it is necessary to limit the range of the PID. When the boundary is exceeded, the boundary value is taken as the adjusted PID. In this paper, the range of Pd is set to [0, 5], and the range of Pa is set to [0, 0.1].

4.2. Action space

The action space is the set of all actions that the environment interacts with the agent. For the problem of AGV automatically adjusting PID, the action space is a series of operation combinations for PID. Each parameter can perform three actions, plus or minus or remain unchanged. In this paper, there are two PID parameters which are Pd and Pa. And there are a total of 9 actions. The adjustment step of Pd and Pa are set to 0.1 and 0.005 respectively. The set action space is shown in Table 1.

Table 1. Action space

Pd	+	+	+	-	-	-	none	none	none
Pa	+	-	none	+	-	none	+	-	none
Action Value	0	1	2	3	4	5	6	7	8

4.3. Reward function and reward discount

Reward function is the rule by which the environment evaluates the action currently selected by the agent. It plays an important role in learning the network model and improving the strategy. Reward function affects how fast and how well the network model converges. We set the reward value as shown in Equation 1. If AGV runs on a given straight line and the sum of the positional deviations (dy) at each moment is less than or equal to the specified threshold 5, it is considered that AGV can run smoothly. The PID set this time is appropriate. Hence a positive reward value of 1 is given. In each round, a PID adjustment is performed. If the AGV running result does not meet the requirements of smooth movement, a negative reward value of -1 will be obtained.

$$R = \begin{cases} 1, & \text{not reach final state}, \ dy \le 0.5 \\ -1, & \text{reach final state}, \quad\ dy > 0.5 \end{cases} \tag{1}$$

The cumulative sum of reward values is the decaying sum of all reward values when sampling from a certain state S_t to the termination state in Markov Decision Process. The total reward value function obtained in each iteration is shown in Equation 2.

$$G_t = R_{t+1} + \gamma R_{t+2} + \ldots = \sum_{k=0}^{\infty} \gamma^k R_{t+k+1} \tag{2}$$

Introducing a reward discount makes the contribution of a subsequent state to the current state harvest less than its reward. When γ=0, it indicates that the harvest in a certain state is the reward obtained by the current state. It does not consider subsequent states. When γ=1, it indicates that all subsequent states will be considered which is a long-term vision behavior. This paper sets the reward discount factor to 0.9.

4.4. Strategy

The policy can guide the agent to choose the optimal action. The greedy strategy can ensure that the behavior selected by an individual in a certain state is the one that can reach the state with the greatest value among all possible subsequent states. ε-greedy strategy [13] is shown in Equation 3. Its basic idea is to ensure continuous exploration. By setting an ε value less than 1, the probability of 1−ε is used to greedily select what is currently considered to be the maximum behavioral value. And randomly select actions from all m alternative actions with probability ε. This paper sets ε to 0.1 and m to 9.

$$\pi(a|s) = \begin{cases} \varepsilon/m + 1 - \varepsilon & , \text{if } a^* = \arg\max_{a \in A} Q(s,a) \\ \varepsilon/m & , \text{else} \end{cases} \tag{3}$$

4.5. The structure of DQN

DQN adopts two networks with the same structure and different parameters. One is the evaluation network which is used to calculate the Q estimated value. The other is the target network which is used to calculate the Q target value. The deep reinforcement learning DQN model consists of an evaluation network, a target network, an experience pool and a loss function. The evaluation network and target network include input layer, hidden layer, fully connected layer and output layer. The connection between the input layer and the hidden layer is performed through the first fully connected layer. We define the forward function to connect the input layer to the hidden layer and use the activation function ReLu to process the value after passing through the hidden layer. Then we set up a second fully connected layer to connect the hidden and output layers. The number of neurons in the input layer is the number of states (2). The number of neurons in the hidden layer is set to 50. The number of neurons in the output layer is the number of actions (9). The input to the network model is the state, that is, a set of PID parameters that are set. The output is an action, which is a value represents a series of adjustments to the PID parameters. The Q evaluation value and the Q target value combined with the loss function are used to update the parameters of the evaluation network. The parameters of the evaluation network are synchronized to the target network for each fixed steps of training. The specific network structure is shown in Figure 2.

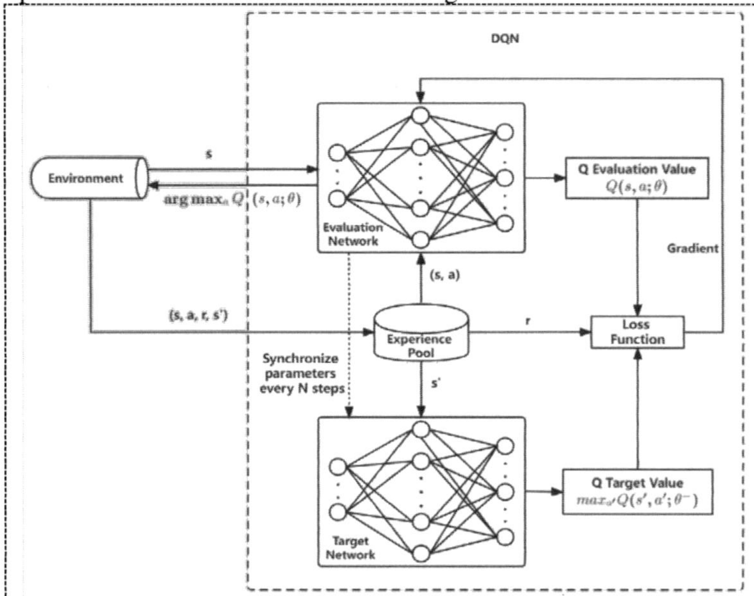

Figure 2. The structure of DQN

4.6. The training method of DQN

Using the DQN network model for training, the flow chart of AGV adjustment PID method is shown in Figure 3.First, initial parameters need to be set. The initial parameters include the number of learning counter, the capacity of the experience pool, the update frequency of target network, the number of samples for network learning, the degree of greed, the reward discount factor, the parameter adjustment step size and the parameter accuracy. Establish DQN evaluation network and target network. Define the action selection function. The input is state s. Adopt ε-greedy decision. Greedy if the randomly generated value is less than the greedy degree. The optimal action is selected through model evaluation. Explore if the randomly generated value is greater than or equal to the greedy degree. The action is chosen randomly and the function returns the chosen action. Define the experience storage function to store the relevant parameters (state s, action a, reward r, next state s') in the experience pool. If the learning times counter reaches a multiple of the target network update frequency value, use the evaluation network parameters to update the target network parameters. Define the learning function. Extract batch data from experience pool. Get the evaluation value and target value. And use loss function and optimizer to update evaluation network parameters.

Initialize the initial state. Randomly generate PID. Actions are chosen according to ε-greedy policy. Get the next state PID. The PID is sent to AGV through the topic communication mechanism of ROS. The AGV runs and obtains information such as x, y, theta, end point coordinates and the running end flag in the running process. Calculate the reward value r. Save the current state, action, reward and next state to the experience pool. If the number of storage is greater than the number of begin training, random samples are taken from the experience pool. Calculate the Q value of the current state s using the evaluation network. Calculate the Q value of the next state s' using the target network. Combining two values and using gradient descent and mean squared loss function to calculate the loss. If the number of learning count reaches the target network update steps at this time, the evaluation network parameters will be updated to the target network. If the number of learning count has not been reached, judge whether the PID set in the current state is the final state or whether the PID has been adjusted 50 times according to the running result of AGV. If not, continue to adjust PID. If so, proceed to the next training loop until the end of the loop.

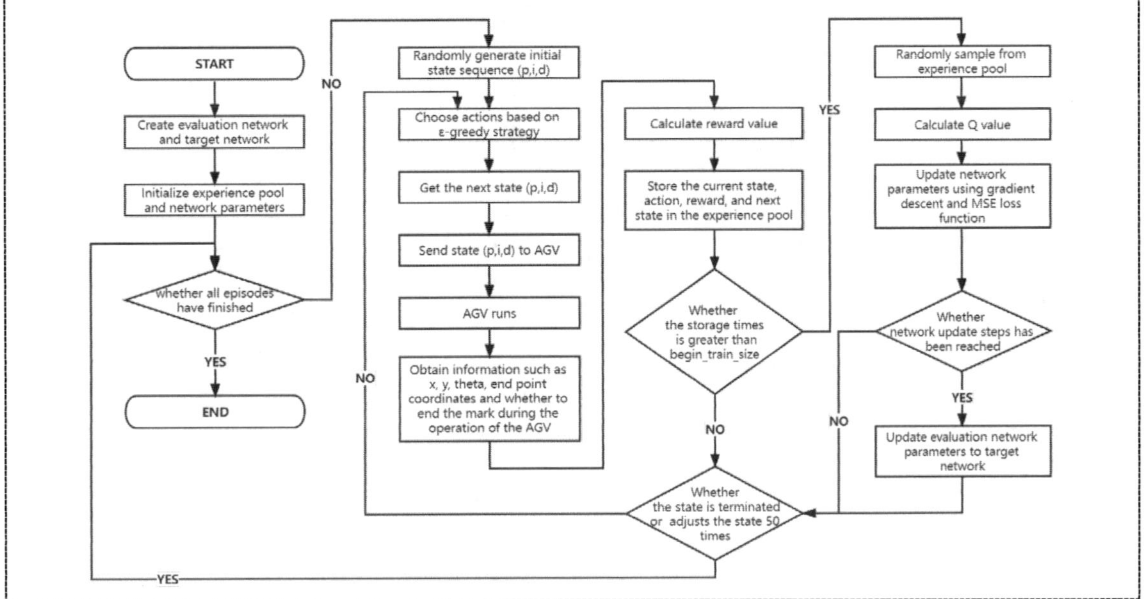

Figure 3. The process of method

5. Simulation experiment and analysis

This paper uses the deep learning framework PyTorch and Python3.6 to build the DQN network model and training algorithm. And we build the simulation environment based on the gazebo physics simulation platform [14] and topic communication mechanism [15] of ROS [16]. The model training parameter settings are shown in Table 2.

Table 2. Parameters for model training

γ	0.9
ε	0.1
learning_rate	0.001
batch_size	32
memory_capacity	10000
target_replace_iter	500
begin_train_size	2000

5.1. Simulation experiments

As shown in Figure 4, there are a total of four paths in the simulation environment. They form a square loop. Each path has the same length (10m) and is a straight line. After each task is completed, it rotates and can run on the next straight line. Finally, it returns to the origin through four straight lines so as to ensure that the movement of AGV in the simulation environment can be seen intuitively in the interface. After each task, the running process needs to be evaluated. The position deviation and angle deviation need to be analyzed.

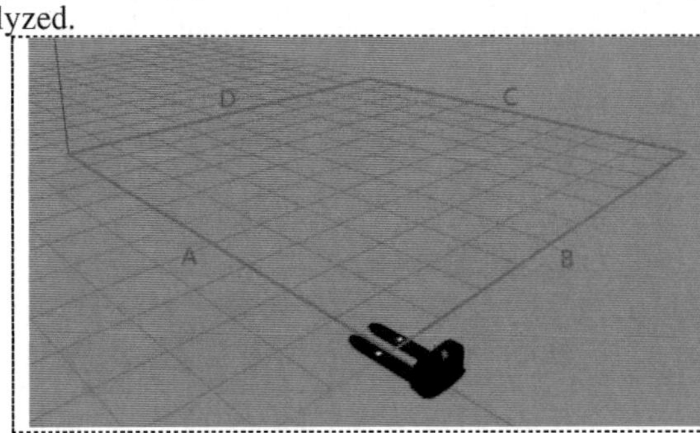

Figure 4. Simulation environment

We set up a training of 5000 episodes and add the test during the training process. The model is tested every 10 times of training. Test process uses the current model for 8 episodes and records the cumulative sum of the average reward value. Store AGV running track graph of each episode during the experiment.

5.2. Experiment results and analysis

Figure 5 is a graph of the cumulative sum of rewards for the training process. Figure 6 is a graph of the cumulative sum of average rewards for the test. It can be seen from the two figures that as the number of training increases, the cumulative sum of reward values gradually increases and the trend gradually converges. It indicates the feasibility and effectiveness of the automatic PID adjustment method based on deep reinforcement learning.

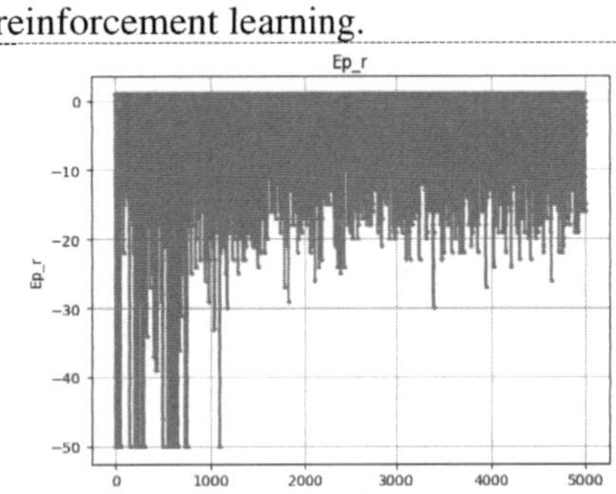

Figure 5. Sum of reward values in training

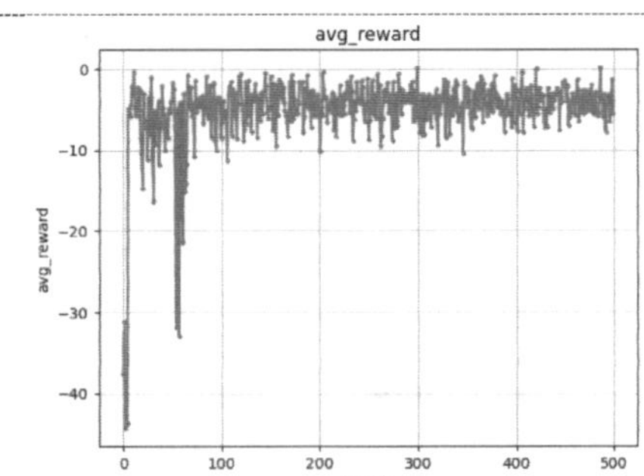

Figure 6. Sum of average reward values in the test

In the early stage of model training, because the model does not know how to make a correct decision, there will be many situations that cannot make the AGV run smoothly. In other words, the appropriate PID has not been adjusted after 50 adjustments in an episode. We record the running process information of AGV after setting the PID in each episode and draw them in a graph for comparison. The AGV running track of some experiments are shown in Figure 7-9.

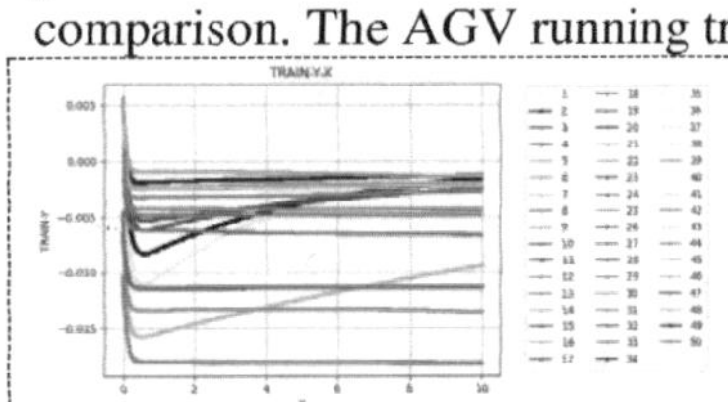

Figure 7. Running track(a)

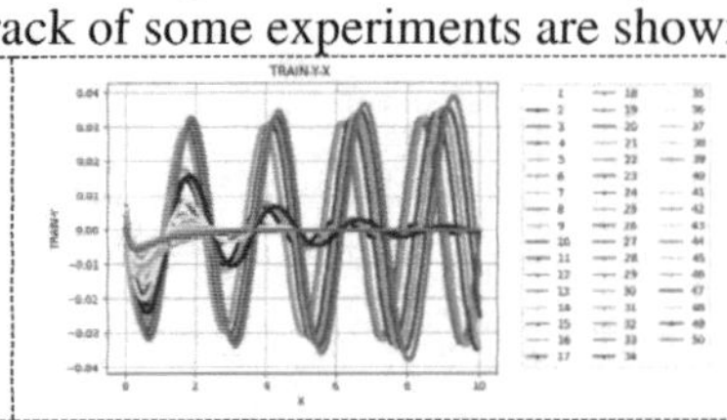

Figure 8. Running track(b)

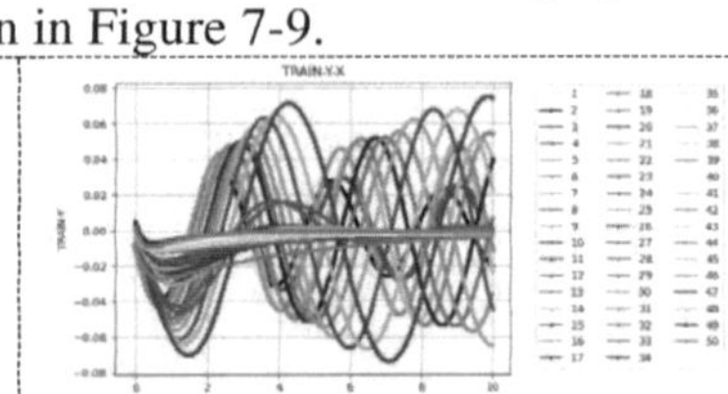

Figure 9. Running track(c)

After the model is trained for a period of time, the model can well adjust the PID to the appropriate level. And AGV can run stably on the specified path. AGV running tracks of some experiments are shown in Figure 10-12.

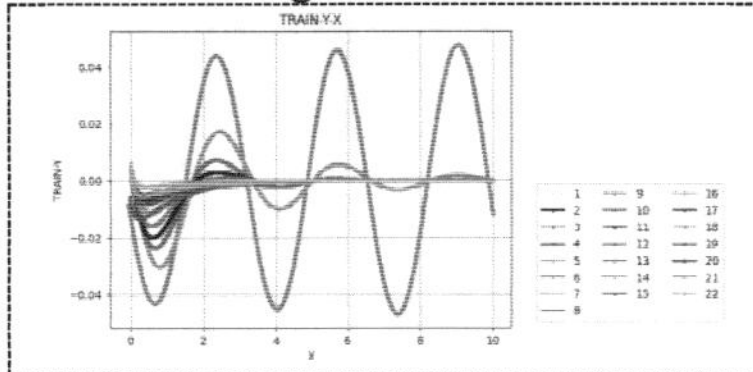

Figure 10. Running track(d)

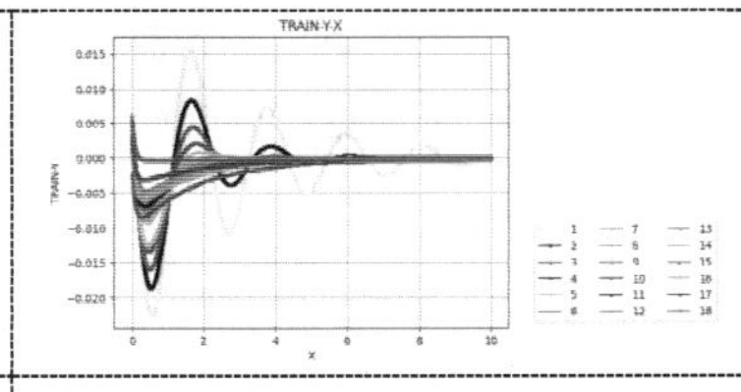

Figure 11. Running track(e)

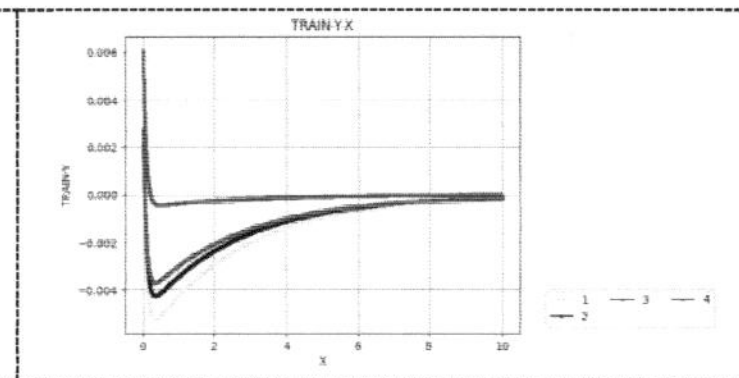

Figure 12. Running track(f)

5.3. Method comparison

In the actual production environment, PID is adjusted manually using the engineering experience method. This method requires the participation of personnel. It takes about 10 hours for technicians to adjust PID of each AGV. But using the artificial intelligence method can release labor, reduce unnecessary time consumption and promote the intelligent development of the production environment. Using the automatic adjustment PID method based on deep reinforcement learning only takes a short time to solve the problem after training the model. PID that enables the AGV to run smoothly is not the only solution. Only a suitable PID can be found in the end while Using the manual method to make adjustments within the known fuzzy range. Since the method proposed in this paper is to randomly generate PID for adjustment and training, multiple suitable PID can be obtained finally.

6. Conclusion

Aiming at the problem of automatic PID adjustment of AGV to achieve smooth operation, this paper proposes a method for AGV to automatically adjust PID based on deep reinforcement learning. Through the ROS simulation environment, the AGV running process information was obtained. The reinforcement learning algorithm was used to build the DQN model. The method was verified and compared with the actual scene. The results show that the method can realize the adjustment of PID without manual participation and can adjust the PID to an appropriate value. In the following research, we will tune the network model parameters of DQN in order to adjust the PID faster and achieve smooth movement of AGV.

7. References

[1] Xue Xu. Research on PID parameter engineering tuning method[J]. *China Chlor-Alkali*, 2016(09):44-46.

[2] Ganesharajah T, Hall N G, Sriskandarajah C. Design and operational issues in AGV-served manufacturing systems[J]. *Annals of Operations Research*, 1998, **76**:109-154.

[3] Sutton R, Barto A. *Reinforcement Learning:An Introduction*[M]. MIT Press, 1998.

[4] Oroojlooyjadid A, Nazari M R, Snyder L V, et al. A Deep Q-Network for the Beer Game: Deep Reinforcement Learning for Inventory Optimization[J]. *Manufacturing & Service Operations Management*, 2021.

[5] Tingfei Huang, Guangquan Cheng, Kuihua Huang, et al. Task allocation method for multi-type interception equipment composite anti-UAV based on DQN[J]. *Control and Decision*, 2022, **37**(1):9.

[6] Xin Deng, Jun Na, Handuo Zhang, Yulin Wang, Bin Zhang. Personalized adjustment method of smart lights based on deep reinforcement learning[J]. *Computer Engineering and Applications*, 2022, **58**(06):264-270.

[7] Hui Liu, Ke Xiao, Jingbo Wang. Multi-AGV path planning method based on multi-agent reinforcement learning[J]. *Automation and Instrumentation*, 2020, **35**(02):84-89.

[8] Xiangyu Gu, Zhiyuan Peng, Kaiyi Xian, Liping Yang, Wei Chen, Juan Zou, Manlu Liu. Planning method of orbital inspection robot based on deep reinforcement learning[J]. *Industrial Instrument and Automation Device*, 2021(01):73 -76.

[9] Shijing Lin, Anjun Xu, Cheng Liu, Kai Feng, Ji Li. The scheduling method of cranes in steelmaking workshop based on deep reinforcement learning[J]. *China Metallurgical*, 2021, **31**(03):37-43.

[10] Watkins, C.J., Dayan, P. sTechnical Note: Q-Learning[J]. *Machine Learning*, 1992, **8**(3-4):279-292.

[11] Csaba Szepesvari. Algorithms for reinforcement learning[M]. *Synthesis lectures on artificial intelligence and machine learning*, 2010, **4**(1):1-103.

[12] Mnih V， Kavukcuoglu K，Silver D，et al. Human level control through deep reinforcement learning[J]. *Nature*, 2015, **518**(7540):529-533.

[13] Gomes E R, Kowalczyk R. Dynamic analysis of multiagent Q-learning with ε-greedy exploration[C]. *ACM*, 2009.

[14] Pingping Hao, Lifeng Yao. Simulation realization of mobile robot path planning based on Player/Gazebo[J]. *Education*, 2010(10):138.

[15] Quigley M, Conley K, Gerkey B P, et al. ROS:an open-source Robot Operating System[C]// *ICRA Workshop on Open Source Software*, 2009.

[16] Wiki. ROS Tutorials[EB/OL]. http://wiki.ros.org/ROS/Tutorials.

Acknowledgments

This work was financially supported by the Project of China's Ministry of Education and supported by Alibaba in 2020 (Grant No. 202002001063). Heartfelt thanks also to the reviewers for their constructive comments to the paper.

Journal of Physics: Conference Series **2320** (2022) 012009 doi:10.1088/1742-6596/2320/1/012009

Magnetic Field Crosstalk Suppression Method on Superconducting Quantum Chips

Yan Ye[1], Jiangtao Liu[2], Xiaoqiang Hou[2], Jianqiong Zhang[3] and ZhiZhuo Huang[3]

[1]Beijing Metro Operation Administration Corporation Limited. Beijing, China
[2]CRRC Qingdao Sifang Co.,Ltd. Qingdao, China
[3]Southwest Jiaotong University. Chengdu,China

13001957600@163.com

Abstract. The arithmetic power of a quantum computer is determined by a combination of the number of quantum bits and the fidelity of a single bit. Superconducting quantum chips can be prepared using a process similar to that of semiconductor chips, and are therefore of interest because of their natural advantages in terms of scalability. However, as the number of quantum bits increases, the z-control lines (magnetic field control lines) on the superconducting quantum chip also increase, leading to increased magnetic field crosstalk between individual control lines, which affects the fidelity of a single bit and further affects the arithmetic power of superconducting quantum computing. To address the above problems, this paper proposes two methods to reduce the magnetic field crosstalk between the control lines by modeling and simulating the superconducting quantum chip using air bridges and inflow structures, and the simulation results show that the magnetic field crosstalk between the control lines is significantly improved after the introduction of air bridges and inflow structures. This work provides a basis for further suppression of magnetic field crosstalk between the z-control lines of superconducting quantum chips.

1. Introduction

Quantum computers have attractive application prospects in public key password cracking and new drug development. As the most promising current implementation, superconducting quantum computers have advantages such as large-scale scalability because their implementation processes are compatible with existing semiconductor processes. In order to improve its performance, the fidelity of the quantum chip needs to be improved on the one hand, and the number of quantum bits on the chip needs to be increased on the other. The number of control lines on a quantum chip increases proportionally with the number of quantum bits, which increases the magnetic interference between control lines and thus leads to a decrease in the fidelity of the quantum chip, i.e., a decrease in the performance of the quantum computer. Therefore, it is important to study the crosstalk between control lines.

In 1980, Paul Benioff at Argonne National Laboratory was the first to throw out the new concept of quantum computing and developed a quantum Turing machine with a classical analog based on this idea, which is usually considered as the prototype of quantum computer[1]. 1982, Feynman continued to enrich Benioff's s theory and proposed the idea that quantum systems could be simulated by designing a quantum computer based on quantum principles, which is unanimously considered by experts and scholars in the field of quantum computing as the earliest idea of quantum computers[2].

In 2017 IBM held a conference to announce the launch of the world's first commercial "universal" quantum computing service system, IBM O. IBM explained that the service is equipped with the ability to be accessed directly over the Internet and that the system has a transformative role in drug development and various scientific research. IBM explained that the service is equipped with direct Internet access and that the system has a transformative impact on drug development and scientific research, and has begun soliciting consumer users. In addition to IBM, other companies, such as Intel, MIT, ETH, Google and Microsoft, are also exploring the field of practical quantum computers [3,4,5].

Google released in 2019 a quantum chip called "Sycamore", which is the first experimental demonstration of the superiority of quantum computers over traditional classical architectures, and this chip containing 54 quantum bits was announced to achieve "quantum hegemony"[6]. Quantum hegemony refers to the computational power possessed by quantum computing that surpasses that of all conventional computers.

On December 4, 2020, Pan Jianwei's team at the University of Science and Technology of China (USTC) announced the design and implementation of a quantum computing prototype containing 76 quantum bits - "Nine Chapters" - and announced the realization of the "quantum hegemony" through a demonstration computational display of the algorithm of Gaussian Boson sampling. "This breakthrough makes China the second country in the world to achieve "quantum hegemony" [7,8].

Superconducting quantum computing has become an international hot topic, but this technology is still in a nascent stage, and there are still many technical problems to be broken through one by one. In a superconducting quantum chip, each superconducting quantum bit is located on a cross capacitor, and the state of the superconducting quantum bit is regulated by a set of chip control lines connected in parallel with it (see Figure 1 for the specific structure). This state modulation consists in two aspects:

1) The first aspect refers to the application of an excitation modulation signal (i.e., xy-control) to the chip control line for excitation to change the state of the superconducting quantum bit.

2) The second aspect refers to the application of a bias flux current signal to the chip control line via a flux modulation line (i.e., z-control) for tuning the frequency of the superconducting quantum bits.

Wherein: said bias flux current signal is usually provided directly with a constant current voltage source and the magnetic flux of the chip control line is varied by an applied constant bias flux current or by mutual inductive coupling with the chip control line, which in turn modulates the frequency of the superconducting quantum bits containing the control line of this superconducting quantum chip[9].

The research in this paper focuses on investigating the flux crosstalk phenomenon on the Z control line and testing the effectiveness of two structural solutions to suppress magnetic field crosstalk.

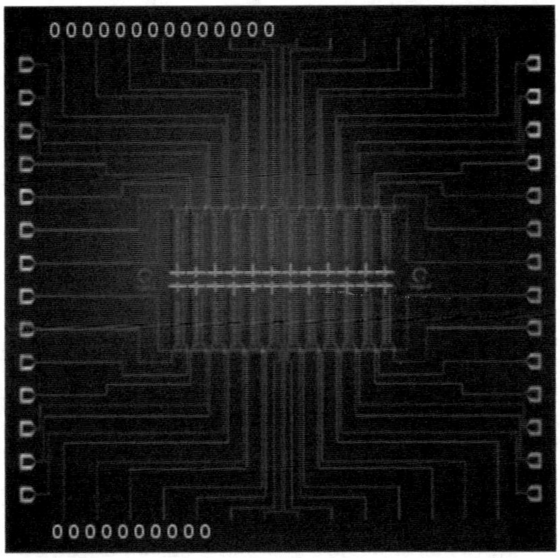

Figure 1. Kuafu KF C24-100 quantum chip

2. Chip control line modeling

2.1. Basic structure

For the time being, the complex structure of the chip and the influence of the working environment are not considered, and only a single control line model structure is established to explore the magnetic flux density distribution near it and verify the correctness of the simulation setup. In this scheme, a control line model of similar size to the actual chip is considered, and then three monitoring lines are inserted at 40um and 2mm from the control line terminal. These three pairs of detection lines are located 10um above the control line, through the middle of the control line, and 10um below the control line, and the low-frequency magnetic field crosstalk phenomenon on the three pairs of monitoring lines is detected by applying a voltage excitation of 1.0nV to the chip.

As shown in Figure 2, the dielectric plate model is established in the studio, and the dielectric plate is divided into two parts: dielectric substrate and superconducting layer, and the dielectric substrate is made of sapphire (excellent electrical insulation performance, hexagonal symmetric structure, stable chemical performance, and excellent superconducting substrate material) with a dielectric constant of 11.5; the superconducting layer material is made of aluminum, which shows superconducting characteristics at 1.196K, and the existing superconducting quantum chip at The operating temperature is about 10mK, which is the most commonly used superconducting material for the preparation of superconducting quantum chips. Theoretically, the superconducting conductivity is infinite and the magnetic permeability is 0. However, since the model simulation will report errors when run under this condition, the conductivity is set to 1×10^10 S/m and the magnetic permeability is 1×10^(-10)H/m to approximate the ideal condition.

The chip control line model is established on the surface of the dielectric plate, the dielectric plate size is 5mm×5mm×0.5mm. The chip control line structure is shown in Figure 2, the surface superconducting layer thickness t is 100nm, the control line single line length is 2060um, a voltage port excitation of 1nV is applied to the control line, and the working environment is set to vacuum environment. The model is shown in Figure 2.

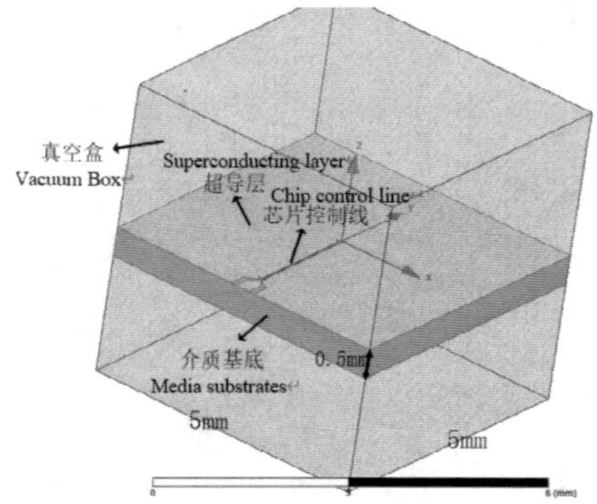

Figure 2. Simple model of single control line

2.2. Single control line result

The magnetic field distributions of the two monitoring lines A2 and B2 that pass through the middle of the control line (z-coordinate t/2) are shown in Fig. 4, with magnetic induction intensity maxima of -175.1316 dB and -175.2354 dB, respectively.The values of magnetic induction intensity measured at the position of the monitoring line passing through the superconducting layer are much lower than the values of magnetic induction intensity inside the control line, which is due to the fact that the superconductor when entering the superconducting state all magnetic fields inside the superconductor are repelled and the magnetic field inside the superconductor always remains zero, a phenomenon known as the Meissner effect[10]. However, since the superconducting layer is not set to the ideal condition in the simulation, the magnetic induction intensity value in the superconducting layer does not become 0. However, the magnetic field in the superconducting layer is much smaller than the magnetic field in the medium, indicating that our simulation setup is correct.

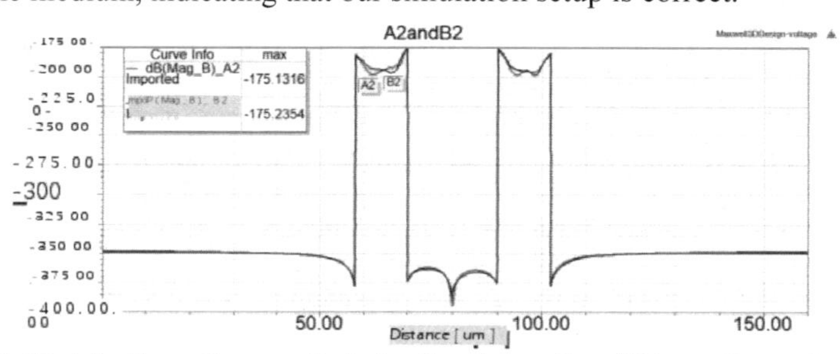

Figure 3. Distribution of magnetic induction intensity of Group 2 monitoring lines

3. The diversion structure suppresses magnetic field crosstalk

Due to the limitation of chip size, the chip control line spacing cannot be increased indefinitely. In order to reduce the crosstalk phenomenon of magnetic field, it is necessary to consider from the perspective of control line structure, so this paper gives a scheme of adding a guide structure between two groups of control lines to form a circulating field between the guide structure and the control line structure, so that the magnetic induction intensity distribution near the Josephson junction on the cross capacitor is more stable.

In order to closely match the structure size of the actual chip control lines, this scheme will set the control line spacing to 500um in the subsequent simulation operation, and set one monitoring line for each group of control line models, which is set on the surface of the cross capacitor, between the superconducting layer and the substrate material. The size and position of the conductive structure are shown in Figure 3, and the thickness of the conductive structure is the same as the thickness of the control line and the cross capacitor, which are both 100nm.

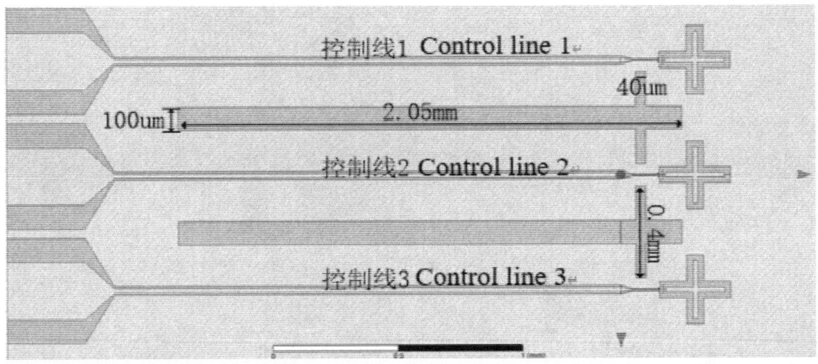

Figure 4. Schematic diagram of diversion structure dimensions

3.1. *The first set of incentive results*

In the case of the first set of excitation without the addition of the current guide structure, the difference in magnetic induction intensity between monitoring lines 1 and 2 is about 15 dB and between monitoring lines 1 and 3 is about 18 dB when the control line spacing is 500 um. After the addition of the current guide structure, the difference in magnetic induction intensity between monitoring lines 1 and 2 is about 25 dB and between monitoring lines 1 and 3 is about 28 dB (See Figure 5). In general, the magnetic induction intensity on monitoring lines 2 and 3 decreased by about 10 dB relative to the result before the addition of the current guide.

The distribution of magnetic inductance on the chip surface in Figure 6 shows that the magnetic inductance near the cross capacitor is more uniform and stable after the addition of the current guide structure, indicating that the current guide structure plays a good role in crosstalk suppression.

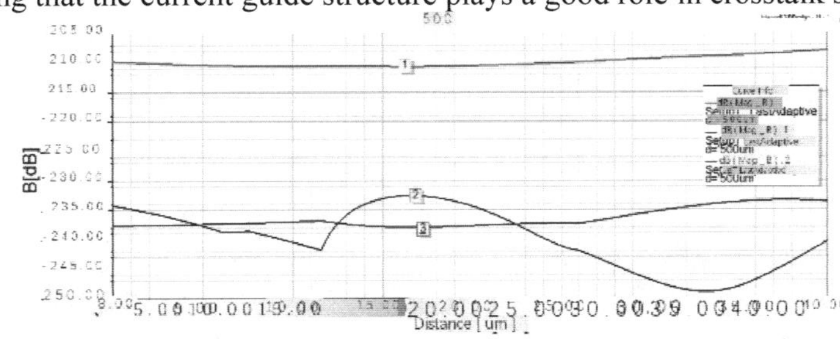

Figure 5. Variation curve of magnetic induction intensity

3.2. *The second set of incentive results*

In the case of the second set of excitation without the current-conducting structure, the difference in magnetic induction intensity between monitoring lines 1 and 2 is about 4.5 dB and between monitoring lines 2 and 3 is about 3 dB when the control line spacing is 500 um. After adding the current-conducting structure, the difference in magnetic induction intensity between monitoring lines 1 and 2 is about 7 dB and between monitoring lines 2 and 3 is about 12 dB (see Fig. 6). The magnetic induction intensity change between monitoring lines 2 and 3 is more obvious due to the fact that the inflow structure creates a circulating flow field between control line 2 and control line 3, which makes the distribution of the excitation flow weaker near the cross capacitor 3, thus reducing the magnetic induction intensity distribution near the cross capacitor 3.

The distribution of magnetic induction intensity on the chip surface in Fig. 7 also shows that the crosstalk of the magnetic field between cross capacitor 2 and cross capacitor 3 is improved by adding the inflow structure.

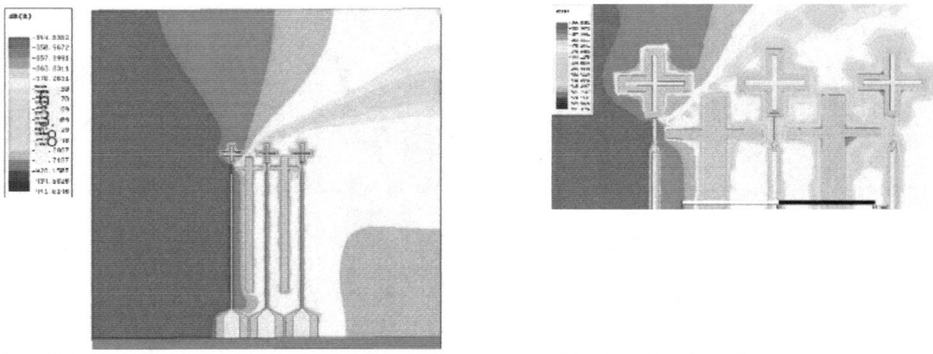

(a) Chip surface distribution map (b) Distribution diagram near cross capacitance

Figure 6. Distribution of magnetic induction intensity of the first group of excitation

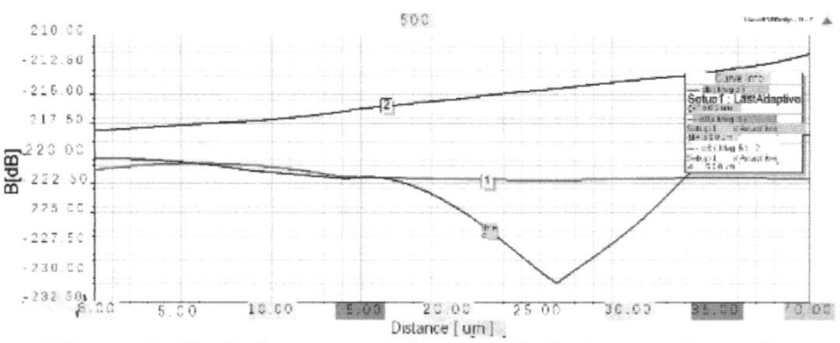

Figure 7. Variation curve of magnetic induction intensity

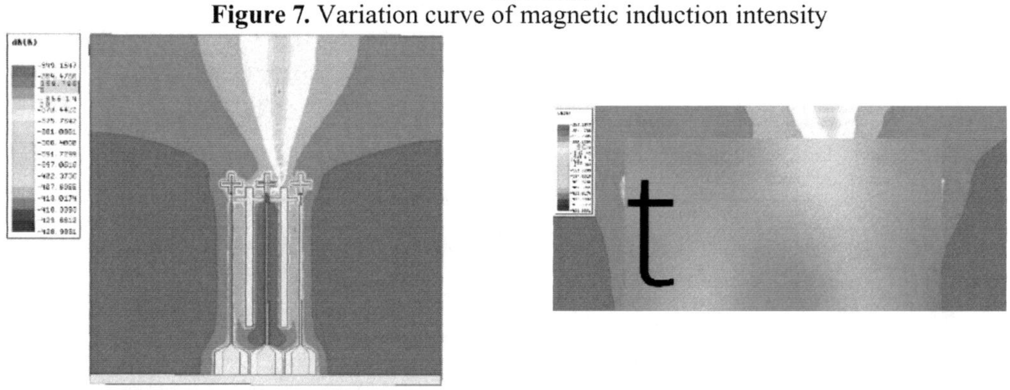

(a) Chip surface distribution map (b) Distribution diagram near cross capacitance

Figure 8. Distribution of magnetic induction intensity of the second group of excitation

3.3. The third group of incentive results

In the case of the third set of excitation without the addition of the current guide structure, the difference in magnetic induction intensity between monitoring lines 1 and 3 is about 12 dB and between monitoring lines 2 and 3 is about 8.5 dB when the control line spacing is 500 um. After the addition of the current guide structure, the difference in magnetic induction intensity between monitoring lines 1 and 3 is about 17 dB and between control lines 2 and 3 is about 20 dB (See Figure 9). The magnetic induction intensity of monitoring line 2 is slightly lower than that of monitoring line 1. In addition to the influence of the inflow structure, the reason also includes the error of gridding, because the number of gridding of the cross capacitor is not enough, which makes the variation curve of magnetic induction intensity on the monitoring line not smooth enough, and there is a certain error in the value, but this error has less influence on the conclusion, and in order to save simulation resources, the grid has not been optimized in this paper In order to save simulation resources, the grid has not been optimized in this paper.

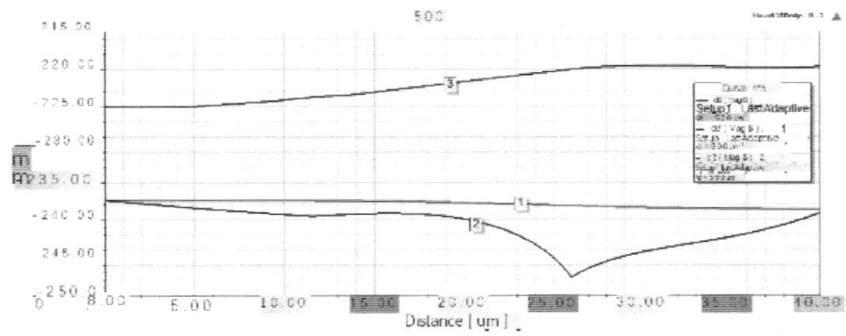

Figure 9. Variation curve of magnetic induction intensity

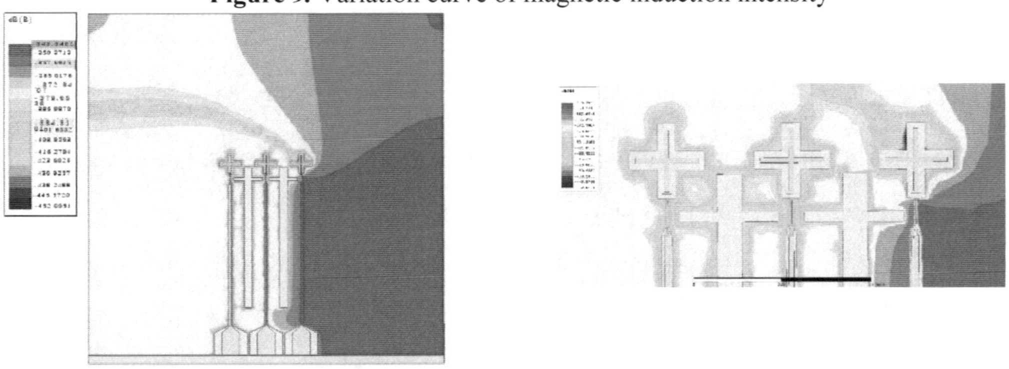

(a) Chip surface distribution map (b) Distribution diagram near cross capacitance

Figure 10. Distribution of magnetic induction intensity of the third group of excitation

4. Air Bridge Suppresses Magnetic field Crosstalk

Considering that the current distribution on both sides of the control line is uneven due to the current guiding effect of the capacitor junction on the excitation current, this paper gives a method to build air bridges on both sides of the capacitor junction of the control line. The air bridges are made of superconducting materials and connected to the superconducting layer, with the purpose of stabilizing the current distribution on both sides of the capacitor junction and forming a set of circulating currents through the air bridges, so as to stabilize the magnetic field distribution and reduce the magnetic field crosstalk. The air bridge is 30um wide, 30um high, and 10um line width, as shown in Figure 11(c). The simulation still uses a single-port excitation to compare the magnetic induction intensity distribution on the cross capacitor before and after adding the air bridge.

Compared with the actual superconducting quantum chip, the size of the air bridge in this simulation model is large, which is difficult to implement in reality, but it is simplified to save simulation resources.

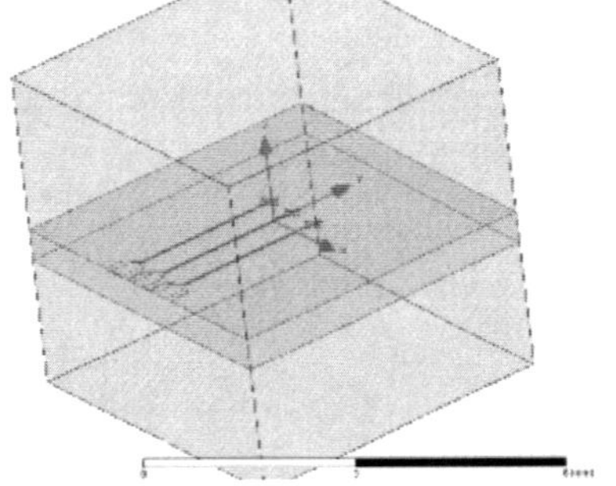

(a) Three sets of actual control line models

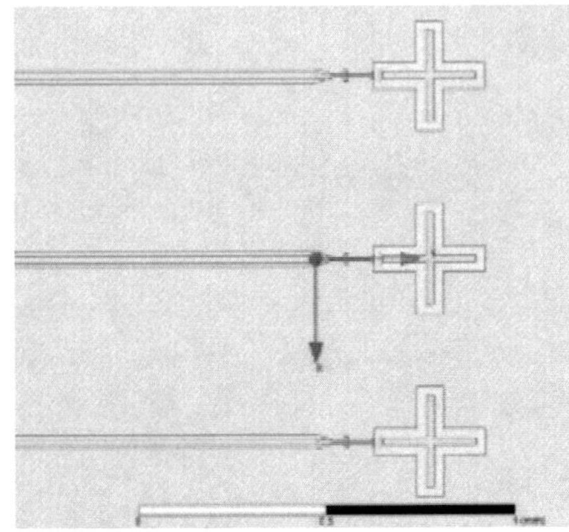

(b) Comparison of air bridge sizes

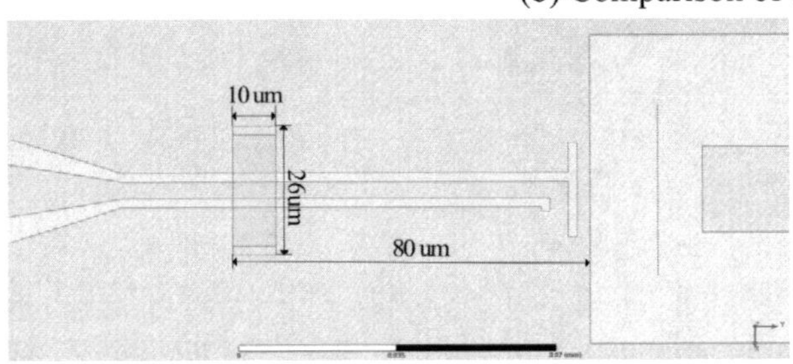

(c) Size of air bridge

Figure 11. Control line model of air bridge chip

4.1. The first set of incentive results

When the air bridge structure is not added, the magnetic induction intensity difference between monitoring lines 1 and 2 is about 15 dB and between monitoring lines 1 and 3 is about 18 dB for the first set of excitation when the control line spacing is 500 um. After the air bridge structure is added, the magnetic induction intensity difference between monitoring lines 1 and 2 is about 36 dB and between monitoring lines 1 and 3 is about 20 dB (see Figure 12). It can be seen that the air bridge has a good suppression effect on the crosstalk of the magnetic induction intensity on the monitoring lines.

Compared with the results of the magnetic induction intensity distribution in the first set of excitation cases with the deflector structure, the air bridge has a better effect on the magnetic field crosstalk suppression on monitoring lines 1 and 2, and the deflector structure has a better effect on the magnetic field crosstalk suppression on monitoring lines 1 and 3.

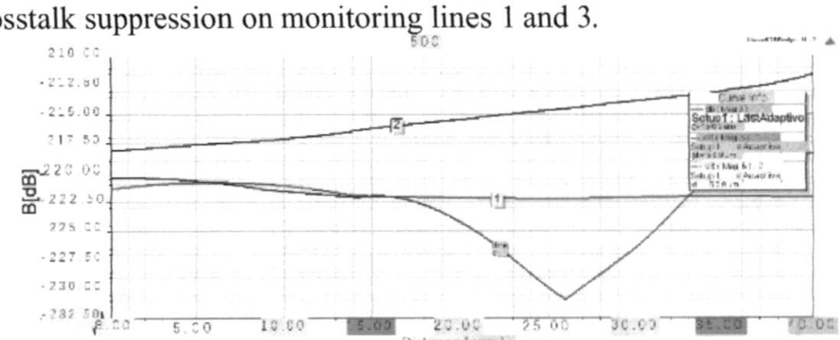

Figure 12. Variation curve of magnetic induction intensity

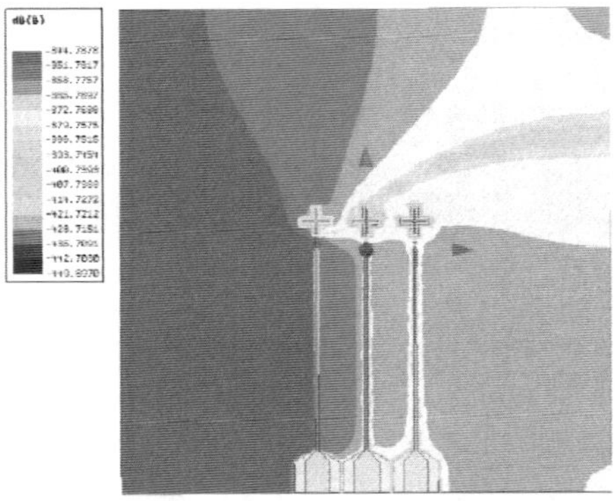

(a) Chip surface distribution map

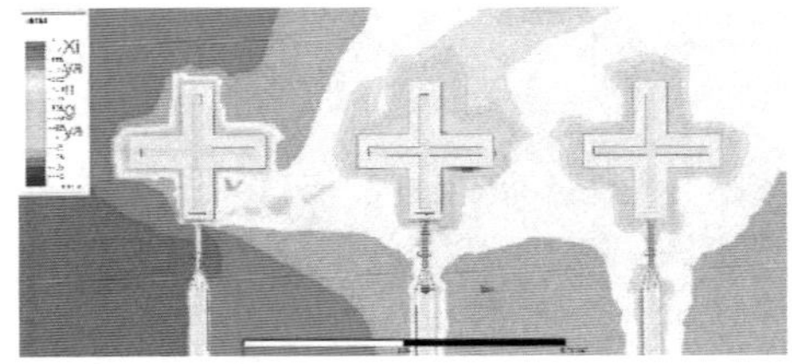

(b) Distribution diagram near cross capacitance

Figure 13. Distribution of magnetic induction intensity of the first group of excitation

4.2. The second group of incentive results

When the air bridge structure is not added, the magnetic induction intensity difference between monitoring lines 1 and 2 is about 4.5 dB and between monitoring lines 2 and 3 is about 3 dB in the second group excitation case when the control line spacing is 500 um. After adding the air bridge structure, the magnetic induction intensity difference between monitoring lines 1 and 2 is about 5 dB and between monitoring lines 2 and 3 is about 8.5 dB (See Figure 14). It can be seen that the air bridge also plays a good role in suppressing the crosstalk phenomenon of the magnetic induction intensity on the monitoring line.

Compared with the results of the magnetic induction intensity distribution in the second set of excitation cases with the inflow structure, the inflow structure has a better effect on the crosstalk suppression of the magnetic field on the three test lines.

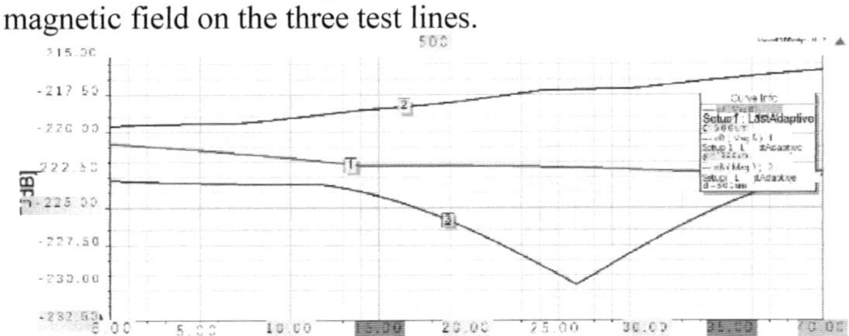

Figure 14. Variation curve of magnetic induction intensity

4.3. The third group of incentive results

When the air bridge structure is not added, the magnetic induction intensity difference between monitoring lines 1 and 3 is about 12 dB and between monitoring lines 2 and 3 is about 8.5 dB for the third group excitation case when the control line spacing is 500 um. After adding the air bridge structure, the magnetic induction intensity difference between monitoring lines 1 and 3 is about 12 dB and between control lines 2 and 3 is about 10 dB (see Figure 16). It can be seen that the air bridge also plays a good role in suppressing the crosstalk phenomenon of magnetic induction intensity on the monitoring line.

In contrast to the magnetic induction intensity distribution results for the third set of excitation cases with the deflector structure, the deflector structure has a better suppression effect on the magnetic field crosstalk on the three test lines.

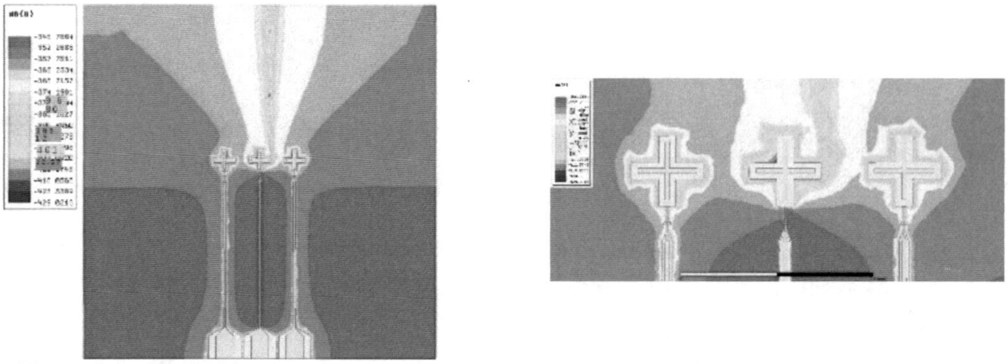

(a) Chip surface distribution map (b) Distribution diagram near cross capacitance

Figure 15. Distribution of magnetic induction intensity of the second group of excitation

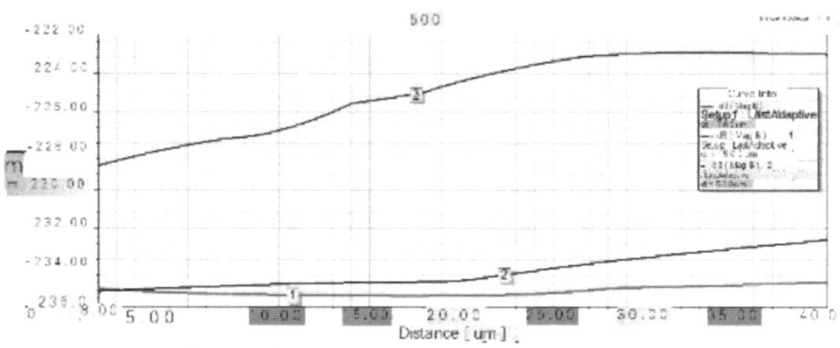

Figure 16. Variation curve of magnetic induction intensity

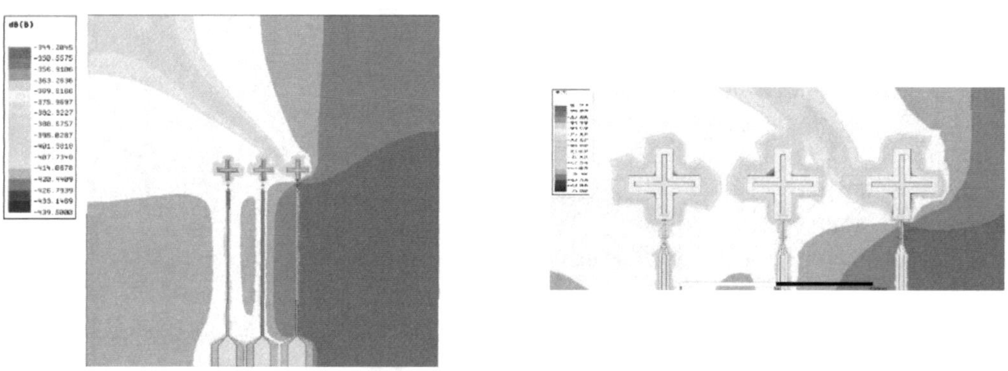

(a) Chip surface distribution map (b) Distribution diagram near cross capacitance

Figure 17. Distribution of magnetic induction intensity of the third group of excitation

5. Conclusion

To address the magnetic field crosstalk on superconducting quantum chips, this paper establishes the analysis method of magnetic field crosstalk on superconducting quantum chips by using electromagnetic simulation software, and further designs the inflow structure and air bridge structure to suppress the magnetic field crosstalk, and its simulation results show that both methods have certain suppression effect on the magnetic field crosstalk. This study is of great significance and reference value for solving the crosstalk on superconducting quantum chips and improving the single-bit fidelity of superconducting quantum chips, thus enhancing the arithmetic power of superconducting quantum computing.

6. References

[1] Wang Jilin, Liu Jianjian, Chen Peiyi. Quantum computing and superconducting quantum computers[J]. Micro and Nanoelectronics, 2009,46(06):321-326+345. Ying Jiang. Modeling and simulation of transponder systems and their applications[D]. Beijing Jiaotong University,2014.

[2] R. P. Feynman. Simulating physics with computers[J]. International Journal of Theoretical Physics,1982, 21: 6-7.

[3] Ren Jirong. Development and application prospects of quantum computers[J]. Electronic World, 2018,(1):97,99.

[4] Han, Zhexin, Gu, Guotai, Xiao, Han. Research and application of quantum computers[J]. Henan Science, 2015,(9):1559-1563.

[5] Zhang Jianfen. Conceptual principles and prospects of quantum computers[J]. Physics Bulletin, 2015,(4):125-128.

[6] Arute F, Arya K, Babbush R, et al. Quantum supremacy using a programmable superconducting processor[J]. Nature, 2019, 574: 505-510.

[7] Wang Jinlan. Chinese scientists achieve a milestone breakthrough in "quantum computing superiority"[J]. Science, 2021,73(01):43+2.

[8] Guo Guangcan, Chen Yipeng, Wang Qin. Advances in quantum computer research[J]. Journal of Nanjing University of Posts and Telecommunications (Natural Science Edition), 2020,40(05):3-10.

[9] Kong, Weicheng. A crosstalk analysis-based method for superconducting quantum bit modulation [P]. Anhui, China: CN110488091A,2019-11-22.

[10] Li Xiaohu. Preparation and performance study of superconducting JosephsonF Suidon junction and structural exploration of superconducting quantum bit circuits[D]. Nanjing University,2015.

Design and Implementation of AGV Scheduling System based on Spring

Ruihang Zhao[1] **and Wei Zhang**[1]

[1] Software College, Northeastern University, Shenyang, China

2171423@stu.neu.edu.cn, Zhangwei@swc.neu.edu.cn

Abstract. This paper designs and implements an AGV scheduling application based on spring framework. This application consists of four modules, intelligent scheduling module, traffic control module, equipment management module and communication protocol module.Through interaction with AGV vehicle-mounted applications, the intelligent scheduling module realizes the function of automatic task allocation and vehicle body accepting tasks autonomously, which solves the problem of low efficiency of existing applications.Traffic control can avoid congestion by announcing node, path and area possession, which solves the problem of congestion that is easy to occur in multi-vehicle operation.The equipment management module realizes the management function of charging station, temporary parking point and other equipment;Communication protocol module to solve the vehicle, equipment, applications between the information encryption, protocol definition and other functions.

Keywords: Automated Guided Vehicle (AGV), task scheduling, robot traffic control, Spring

1. Introduction

The booming development of the Internet of Things at home and abroad, in many industries, such as warehousing, logistics and many other places, such as post offices, docks, airports and many other places people's demand for autonomous mobile cars is increasing[1].AGV serves as the intersection of Internet of Things technology and Internet technology[2], has good prospects for development, breeds a huge market value, has a great role in promoting all walks of life, but also there are the following problems, Due to the dense population density and rich and unique material resources in China, there is a huge demand for logistics engineering. However, the current application of AGV is more special case development for small places, which cannot meet the growing demand of large logistics and handling bases; Due to site-specific development, AGV applications of existing companies can only be used for their own sites, with weak portability and cannot be widely promoted[3];At present, most applications are directly connected with the AGV body itself for development, which is limited by the body type and inconvenient for demonstration effect[4], the car body loss, time and space are inconvenient problems.AGV simulation application is a good way to solve the above problem, will first need to be carried out in accordance with the topological structure has been the site of the original survey, robot programming, application realization partitioned into multiple tasks at the same time the task steps to speed up the development time and guarantee the reusability of the program, can according to different scenarios, car body and user needs to carry on the secondary development, And because of the characteristics of simulation application, on the one hand, it can simulate large demand sites[5]On the other hand, it is convenient to demonstrate and obtain user requirements. In the whole project development process, it can better cope with the change of requirements and use a variety of algorithms to deal with problems flexibly[6]The terminals use TCP connections[7], easy to program itself iteration[8]Changes can be made to meet functional incremental requirements.

Spring is an open source framework created by Rod Johnson[9]It was created to address the complexity of enterprise development by using basic Javabeans for tasks that were previously only possible with EJBs.

The current Spring framework has the following characteristics:

Light,Spring as a framework, a major advantage is its portability, no matter from the aspects of memory, how application or framework used in the process of the cause of the overhead of ways are portable, a complete Spring framework jar package release can't more than 2 MB, and the entire framework caused in the running costs are negligible.

Inversion of control, the decoupling between modules is completed. When a module needs the classes of other modules during the program running, the object will pass the object in a passive way[10]; The aspect oriented,in the development process, each object only need to do their own business logic, without the need to responsible for other functional requirements, the application level of object itself is not even aware of the existence of the function of application level, and these functions by aspect oriented, the application functions of cohesion in the module development.

2. Application design

The application mainly adopts the way of front and back end separation. The front end uses VUE framework and interacts with the back end through Ajax technology. The back end uses Springboot framework to expose the interface, and additionally connects to AGV vehicle-mounted applications and monitoring application pages. The user interacts with the information from the front end to the back end, which encrypts and decrypts the messages and then interacts with the database.

The main users of this application are operators and AGV. The main functions of operators need to select task scheduling mode, task management, device management, and protocol management mode. Relying on the intelligent algorithm of the application itself, tasks can be assigned to the appropriate robot at the appropriate time according to their own needs.

AGV are mainly responsible for autonomous state reporting, task acceptance and action reporting to ensure that the robot itself can apply for tasks independently in idle state. When applying for tasks, the position and model of the robot body are reported to ensure that it is assigned to suitable tasks.

AGV scheduling application uses MVC architecture. Various operators in the client use browsers (or materials tablets) to access all scheduling monitoring applications. The interface uses HTML pages to dynamically interact with Ajax data access, and the Tomcat server built in Spring is used as the container of the application. After the version is updated, there is no need to patch or install software. Refresh the browser on PC or mobile terminal to access the interface after the latest change and use the changed service interface, etc. No other deployment or operation is required.

The backend server is developed by SpringMVC framework, Mybatis and JDBC are used to develop the company's own database call interface, and NavICat is used for database visualization. The whole back end is divided into controller layer, Service layer and Mapper layer. The Controler layer is responsible for monitoring the access of the front end. The Service layer is responsible for completing the corresponding services. Mapper layer is responsible for the whole application and database of the bottom data interaction and operation, in the whole process of program development encountered the need for JAR package, through SpringBoot and Maven with annotations to inject download, such as log operation is through the import of LOG4j JAR package finally realized, Finally, the whole project will be packaged into a JAR package and run on the cloud server to complete the project online.

This paper mainly divides the application structure into three layers, which are presentation layer, logic layer and data access layer respectively. The main task of the business presentation layer is to be responsible for the direct human-computer interaction between applications and users. All interface design and display, interface operation and feedback should be completed in this layer. The logical layer mainly serves as a link between the preceding and the following for data exchange. The overall architecture is shown in Figure 1.

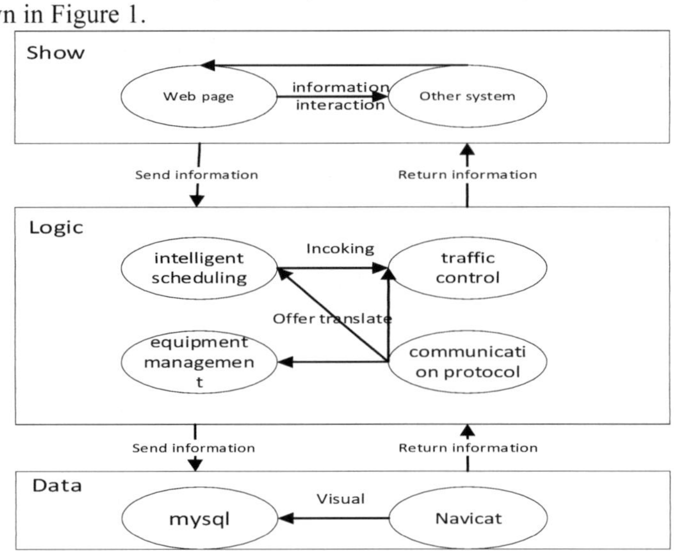

Figure1. Architecture of system layered

In the intelligent scheduling module, operators can assign tasks in the application according to their own handling requirements, which can be divided into common task cyclic task, mandatory task, etc., and select the working mode of the task,The longest priority, the shortest priority, according to the right of priority, the backend application will assign tasks to receive according to the algorithm of AGV car body, at the time of deployment of tasks need to consider the weight of the task, consumption, the car body limit, receiving vehicle power, body type, rational and efficient allocation of tasks, the tasks are completed at the same time, to ensure the minimum loss, The fastest, the most efficient.

Traffic control module is responsible for dealing with traffic control problems in the operation of the program, such as AGV car need to access resources cannot be Shared at the same time, you need to traffic control, reasonable allocate resources, ensuring the security of the tasks can be issued and ensure the completion of the task, also need to achieve high efficiency, low wait, independently by AGV vehicle detection and reporting, The traffic control module class diagram is shown in Figure 2.

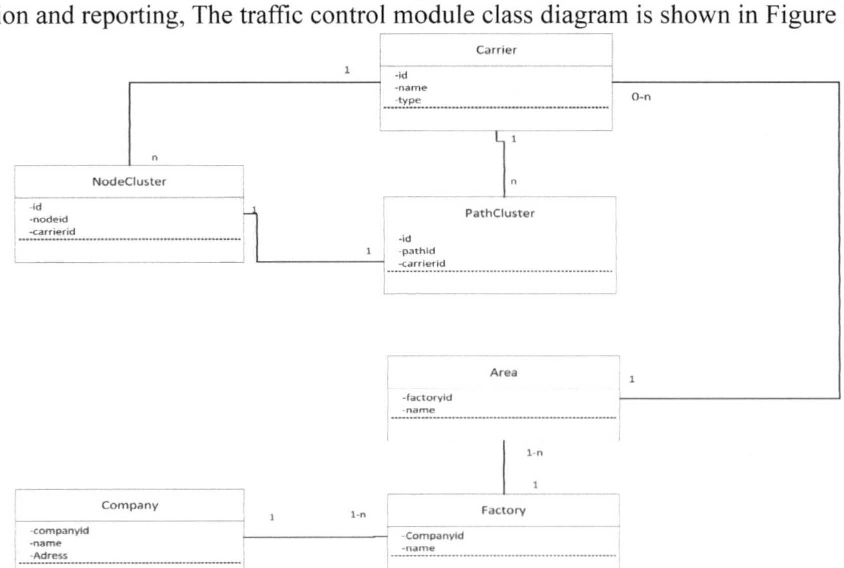

Figure2. Diagram of traffic control class

This module is responsible for adding, deleting, checking and parameter configuration of company, factory, region and other external equipment.Communication protocol module is mainly responsible for the interaction between application and car body, charging station and application that cannot use HTTP request, data protocol encryption and translation reading.

3. Implementation process

Intelligent scheduling module is used to realize and operate the tasks delivered to AGV vehicles. Users can add and modify the tasks to be delivered in this module, and cancel and enforce the delivered tasks. Click Task Management to get his expansion module. This module contains the function modules of task editing, standard task, circular task, forced task, cancel task and forced completion. The following three scheduling modes are realized in the scheduling mode, The longest priority, the shortest priority and according to the right of scheduling, the scheduling changes on the architecture of most of the existing pool model is only a single task, a new car pool to ensure the independent report of the body itself, when the body in the working state will be his registration to car pool, to adopt the method of traverse to accept the task, so as to realize the intelligent robot functions.The task allocation flow chart is shown in Fig3.

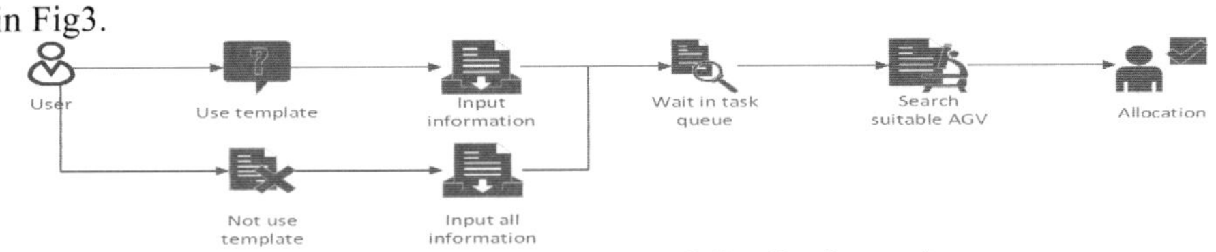

Figure3. Flow chart of distributing task

Traffic control module is automatically run application background, not on the front end, the final implementation of the specific process, announced after a car good path planning first take up the route, and traffic control direction to demand direction, follow-up bodywork is required for a search path to avoid take position, if they happen to find the position of has been preempted and myself in the same direction and need through the announced to grab at the same time, There are three kinds of preemption problems to be solved, intersection problem, encounter problem and pursuit problem. By solving these three problems, collision can be fundamentally avoided.

The collision avoidance method originally used by the app is to avoid the collision from the root by dividing different maps into controlled areas at the beginning of the map setting, or directly ignore the collision. When the vehicle body will collide, deadlock will occur, and the app will be reset artificially.This application wants to use a general solution, that is, in the process of running the car body, the collision is solved when the collision occurs, which is also the way of collision occupation, but it is dynamic occupation. This application mainly divides the collision situation into co-direction collision, opposite direction collision and intersection collision.

Co-direction collision can be solved by collision detection and speed avoidance of car body itself. The direction and intersection need to be solved by preemption mechanism. Preemption can be divided into two modes. One is to submit occupation information to the database at this section. Contra-directional collision divides the path into two passable conditions, positive direction and opposite direction. The intersection is divided into four junctions 1,2,3 and 4 according to the way of traffic lights, and only one direction can pass through the other three directions.

The work area module is mainly used to edit the information of the existing company, workshop and work area or add the new company and its related workshop and work area. This module mainly includes company setting, workshop setting and work area setting.

The communication protocol module mainly defines the message and connection request. The main structure of the protocol is as follows, Frame header,takes a fixed value;The version number must be consistent between the communication parties;The sender address of the message frame;The receiver address of the message frame, and the receiver needs to perform address verification;The type of message frame, different types of frame, the data part is different;The data of the message frame is different;Message frame ID, external application maintenance increment, AGVServer returns the same value; Message frame verification;Endframe, fixed value.

The flow chart of determining whether the next running destination can pass the traffic control is shown in FIG4.

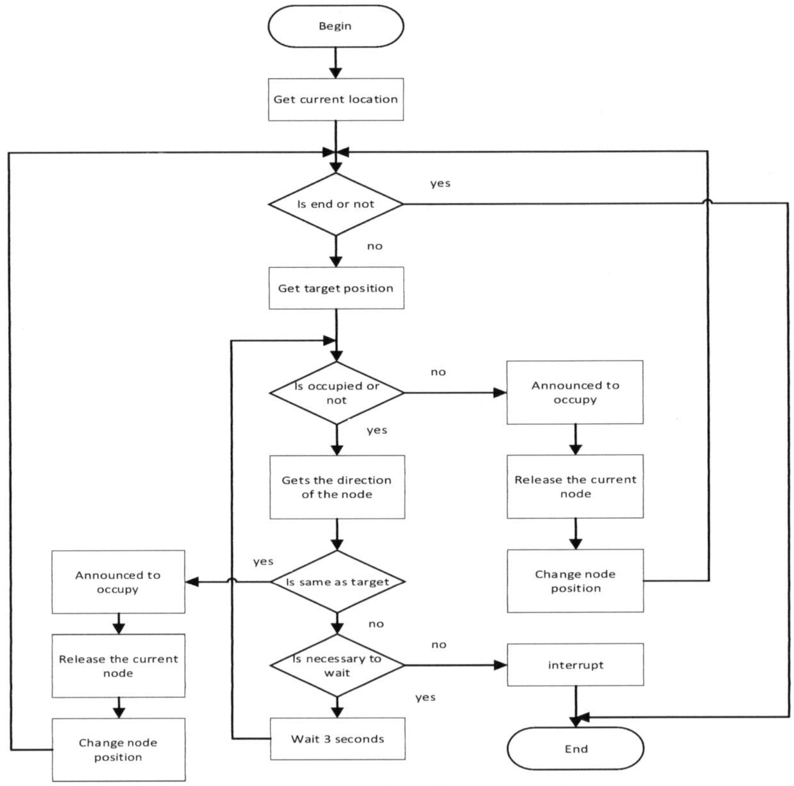

Figure4. Chart of traffic control flow

4. Simulation results

The task module is used to display and operate the tasks issued to AGV vehicles. Users can add and modify the tasks to be delivered in this module, and cancel and enforce the delivered tasks. Click Task Management to obtain his expansion module. This module includes task editing, standard task, circular task, mandatory task, cancel task and forced completion of these functional modules. The overall schematic diagram of the application of these functional modules is shown in Figure 5, which can intuitively display the task scheduling and vehicle operation in the factory.

The traffic control area setting component uses one or more user-set nodes (on the map) to form a virtual area.

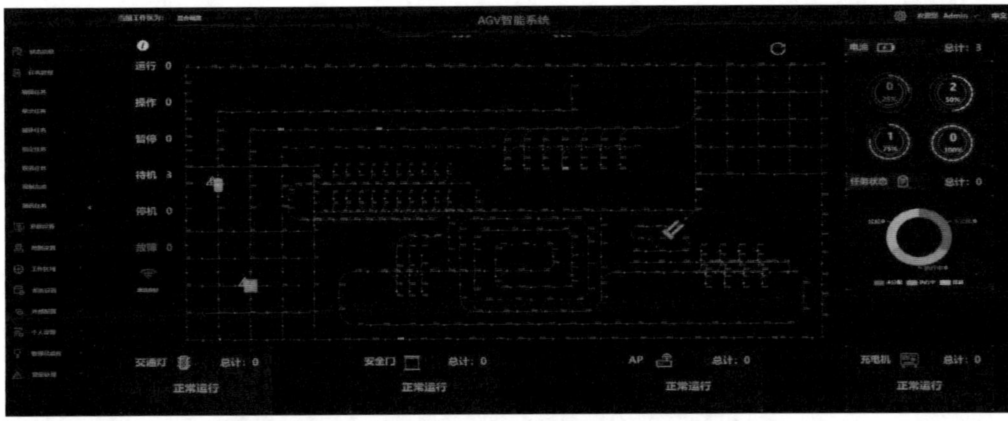

Figure5. Overall schematic diagram of application

5. Conclusion

AGV scheduling application is based on AGV simulation scheduling application module, so that the company, the robot and the operator through intelligent scheduling, traffic control and other main functions to form a good organic cooperation.

This paper divides the whole application into four modules,intelligent scheduling, traffic control, equipment management, communication protocol, and carries out the work according to each module in each stage.Finally, it is concluded that the application is suitable for multiple use scenarios with current logistics and transportation requirements, which can well solve the problems mentioned above,Existing AGV management applications have problems such as low task allocation efficiency, easy blockage in multi-car body operation, narrow application area, and serious wear and consumption caused by direct docking robot test, which can form a more convenient logistics and transportation environment and is an effective application to promote the vigorous development of logistics and transportation industry.

6. References

[1] Zhou Wei, ZENG Yuhang. Overall Architecture Design of AN AGV System [J]. *Computer Integrated Manufacturing Systems* **2020, No. 2** P402-409 1006-5911.

[2] Guorong Wu 1 ; Xuan Sun 1. AGV Task Distribution Study**[J].** *Journal of Physics: Conference Series* **2020 Vol.1486** P072016 1742-6588;1742-6596.

[3] Yuan Peng, Zhou Jun, Yang Zi-bing, Wu Di, Huang Pan-ling. Research on AGV Path Planning and Deviation Correction [J]. *Modern Manufacturing Engineering*, **2021, 4th** Issue P26-32 1671-3133.

[4] Hu Weimin, Jin Wenzhou. Improved Smoothing A* Algorithm for Multi-AGV Path Planning [J]. *Computer Engineering and Applications* **2020, 56 (16)** p204-210 1002-8331.

[5] Xiao Haining, WU Xing, Zou Ting, Zhai Jingjing. A Deadlock-avoidance Scheduling Method for Transportation Systems Based on Multi-load AGV [J]. *Transactions of Nanjing University of Aeronautics and Astronautics* **2021 Volume 38, No. 1** P181-192 1005-1120

[6] Sun Bo, JIANG Ping, ZHOU Genrong, DONG Dianyong. AGV Path Planning Based on Improved Genetic Algorithm [J]. *Computer Engineering and Design* **2020 no. 2** P550-556 1000-7024.

[7] Lingfeng Guo 1 ;Jack Y. B. Lee 2. TCP-FLASH -A Fast Reacting TCP for Modern Networks[J].*IEEE Access* **2021 P1** .

[8] James A.Senn. *Analysis and design of Electronic prescription* [M]. Beijing：McGraw-Hill, **2007**: 8-342.

[9] Qimei Chen,Hongbo Shi. Design and implementation of domain knowledge service platform based on spring framework[J]. *Proceedings of the Association for Information Science and Technology*,**2017,54(1)**.

[10] Priyanka Sarode,R. N. Jugele. Facets of Aspect Oriented Programming[J]. *International Journal*

of Innovative Technology and Exploring Engineering (IJITEE),**2020,9(4)**.

Acknowledgement

This work was financially supported by the Project of China's Ministry of Education and supported by Alibaba in 2020 (Grant No. 202002001063). Heartfelt thanks also to the reviewers for their constructive comments to the paper.

Redundancy Strategy of Cascaded Energy Storage System Based on Model Prediction Control

Bo Yang[1,2*], Wencheng Tang[1]

[1] School of Mechanical Engineering, Southeast University, Nanjing 210009, China;
[2] China Electric Power Research Institute Company Ltd., Nanjing 210003, China;

yangbo@epri.sgcc.com.cn

Abstract. The cascaded energy storage system is difficult to achieve global optimal control, because its controlled object includes not only external variables such as output power, voltage and current, but also various internal variables such as power module capacitor voltage and battery cluster charge state. Redundancy strategy is proposed in this paper based on model prediction control. Firstly, the predictive model of the system in discrete domain is constructed, and the fast model predictive control method of the cascaded energy storage system is studied to realize the global optimal control and improve the dynamic response performance of the cascaded energy storage system. Secondly, in order to realize the continuous and reliable operation of the cascaded energy storage system under the fault state, the redundancy control method of the cascaded energy storage system is studied. Finally, the validity and effectiveness have been testified by relevant simulations.

1. Introduction

Due to the soaring consumption of traditional fossil energy, environmental problems such as greenhouse effect and air and water pollution have been triggered. The sustainable development energy path of replacing traditional fossil energy with renewable energy has become the consensus of the world's energy transformation mode. How to enhance the strong and stable operation capacity of the power grid and effectively promote the safe grid connection and consumption of renewable energy is a key issue to be tackled in the future power grid construction [1]-[3].

The energy storage system has the characteristics of bi-directional power regulation, fast response rate and flexible charge and discharge. Therefore, compared with the conventional power supply in the power system, the energy storage system has faster climbing speed, larger operating range and more flexible start-up and shutdown capacity. It can not only provide power support for new energy power plants, stabilize fluctuations, delay transmission and distribution line congestion, but also serve as a flexible power supply to improve the frequency characteristics of the power system [3]-[5]. The cascaded energy storage system adopts a modular cascade topology, which improves the voltage level of the system port through the series connection of power units. It does not need a transformer to connect directly to the power grid. It has the characteristics of high safety, high reliability, high efficiency and long service life. At the same time, it has the technical advantages of large single machine capacity, fast dynamic response, high power utilization and good output quality.

As the core equipment of the energy storage system, it can be divided into low voltage and medium/high voltage energy storage converter according to the access voltage level. Low voltage energy storage converters mostly adopt diode clamped converter topology, of which the two-level or three-level topology is the most widely used and the most mature technology.

Due to the access voltage of 10kV and above, the medium and high voltage energy storage converter contains many switching devices in order to improve the withstand voltage level of the system. The modular cascade electrical topology is mostly used, and the transformer is not directly connected to the power grid. It can be called direct mounted energy storage system or cascade energy storage system [6]-[8]. The cascaded energy storage system is subdivided according to the type of electrical topology, mainly including cascaded H Bridge (CHB) [9]-[13] and modular multilevel converter (MMC) [14]-[19].

However, it is difficult for cascaded energy storage system to achieve global optimal control, owing to its controlled object includes not only external variables such as output power, voltage and current, but also various internal variables such as power module capacitor voltage and battery cluster charge state. Thus this paper proposed a redundancy strategy based on model prediction control. Through the effective prediction of the expected output voltage, the high-precision and effective control of the output power and current of the system can be realized. Besides, the continuous and reliable operation of the cascaded energy storage system can be realized under the fault state.

2. Establishment of Discrete Domain Mathematical Model

The continuous domain mathematical model of the cascaded energy storage system is shown in (1). The output voltage at the AC side of the direct mounted energy storage converter can be simulated as a continuous output AC voltage source in the form of duty cycle. However, the cascaded energy storage system should be a continuous discrete hybrid electrical model.

$$\begin{cases} e_a = v_{sa} + L\dfrac{di_{sa}}{dt} + Ri_{sa} \\ e_b = v_{sb} + L\dfrac{di_{sb}}{dt} + Ri_{sb} \\ e_c = v_{sc} + L\dfrac{di_{sc}}{dt} + Ri_{sc} \end{cases} \tag{1}$$

The AC side of the cascaded energy storage system is a continuously changing power grid voltage and current variable and an inductance with closed loop attribute, both of which have typical continuous domain mathematical model characteristics. The direct mounted energy storage converter side controls the on-off of the power module IGBT to realize the input and removal in the unit control cycle, However, in the unit control cycle, the output voltage of the power module and the output voltage of the energy storage converter (excluding the inductive voltage drop) remain unchanged, meeting the zero order hold (ZOH) characteristic. Therefore, the relationship between the cascaded energy storage system and the AC network can be expressed as

$$\begin{cases} \sum_{x=1}^{N} S_{ax} \cdot V_{cax} = e_a - L\dfrac{di_{sa}}{dt} - Ri_{sa} \\ \sum_{x=1}^{N} S_{bx} \cdot V_{cbx} = e_b - L\dfrac{di_{sb}}{dt} - Ri_{sb} \\ \sum_{x=1}^{N} S_{cx} \cdot V_{ccx} = e_c - L\dfrac{di_{sc}}{dt} - Ri_{sc} \end{cases} \tag{2}$$

There are some differences among the parameters of each power module and battery cluster contained in each phase, but with the introduction of strategies such as energy balance, this difference can converge to a certain range. At the same time, in order to reduce the complexity of the cascaded energy storage system, (2) can be simplified to obtain the relationship between the switching number of sub modules and network side variables, as shown in equation (3)

$$\begin{cases} S_{Na} \cdot \overline{V}_{ca} = e_a - L\dfrac{di_{sa}}{dt} - Ri_{sa} \\ S_{Nb} \cdot \overline{V}_{cb} = e_b - L\dfrac{di_{sb}}{dt} - Ri_{sb} \\ S_{Nc} \cdot \overline{V}_{cc} = e_c - L\dfrac{di_{sc}}{dt} - Ri_{sc} \end{cases} \tag{3}$$

where S_{Na}, S_{Nb} and S_{Nc} respectively represent the switching number and switching state of each phase sub module, and V_{ca}, V_{cb} and V_{cc} respectively represent the average value of DC side voltage of each phase power module.

Then, (3) can be discretized as

$$\begin{cases} S_{Na}(k) \cdot \overline{V}_{ca}(k) = e_a(k) - \dfrac{L}{T}(i_{sa}(k) - i_{sa}(k-1)) - Ri_{sa}(k) \\ S_{Nb}(k) \cdot \overline{V}_{cb}(k) = e_b(k) - \dfrac{L}{T}(i_{sb}(k) - i_{sb}(k-1)) - Ri_{sb}(k) \\ S_{Nc}(k) \cdot \overline{V}_{cc}(k) = e_a(k) - \dfrac{L}{T}(i_{sc}(k) - i_{sc}(k-1)) - Ri_{sc}(k) \end{cases} \tag{4}$$

In order to realize the short-term prediction of output current or power of the cascaded energy storage system, the unit period derivation of (4) can obtain the predicted value of AC side current of the cascaded energy storage system.

$$\begin{cases} i_{sa}(k+1) = \dfrac{T}{L-RT}(e_a(k+1) - S_{Na}(k+1) \cdot \overline{V}_{ca}(k+1)) + \dfrac{L}{L-RT}i_{sa}(k) \\ i_{sb}(k+1) = \dfrac{T}{L-RT}(e_b(k+1) - S_{Nb}(k+1) \cdot \overline{V}_{cb}(k+1)) + \dfrac{L}{L-RT}i_{sb}(k) \\ i_{sc}(k+1) = \dfrac{T}{L-RT}(e_c(k+1) - S_{Nc}(k+1) \cdot \overline{V}_{cc}(k+1)) + \dfrac{L}{L-RT}i_{sc}(k) \end{cases} \tag{5}$$

Considering that the sampling frequency and control frequency of the digital processor can reach us level, the voltage change at the AC and DC sides in the unit cycle is small and can be ignored, so the current prediction function can be simplified as follows

$$\begin{cases} i_{sa}(k+1) = \dfrac{T}{L-RT}(e_a(k) - S_{Na}(k+1) \cdot \overline{V}_{ca}(k)) + \dfrac{L}{L-RT}i_{sa}(k) \\ i_{sb}(k+1) = \dfrac{T}{L-RT}(e_b(k) - S_{Nb}(k+1) \cdot \overline{V}_{cb}(k)) + \dfrac{L}{L-RT}i_{sb}(k) \\ i_{sc}(k+1) = \dfrac{T}{L-RT}(e_c(k) - S_{Nc}(k+1) \cdot \overline{V}_{cc}(k)) + \dfrac{L}{L-RT}i_{sc}(k) \end{cases} \tag{6}$$

Based on (6), the output current and power of the cascaded energy storage system can be predicted in a short time, so as to obtain a faster control and tracking effect.

3. Optimization of Fast Power Current Predictive Control Strategy

3.1. Establishment of Value Function

Model predictive control (MPC) applied in the field of power electronics is based on traversing the finite set of switch states or searching for optimization in other ways, so as to realize the short-term prediction of the converter, screen out the optimal switch states and act on the system to control them. Therefore, this method can be called finite control set model predictive control (FCS-MPC). The discrete domain mathematical model of direct coupled energy storage system is established, which is the basis of power and current predictive control. The establishment of value function is another important part of model predictive control, and it is also the optimization process of model predictive control based on value function. The establishment of the value function is to transform the control problem into an online optimization problem of the function. In the current prediction link, the three-phase current can be optimized in parallel to obtain the extreme value. Therefore, the current control value function of the cascaded energy storage system can be constructed as follows.

$$G(k+1)=(i_{sa}^*(k+1)-i_{sa}(k+1))^2+(i_{sb}^*(k+1)-i_{sb}(k+1))^2 \\ +(i_{sc}^*(k+1)-i_{sc}(k+1))^2 \tag{7}$$

The minimum value of (7) can be obtained to obtain the combination of the predicted current of the next cycle and the minimum power module of the given deviation.

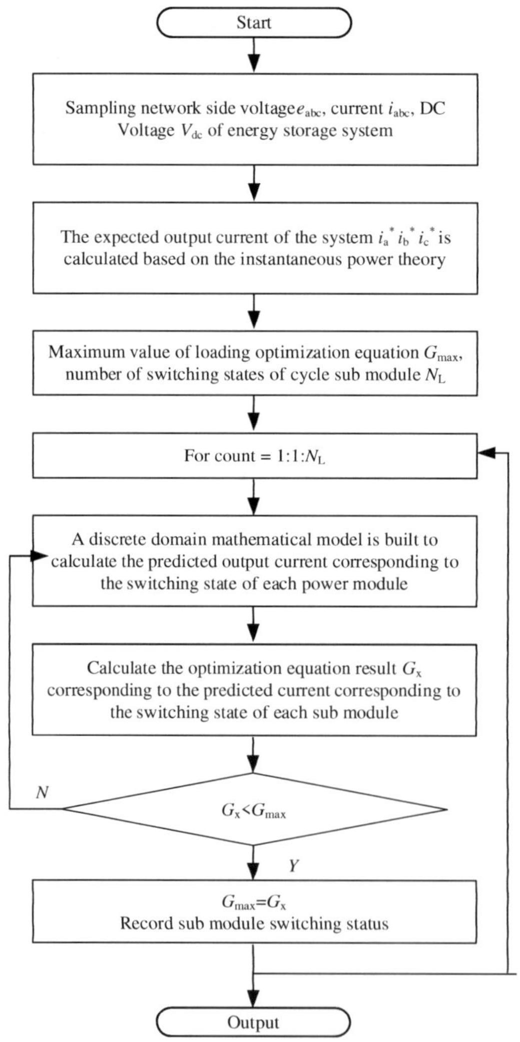

Figure 1. Flow chart of finite set model predictive control for cascaded energy storage system.

The control process of FCS-MPC based cascaded energy storage system is mainly divided into the following steps.

Step1, the AC side voltage $e_{abc}(k)$ and current $i_{abc}(k)$ of the cascaded energy storage system connected to the power grid are sampled, and the DC Side Voltage V_{dc} of the power module, battery SOC and other internal variables are collected.
Step2, based on the instantaneous power theory, the output setting of the system is inversely derived, and the expected output current of the cascaded energy storage system is obtained.
Step3, the mathematical model of the cascaded energy storage system is discretized to predict the AC side current equivalent $i_a(k+1)$, $i_b(k+1)$ and $i_c(k+1)$ of the system in the future control cycle.
Step4, calculate the established current control optimization equation.
Step5, according to the combination number of sub module switching states, the traversal method is used for rolling optimization, and the online calculated G_x is constantly used to refresh G_{max}.
Step6, obtain the optimal sub module switching state and apply it to the cascaded energy storage system.

4. Redundancy Control of Direct Coupled Energy Storage System Based on Model Prediction

In order to improve the safe and reliable operation capability of the cascaded energy storage system, it is not only necessary to fully consider the objectives and constraints in the control system design, but also to give full play to the modular topology characteristics of the cascaded energy storage system, introduce redundant modules, and improve the continuous and reliable operation capability of the system in the electrical topology.

In order to improve the input and bypass speed of the redundant module of the cascaded energy storage system, this paper uses the fast bypass switch composed of IGBT to replace the traditional contactor in the design process of the redundant module, as shown in Figure 2, so as to reduce the input or bypass speed of the redundant module from about 10s to 10~20ms, and avoid the short-term power shortage caused by the action of the system redundant module when the power module or battery pack fails [19].

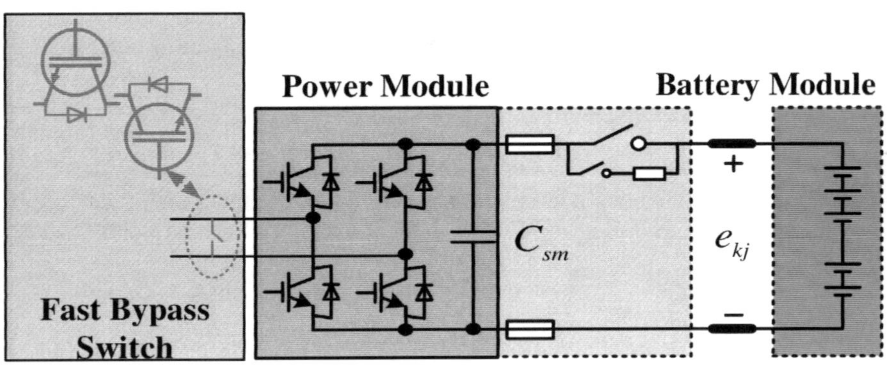

Figure 2. Schematic diagram of electrical topology of redundant module of cascaded energy storage system.

In the process of redundancy control, the cascaded energy storage system shall not only perform electrical action and isolation, but also introduce corresponding control strategies to readjust and distribute the power and current output of each power module, so as to achieve short-term fault transition of the cascaded energy storage system, ensure the output of rated power and improve the reliable operation capacity of the system [19].

4.1. Principle of Redundancy Control Strategy for Cascaded energy storage system

The energy storage system with cascade H-bridge structure is modular. If a certain number of links are reserved, the system can have the function of redundant control. When a link of the system fails, the fault link is bypassed to improve the modulation ratio of the non-fault H-bridge power unit PWM signal in the fault phase, which can maintain the stability and balance of the output voltage at the AC side, so that the system can continue to operate for a period of time. The specific working principle is shown in Figure 3.

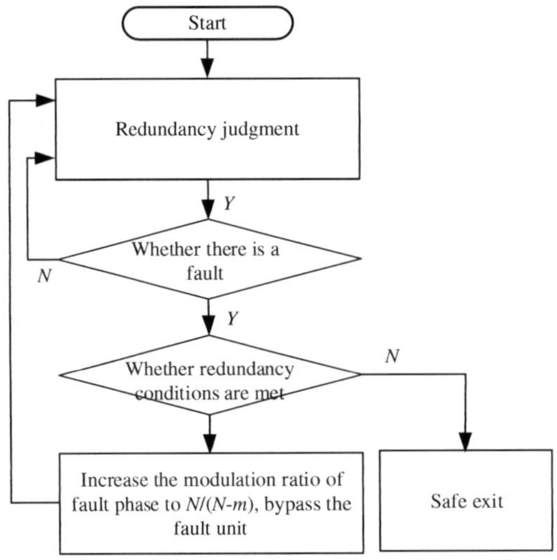

Figure 3. Redundancy control schematic diagram of energy storage system.

When a fault occurs, the system identifies the faulty link and bypasses the faulty power unit. The system judges whether the redundancy conditions are met. If the redundancy conditions are met, bypass the fault power unit and maintain the voltage balance at the AC side to ensure the continuous operation of the system. If the redundancy conditions are not met, the system shall exit safely. Maintain the AC side voltage by increasing the modulation ratio of the non-fault H-bridge power unit PWM signal in the fault phase.

Assuming that M links of phase a are faulted, if the redundancy condition is met, the output voltage modulation ratio of the non-fault link power unit in phase A is $n/(n-m)$ times of the original value, then the AC side input line voltage balance can be maintained in case of fault.

During the simulation, it is set that the first unit of phase a sends out fault, and there are 13 links in phase A. at this time, the output voltage modulation ratio of the non-fault power unit in phase A is changed to 13/12 times of the original.

4.2. Redundancy Control Strategy Based on Model Prediction

The power and current of the cascaded energy storage system can be controlled by fast voltage prediction. In order to improve the voltage output accuracy and rapidity of the cascaded energy storage system, this paper proposes a redundancy control strategy based on model prediction, and its control process is shown in the Figure 4.

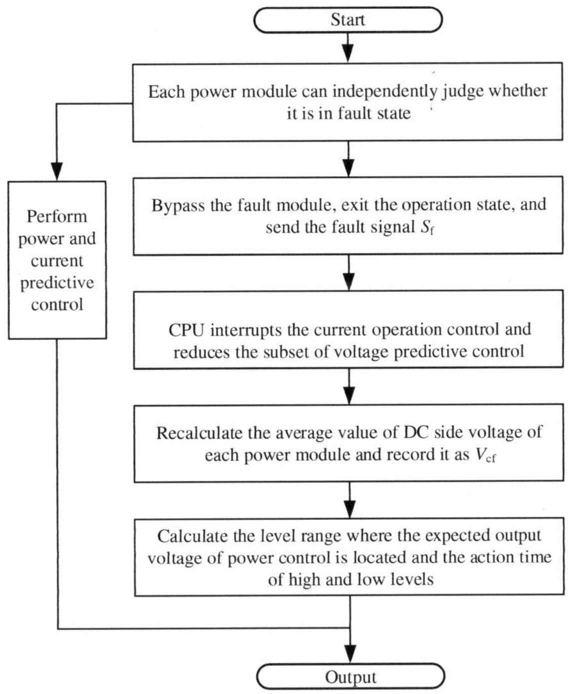

Figure 4. Redundancy control flow chart of cascaded energy storage system based on model prediction.

When a phase power module fault occurs in the system, the power unit first judges the fault information independently, bypasses the module, and sends the fault signal SF and bypass signal to the central processing unit;

After receiving the fault signal, the central processing unit compresses the voltage prediction subset of each phase from 2*N*+1 level range to 2*N*-1 range, interrupts the voltage prediction control strategy in this control cycle, recalculates the average voltage of each phase, and the average DC side voltage obtained can be recorded as VCF [19], which is represented as

$$V_{cf} = \frac{N}{N - N_f} V_c \tag{8}$$

where N_f refers to the number of fault barrier sub modules of each phase of the cascaded energy storage system, and V_{cf} refers to the average DC side voltage of each phase sub module after the fault.

Recalculate the level range of the expected output voltage of the system and the action time of the high and low levels. The formula is as follows:

$$v_{ex}(k+1) = M \cdot V_{dc} \cdot T_{sM} + (M+1) \cdot V_{dc} \cdot T_{sM+1} \tag{9}$$

where T_{sM} and T_{sM} +1 respectively represent the action time of high and low level of the system after the power module fault occurs.

5. Simulation Verification and Analysis

In order to verify the correctness and effectiveness of the proposed safety constrained fast voltage predictive control strategy for the cascaded energy storage system, a Simulink simulation model is built. The simulation environment of this part is consistent with the previous parameters. In order to ensure the safe and stable operation of the cascaded energy storage system, the first step is the design of safety constraints. Figure 5(a) shows the dynamic response characteristics of the fast voltage predictive control of the cascaded energy storage system. As shown in the figure, when the polarity of the system output power is reversed, the system quickly predicts the expected output port voltage and jumps to the target value step by step within the unit control cycle.

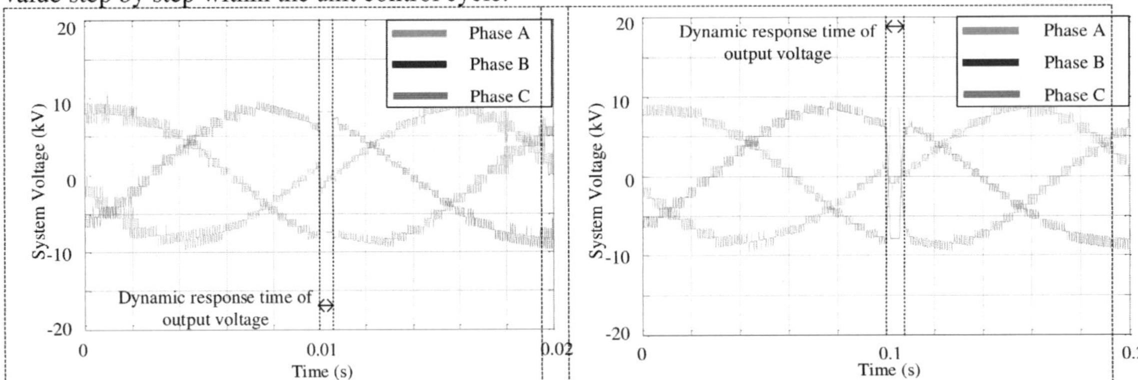

(a) Output voltage waveform of cascaded energy storage converter (FCS-MPC).

(b) Output voltage waveform of cascaded energy storage converter (SS-FVMPC).

Figure 5. The dynamic process comparison simulation curves of output port voltage with different control strategies.

Based on SC-FVMPC, as shown in Figure 5(b), if the output power of the cascaded energy storage system is given a large step, the SC-FVMPC strategy first predicts the system output port voltage, online judges the serial number of the expected output voltage in the level interval, and judges whether the difference between the level interval and the level interval of the previous control cycle is greater than the present threshold (voltage jump threshold, VJT). Its control flow is shown in Figure 4. According to Figure 5(b) dynamic simulation waveform of output port voltage of cascaded energy storage system based on SC-FVMPC, based on this strategy, the system output voltage changes step by step in the dynamic process. Although the dynamic response time is extended, the increase time is controlled within 0.2ms, and the dynamic response time of the system is controlled at about 1ms, which ensures the system safety and excellent dynamic response performance of the system.

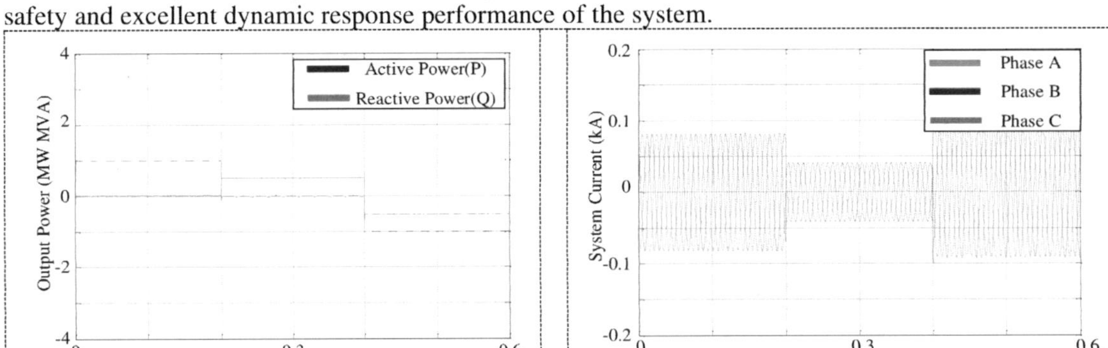

(a) System output active power and reactive power waveform.

(b) Three phase output current waveform of cascaded energy storage converter.

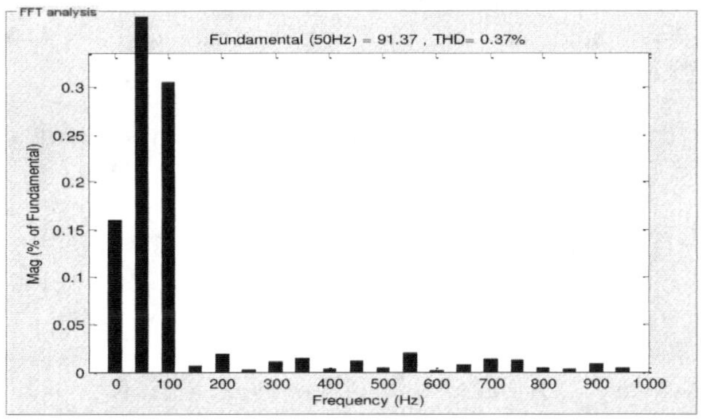

(c) Output current THD of cascaded energy storage converter.

Figure 6. Dynamic response waveform of output power of direct mounted energy storage based on SC-FVMPC.

Figure 6 shows the dynamic waveform of active and reactive power regulation of the cascaded energy storage system based on SC-FVMPC. As shown in the figure, in the initial stage, the cascaded energy storage system operates with 1000kW output active power. At 0.2S, the system output active power steps to 0, the reactive power rises to 500KVA, and at 0.4s, the reactive power steps to -500kw and the active power steps to -1000kw. It can be seen from Figure 6(b) and 6(c) that the cascaded energy storage system based on SC-FVMPC strategy indirectly controls the system output current by quickly predicting the system output port voltage. As shown in the figure, the SC-FVMPC strategy can realize the rapid control of active and reactive power. At the same time, the system output current is three-phase symmetrical, and the harmonic content of the output current is less than 0.4%, which has good output quality.

6. Conclusions

Based on the continuous domain mathematical model of the cascaded energy storage system, a prediction model in the discrete domain of the system is constructed. The problems of the application of finite set model predictive control to the cascaded energy storage system are studied, such as the operation load is too heavy, the concentration and the jump of a single control cycle are too large. An innovative safety constrained fast voltage prediction method for the cascaded energy storage system is proposed, Through the effective prediction of the expected output voltage, the high-precision and effective control of the output power and current of the system can be realized. At the same time, the system can quickly track the given value within 0.7ms, which has excellent dynamic characteristics.

Secondly, in order to realize the continuous and reliable operation of the cascaded energy storage system under the fault state, connect with the fast voltage predictive control technology mentioned above, and improve the dynamic performance of the system redundancy control, a redundancy control method of the cascaded energy storage system based on model prediction is proposed. After the system fault occurs, the central processing unit receives the fault signal quickly, the expected average voltage can be optimized online to complete the whole redundancy process.

Finally, through the theoretical simulation results, it can be seen that within the unit control cycle (< 1ms), the cascaded energy storage system quickly enters the redundant operation mode, the system output power has no shortage, and the output current is continuous, which verifies the effectiveness of the proposed redundant control method.

7. References

[1] Li X, Dong C, Jiang W, et al. "An improved coordination control for a novel hybrid AC/DC microgrid architecture with combined energy storage system." Applied Energy 292.4(2021):116824.

[2] L. Xiong, X. Liu, Y. Liu and F. Zhuo, "Modeling and stability issues of voltage-source converter dominated power systems: a review," in CSEE Journal of Power and Energy Systems, doi: 10.17775/CSEEJPES.2020.03590. (Early Access)

[3] Hajiaghasi, S., A. Salemnia, and M. Hamzeh. "Hybrid energy storage system for microgrids applications: A review." The Journal of Energy Storage 21.FEB.(2019):543-570.

[4] Huang M, He W, Incecik A, et al. "Renewable energy storage and sustainable design of hybrid energy powered ships: A case study." Journal of Energy Storage, 2021, 43: 103266.

[5] L. Xiong, X. Liu, H. Liu and Y. Liu, "Performance comparison of typical frequency response strategies for power systems with high penetration of renewable energy sources," in IEEE Journal on Emerging and Selected Topics in Circuits and Systems, vol. 12, no. 1, pp. 41-47, March 2022.

[6] Cajigal-Núñez J M, Winter P, Wrede H, et al. "Design and Configuration of a Suitable Electrical Energy Storage Device for an MMC-STATCOM with Voltage Source Behavior." 2020 IEEE 11th International Symposium on Power Electronics for Distributed Generation Systems (PEDG) IEEE, 2020.

[7] Zeng W, Li R, Huang L, et al. "Approach to Inertial Compensation of HVDC Offshore Wind Farms by MMC with Ultracapacitor Energy Storage Integration. " IEEE Transactions on Industrial Electronics, 2021.

[8] L. Xiong, X. Liu, L. Liu, and Y. Liu, "Amplitude-phase detection for power converters tied to unbalanced grids with large X/R ratios," IEEE Transactions on Power Electronics, vol.37, no.2, pp. 2100-2112, February 2022.

[9] Gómez P J, Galván L, Galván E, et al. "Energy Storage Systems Current Ripple Reduction for DC-Link Balancing Method in Hybrid CHB Topology." IECON 2020 - 46th Annual Conference of the IEEE Industrial Electronics Society IEEE, 2020.

[10] Rodriguez E, Leyva R, Liang G, et al. "Passivity Control in Modular Battery Energy Storage Systems." 2021 IEEE Energy Conversion Congress and Exposition (ECCE). IEEE, 2021: 148-152.

[11] Pirooz A, Firouz Y, Van Mierlo J, et al. "Voltage vector redundancy exploitation for battery balancing in three-phase CHB-based modular energy storage systems. " IEEE Transactions on Industrial Electronics, 2021.

[12] Maharjan L, Inoue S, Akagi H, et al. "State-of-Charge (SOC)-Balancing Control of a Battery Energy Storage System Based on a Cascade PWM Converter." IEEE Transactions on Power Electronics 24.6(2009):1628-1636.

[13] L. Xiong, F. Zhuo, "Chopper controller based DC voltage control strategy for cascaded multilevel STATCOM," in Journal of Electrical Engineering & Technology, vol.9, no.2, pp.576-588, Mar. 2014.

[14] Easley M, Shadmand M B, Abu-Rub H. "Computationally-efficient optimal control of cascaded multilevel inverters with power balance for energy storage systems." IEEE Transactions on Industrial Electronics, 2021, 68(12): 12285-12295.

[15] Li N, Gao F, Hao T, et al. "SOH Balancing Control Method for MMC Battery Energy Storage System." IEEE Transactions on Industrial Electronics, 2017:1-1.

[16] Bayat H, Yazdani A. "A Hybrid MMC-Based Photovoltaic and Battery Energy Storage System." IEEE Power & Energy Technology Systems Journal, 2019:1-1.

[17] Gan C, Sun Q, Wu J, et al. "MMC-Based SRM Drives with Decentralized Battery Energy Storage System for Hybrid Electric Vehicles." IEEE Transactions on Power Electronics, 2018, PP(99):1-1.

[18] Danqing L, Guangzhu W, Zhujian O, et al. "A Control Strategy of MMC Battery Energy Storage System Based on Arm Current Control." 2018 International Power Electronics Conference (IPEC-Niigata 2018 –ECCE Asia). 2018.

[19] L. Xiong, F. Zhuo, X. Liu, et al, "Fault-tolerant control of CPS-PWM based cascaded multilevel inverter with faulty units," in IEEE Journal of Emerging and Selected Topics in Power Electronics, vol. 7, no. 4, pp. 2486-2497, Dec. 2019.

Acknowledgments
This work was supported by the State Grid Jiangsu Electric Power Co., Ltd. Science and Technology Project, i.e., Research on Redundancy Control and Condition Evaluation Technology of Medium Voltage Cascaded energy storage system (J2021017).

Study of Electromagnetic Characteristics of Silicon Steel Sheet and Transformer Vibration Under Different Tension/Compression Stress

Chu Houguang, Li Longfei, Liu Lei, Zhang Fei, Zhu Sirui, Zhou Libing, Du Zhaoguang and Wen Jintao

State Grid Xinjiang Electric Power Co., Ltd. Electric Power Science Research Institute, Wulumuqi, 830011

chuhouguang19840409@126. com

Abstract. In order to study the electromagnetic and vibration performance of transformer under different tensile and compressive stresses. This paper firstly studies the magnetic properties of core silicon steel sheets for transformers under different tensile and compressive stresses, the B-H curves, B-P curves, and permeability curves of silicon steel sheets under different tensile and compressive stresses are given. Secondly, based on the test results of magnetic characteristics of the silicon steel sheet, the electromagnetic performance and vibration of a 3kVA transformer are simulated and analyzed. In the simulation process, the influence of different tensile and compressive stresses on the performance of the silicon steel sheet core of the transformer is considered. It provides a new idea for the electromagnetic-vibration analysis of transformers.

1. Introduction

With the continuous development of China's industry and the increasing improvement of people's living standards, people in daily life not only have a stronger dependence on the demand for electricity, but also put forward higher requirements for the vibration and noise pollution caused by the transformer which can adjust the voltage level.The vibration of transformer body (winding, core, oil tank) and cooling device is the source of noise, and the vibration of core is mainly caused by magnetostriction of silicon steel sheet during operation.Under the action of stress, the main domain and sub domain in silicon steel sheet will be deformed, which will change the magnetic properties and magnetostrictive properties, resulting in the increase of core vibration and noise[1-3].

Magnetostriction refers to the size change of ferromagnetic materials during magnetization.Magnetostriction can be divided into spontaneous magnetostriction and field magnetostriction. The former is related to the internal magnetic field of ferromagnetic materials, and the latter is caused by the interaction between the external magnetic field and the sample[4].The magnetostrictive properties of oriented silicon steel sheet and non oriented silicon steel sheet are very different under the excitation of alternating magnetic field, rotating magnetic field and alternating magnetic field with DC bias, respectively[4-6].The magnetostrictive phenomenon of silicon steel sheet is one of the main reasons for the vibration and noise of power transformer, motor and other electromagnetic equipment, and the measurement of magnetostrictive characteristics of silicon steel sheet is the premise and basis for the study of vibration and noise of electrical equipment.Foreign research on magnetostrictive measurement technology started earlier, and gradually formed an advanced non-contact magnetostrictive measurement technology[7-9].In 2010, Japanese enokizono team obtained the relationship between vector magnetic characteristics and applied stress through stress measurement system[10-11], In 2011, the effect of mechanical stress on the magnetic properties of electrical silicon steel sheet under alternating magnetic flux was studied, and the different effects of tensile and compressive stress on permeability and power loss were obtained[12-13].

In this paper, a set of magnetostrictive measuring equipment and a stress effect measuring system are used,the magnetic and magnetostrictive properties of grain oriented silicon steel (grade: 27sqgd090) sheet under different tensile and compressive stresses along the rolling direction were tested and studied. The B-H curve, B-P curve, permeability curve and single value curve of silicon steel sheet under different tensile and compressive stresses were given.Secondly, based on the above measurement results, the model material library is established, and the finite element analysis of a 3KVA transformer under different tension and compression stress conditions is carried out based on COMSOL finite element analysis software.Finally, an experimental study is carried out to verify the correctness of the proposed scheme, which provides a reference for the study of power transformer vibration and noise under different tension and compression stresses.

2. Analysis of magnetic properties of silicon steel sheet under different tensile and compressive stresses

2.1. Measuring device

The main equipment of the magnetostrictive experimental measurement platform is a German brockhous mst500 one-dimensional single-chip magnetostrictive measurement system, as shown in Figure 1.The measurement system consists of three parts: host, magnetic circuit device and laser transmitting and receiving device. In order to prevent the measurement error caused by uneven ground and vibration, the whole magnetic circuit device and laser transmitting and receiving device are placed on a suspended platform supported by air compressor. Before the experiment, air pump is used to inflate them.The effective size of the sample to be measured is 500mm × 100mm. During the measurement, one section of the sample is clamped, and the other end can stretch freely. The shape variable of the silicon steel sheet is calculated by detecting the distance between the laser and the reflector. The working frequency of the laser is 100Hz, and the resolution is 10nm / m.

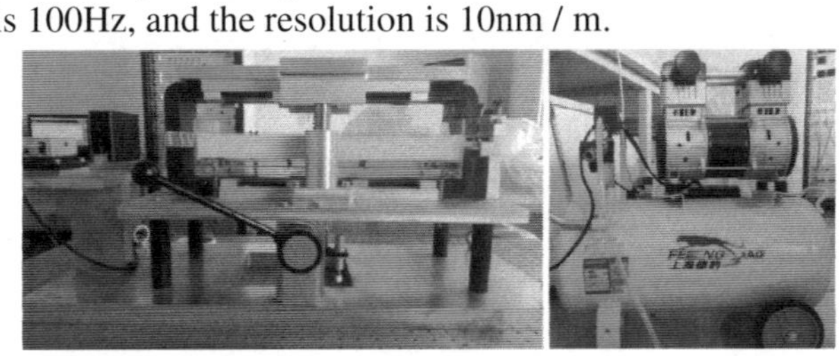

Figure 1. Equipment for measuring magnetostrictive

2.2. F Measurement results and analysis

The magnetic properties and magnetostrictive properties of 27sqgd090 oriented silicon steel sheet along the rolling direction under 50 Hz sinusoidal excitation were measured. In order to prevent the silicon steel sheet from bending, the applied stress range is - 8 MPa (compressive stress) to 8 MPa (tensile stress).The magnetization curve of silicon steel sheet under different stress is shown in Fig. 2. It can be seen from the figure that before the sample sheet enters the magnetic saturation, the magnetization curve begins to shift to the right with the increase of external pressure stress, that is to say, the H value of the same B in the sample sheet increases with the increase of stress in the magnetization process, and the effect is particularly obvious.When H reaches 88A/m without external force, the sample begins to saturate, and the inner flux density B is about 1.6T. When the external force is - 8Mpa and 8Mpa, the H values are 310A/m and 41A/m respectively. When the sample goes into saturation state and continues to increase the value of H, the increase speed of magnetic flux density B is very slow, basically does not change, and the maximum can reach about 1.9T.

a. Tensile stress b. Compressive stress

Figure 2. Effect of stress on core loss curve in rolling direction of oriented silicon steel sheet

Figure 3 shows the relative permeability curve of silicon steel sheet under different stress. It can be seen from the figure that the relative permeability curve moves down with the increase of stress, no matter tensile stress or compressive stress, which reduces the magnetic conductivity of the sample sheet.Figure 3 (a) shows the pull-out curve.It can be seen that with the increase of tensile force, the relative permeability decreases in turn, but does not change the shape of the curve. The peak permeability of each curve appears when the flux density B is equal to 1.2T, indicating that the tensile stress does not change the saturation point and other magnetic properties of the sample. When B is greater than 1.2T, the silicon steel sheet enters the magnetic saturation, and the effect of tensile stress on the permeability becomes smaller When B is greater than 1.75T, the values of relative permeability under different tensile stress are close to each other, which indicates that tensile stress has little effect on the magnetic conductivity of silicon steel sheet when its depth is saturated.
Fig. 3 (b) shows the law of the influence of the compressive stress on the sample. It can be seen that the pressure has a great influence on the relative permeability. With the increase of the pressure, the curve moves down obviously and the shape changes greatly. The permeability peak points appear at different b values and advance continuously. When the force increases from 0Mpa to - 8Mpa, the b value at the relative permeability peak point also changes from 1.2T to 0.6T, indicating that the compressive stress increases

a. Tensile stress b. Compressive stress

Figure 3. Effect of stress on relative permeability in rolling direction of oriented silicon steel sheet

Figure 4 (a) shows the effect of tensile stress on the iron loss curve of silicon steel sheet. It can be seen that when the tensile stress is small, the iron loss decreases. When the force increases from 0MPa to 2MPa, P1.7/50 increases from 0.829W/kg to 0.785W/kg, and the iron loss is reduced by 0.044W/kg. As the tensile stress continues to increase, the iron loss gradually increases, and is greater than that without stress. When the tensile stress is 4MPa,P1.7/50 is 0.89W/kg, increased by 0.061W/kg; when the tensile stress is 6Mpa, P1.7/50 is 0.98W/kg, increased by 0.151W/kg; when the tensile stress is 8Mpa,P1.7/50 is 1.099W/kg, increased by 0.27W/kg.

Figure 4 (b) shows the influence of compressive stress on the iron loss curve of silicon steel sheet. It can be seen that the iron loss P1.7/50 increases significantly with the increase of compressive stress.

When the compressive stress increases from 0 MPa to 8 MPa, P1.7/50 increases from 0.829 W/kg to 1.393 W/kg, with an increase of 0.564 W/kg.

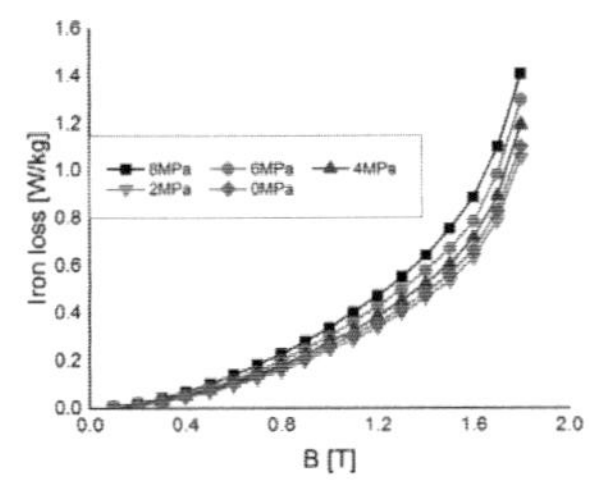

c. Tensile stress

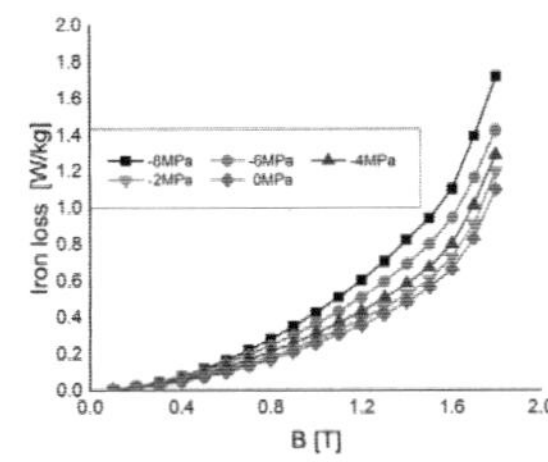

d. Compressive stress

Figure 4. Effect of stress on core loss curve in rolling direction of oriented silicon steel sheet

Figure 5(a) shows the peak to peak curve of magnetostriction of silicon steel sheet under different stresses. It can be seen that the effects of tensile stress and compressive stress on magnetostriction are completely opposite. The compressive stress increases the magnetostriction, while the tensile stress decreases the magnetostriction, and the magnetostriction is especially important for the compressive stress sensitive.Fig.5(b) shows the influence of different stresses on the magnetostrictive peak when the flux density B is equal to 1.0T, 1.3T and 1.5T. It can be seen that the magnetostrictive peak decreases with the increase of applied stress.When the compressive stress is 6Mpa, the peak values of magnetostriction are 5000.03nm/m, 3801.61nm/m and 1999.73nm/m respectively, which are 2.86 times, 3.46 times and 5 times of those without stress;When the compressive stress is 8 MPa, the peak values of magnetostriction are 6630.66 nm/m, 4998.6 nm/m and 2801.12 nm/m respectively, which are 3.79 times, 4.55 times and 7 times of those without stress;When the tensile stress is 6MPa, the peak values of magnetostriction are 1499.87nm/m, 898.95nm/m and 289.6nm/m respectively, which are 0.86 times, 0.82 times and 0.724 times of those without stress;When the tensile stress is 8 MPa, the peak values of magnetostriction are 1099.89 nm/m, 629.76 nm/m and 210 nm/m respectively, which are 0.63 times, 0.573 times and 0.525 times of those without stress.It can be concluded that appropriate applied stress can reduce the magnetostriction of the core silicon steel sheet, so as to reduce the vibration of the transformer.

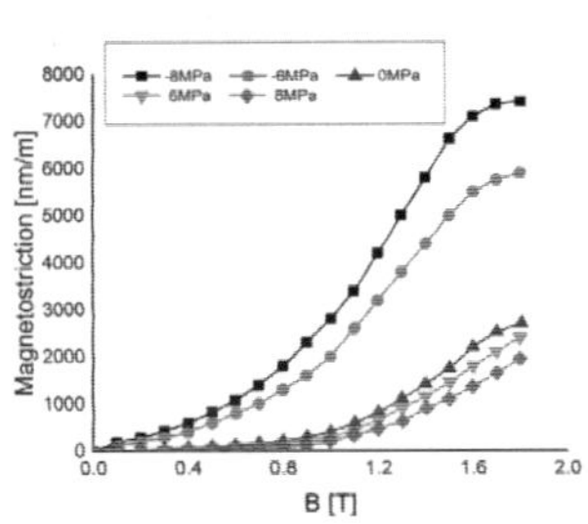

a. Magnetostriction peak to peak under different stresses

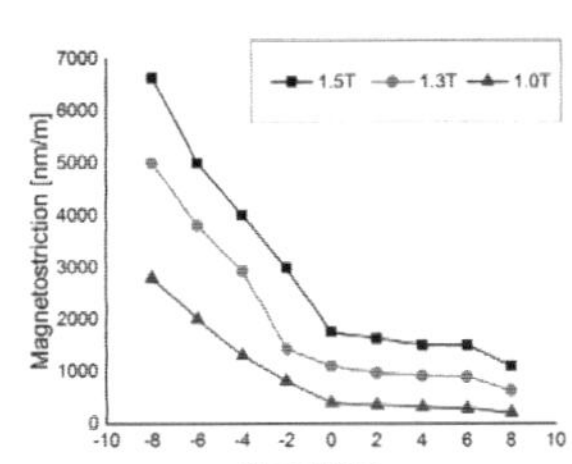

b. B = 1.0T, 1.3t, 1.5T, magnetostriction peak to peak value versus stress curve

Figure 5. Effect of orientation on rolling direction of silicon steel

3. Performance simulation of 3KVA power transformer under different tension and compression stresses

According to the experimental data obtained from the above tests, the electromagnetic vibration characteristics of a 3KVA transformer under different tension and compression stresses are simulated and analyzed based on COMSOL finite element analysis software. Table 1 shows the basic parameters of the transformer.

In Figure 6, μ_0 is the vacuum permeability; μ_r is the relative permeability; A is the vector magnetic potential; J is the current density; E is the electric field strength; M is the quality matrix; C is the damping matrix; K is the stiffness matrix; ξ is the damping ratio; Δ is the partial derivative of displacement to time; λ is a shape variable; σ is the stress; I_r is energy functional.

Table 1. Basic parameters of transformer

Project	Primary side	Secondary side
Rated capacity	3kVA	3kVA
Rated voltage	220V	110V
Rated current	13.6A	27.3A
Number of turns	110	55

Figure 7 shows the magnetic field distribution of power transformer under different stresses. With the increase of applied stress, the internal magnetic density of power transformer will increase. The maximum magnetic density is 1.49T at 0MPa, and 1.5T and 1.53T at 8MPa.

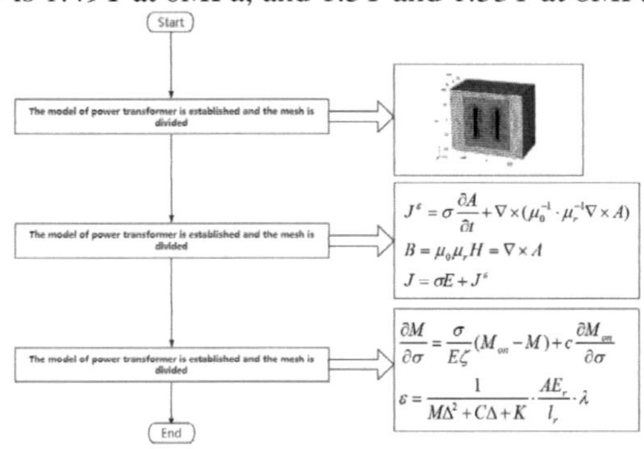

Figure 6. Calculation process of the vibration for transformer

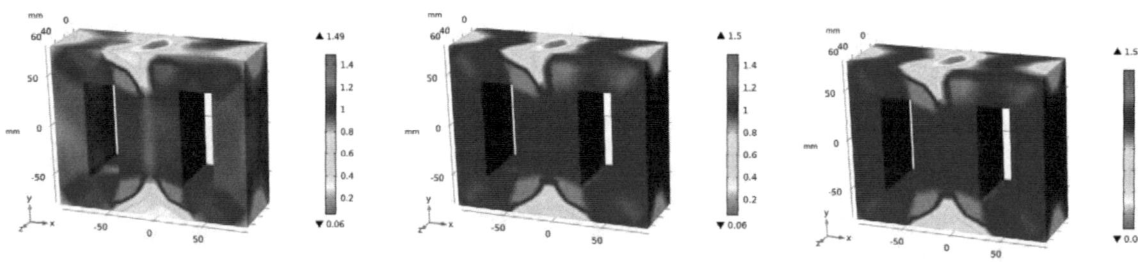

a. 0MPa magnetic field distribution b. 8MPa magnetic field distribution c. -8MPa magnetic field distribution

Figure 7. Magnetic field distribution of power transformer under different stresses

Fig. 8 show the vibration displacement diagram and stress distribution diagram of power transformer core when no stress and compressive stress are 8 MPa. The maximum vibration displacement is 0.46 μm and the vibration stress is 7.79 $\times 10^5$ N/mm^2, while the maximum vibration displacement is 0.5 μm and the vibration stress is 7.93 $\times 10^5$ N/mm^2. Therefore, it can be found that the applied compressive stress intensifies the vibration of the power transformer core, but from the analysis of the calculation results, the compressive stress only changes the size of the vibration, and does not change the vibration form.

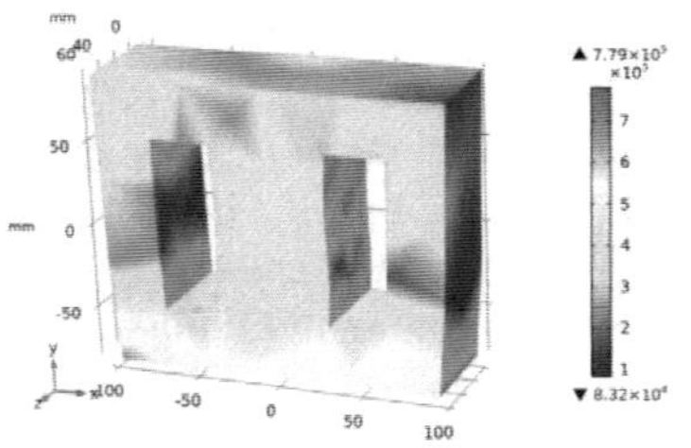

a. 0MPa stress field distribution

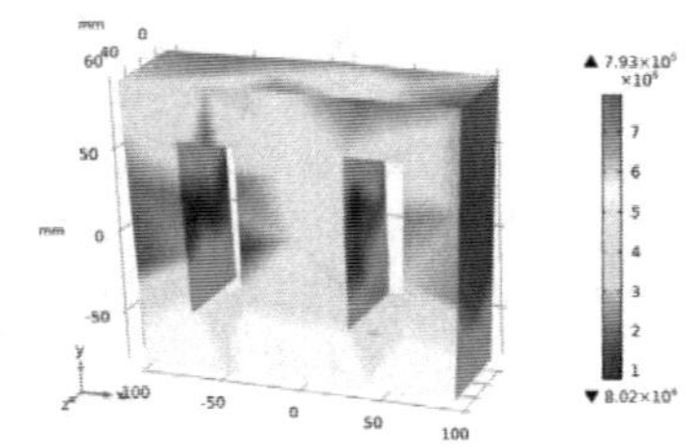

b. -8MPa stress field distribution

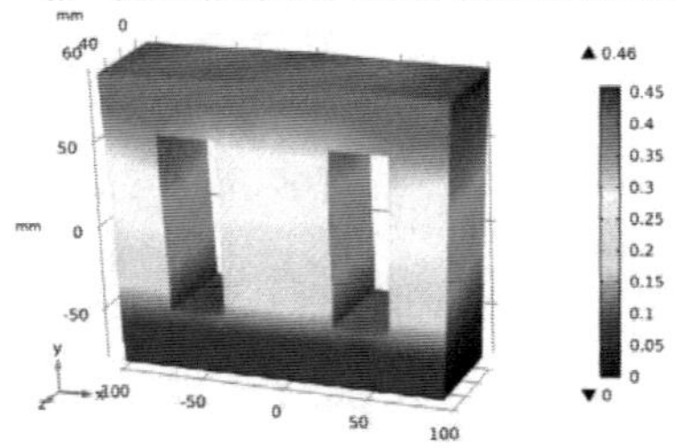

c. 0MPa vibration displacement diagram of iron core

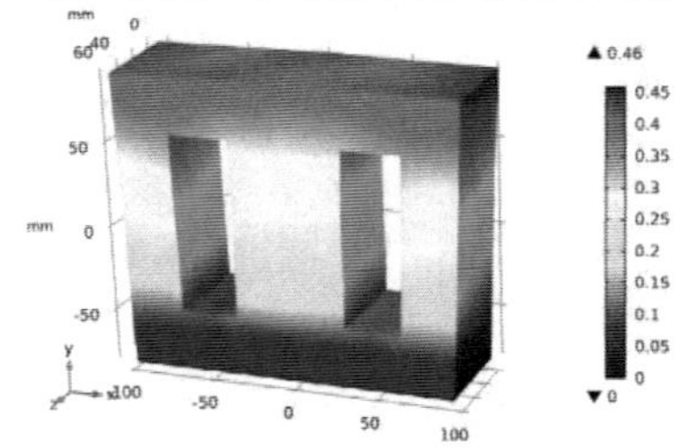

d. -8MPa vibration displacement diagram of iron core

Figure 8. Stress field distribution and vibration displacement diagram of power transformer core

Fig. 9 takes the calculation of power transformer vertex as an example to extract the vibration time domain and frequency domain waveforms of this point when there is no stress. It can be seen from Fig. 9 (a) that the period of vibration time domain signal is 0.01s, which is half of the period of power supply voltage, so its frequency is twice of the power supply frequency, that is, 100Hz, which is consistent with the result of spectrum calculation in FIG. 9 (b), and the spectrum energy is mainly concentrated at the frequency point of 100Hz.

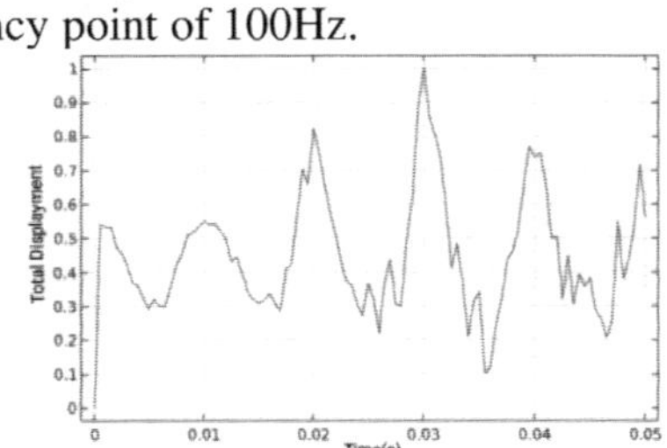

a. Time domain waveform

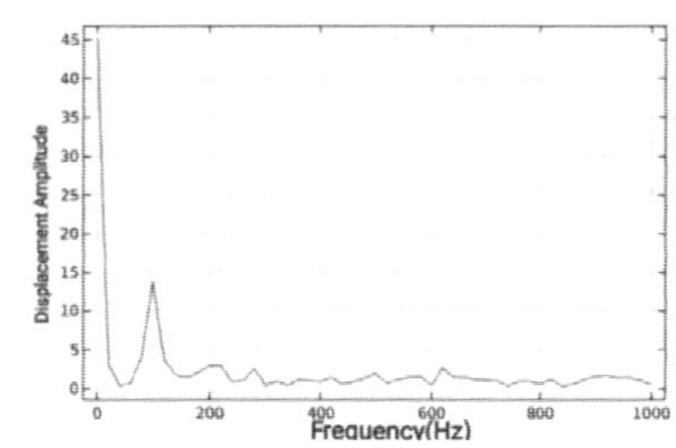

b. Frequency domain waveform

Figure 9. Time domain and frequency spectrum waveform of power transformer vibration without stress

4. Conclusion

In this paper, the magnetic characteristics of a single oriented silicon steel sheet under different tension and compression stresses are studied, and the simulation analysis and Research on the magnetic characteristics and vibration characteristics of a 3KV power transformer are carried out.

(a) The experimental results show that the magnetostriction decreases when the tensile stress is applied along the rolling direction, and increases when the compressive stress is applied, especially when the compressive stress is applied. At the same time, the stress increases the loss value of silicon steel sheet.

(b) The experimental results show that the stress reduces the magnetic conductivity along the rolling direction. Among them, the tensile stress only reduces the permeability value, and does not change the saturation point and other magnetic properties of the silicon steel sheet, and when the silicon steel sheet is deeply saturated, the tensile stress has little effect on its magnetic

conductivity; the compressive stress greatly reduces the magnetic conductivity of the oriented silicon steel sheet, and with the increase of the pressure, the permeability peak point gradually advances, at the same time, the compressive stress changes the oriented silicon steel sheet The time for the steel sheet to enter magnetic saturation.

(c) The simulation results show that for a 3KVA power transformer, when the external pressure stress is 8Mpa, the core vibration displacement increases by 8.69%, and the vibration stress increases by 1.79%.

5. References

[1] Zhu Lihua, Yang Qingxin, Yan Rongge, Zhang Xian. Research on vibration and noise of power *transformer cores including magnetostriction effects*. Transactions of China Electrotechnical Society, 2013, 28(4): 1-6.

[2] Mizokami M, Yabumoto M, Okazaki Y. *Vibration analysis of a 3-phase model transformer core*. Electrical Engineering in Japan, 1997, 119(1): 1-8.

[3] Li Hongzhi, Cui Xiang, Liu Dongsheng, Lu Tiebing, Cheng Zhiguang. *Influence on three phase power transformer by DC bias excitation*. Transactions of China Electrotechnical Society, 2010, 25(5)： 88-96.

[4] S.Somkun, A. J. Moses, P. I. Anderson. *Measurement and modeling of 2-D magnetostriction of nonoriented electrical stee*l. IEEE Transactions on Magnetics, 2012, 48(2): 711-714.

[5] Y.Kai, T. Todaka, M. Enokizono, et al. *Measurement of the two-dimensional magnetostriction and the vector magnetic property for a non-oriented electrical steel sheet under stress*. Journal of Applied Physics, 2012, 111: 07E320.

[6] S. Somkun, A. J. Moses, P. I. Anderson, P. Klimczyk. *Magneto-striction anisotropy and rotational magnetostriction of nonoriented electrical steel*. IEEE Transactions on Magnetics, 2010, 46(2): 302-305.

[7] P. I. Anderson, A. J. Moses, H. J. Stanbury. *Assessment of the stress sensitivity of magnetostriction in grain-oriented silicon steel.* IEEE Transactions on Magnetics, 2007, 43(8): 3467-3476.

[8] P. I. Anderson. Measurement of the stress sensitivity of magnetostriction in electrical steels under distorted waveform conditions. Journal of Magnetism and Magnetic Materials, 2008, 320 (20): 583-588.

[9] B. Weiser, H. Pfützner, J. Anger. *Relevance of magnetostriction and forces for the generation of audible noise of transformer cores*. IEEE Transactions on Magnetics, 2000, 36(5): 3759-3777.

[10] Kai Y, Tsuchida Y, Todaka T, et al. *Effect of local residual stress in rotating machine core on vector magnetic property.* The XIX International Conference on Electrical Machines-ICEM 2010.2010.

[11] Gao Yanhui ， Muramatsu Kazuhiro ， Fujiwara Koji ， et al. *Vibration analysis of a reactor driven by an inverter power supply considering electromagnetism and magnetostriction*. IEEE Transactions on Magnetics， 2009， 45(10)： 4789－4792.

[12] Kai Y, Tsuchida Y, Todaka T, et al. *Influence of stress on vector magnetic property under alternating magnetic flux conditions*. IEEE Transactions on Magnetics, 2011, 47(10): 4344-4347.

[13] Gao Yanhui， Nagata Masahide， Muramatsu Kazuhiro， et al. *Noise reduction of a three-phase reactor by optimization of gaps between cores considering electromagnetism and magnetostriction*. IEEE Transactions on Magnetics， 2011， 47（10）： 2772－2775.

Acknowledge

This research was supported by the science and technology project of State Grid Corporation (SGXJDK00PJJS2100096).

Development and Application of Infrared Video Intelligent Diagnosis Device for Substation Equipment Based on Edge Calculation

Guangwei Liu[1,2], Ka Mu[1,2], Boyan Liu[1,2], Xinsheng Ma[1,2], Zhen Hao[1,2] and Jixian Ma[1,2]

[1]State Grid Jibei Electric Power Co., Ltd, Beijing, China
[2]North China Electric Power Research Institute Co., Ltd, Beijing, China

gwliu2016@qq.com

Abstract. In recent years, infrared detection technology is widely used in the field of state detection of power equipment. It can timely and effectively find equipment defects and avoid the further development of equipment defects. However, the conventional manual inspection has some problems, such as heavy workload, long time-consuming and low efficiency. Based on the research of infrared image intelligent detection technology of substation equipment based on deep learning algorithm and digital image lossless compression technology, a set of infrared video intelligent diagnosis device of substation equipment based on edge calculation is developed and applied in practice. The device realizes local real-time rapid processing and timely diagnosis of equipment defects, and the accuracy of defect identification is about 94.44%, which greatly improves the safe and stable operation level and power supply reliability of substation equipment.

1. Introduction

In recent years, infrared detection technology is widely used in the field of state detection of power equipment when it is live. It can timely and effectively find equipment defects and avoid the further development of equipment defects. However, the conventional manual inspection has some problems, such as heavy workload, long time-consuming and low efficiency. The improvement of computer computing ability and the breakthrough of artificial intelligence technology provide conditions for infrared video intelligent detection of substation equipment.

Scholars at home and abroad have carried out in-depth research on front-end intelligent technology based on edge computing and neural network acceleration technology [1-4]. Reference [5] uses wavelet transform and support vector machine to identify insulator damage in distribution network video monitoring. On the basis of SLIC, HSV, MSER and other segmentation algorithms, scholars from State Grid Research Institute and Tianjin University of technology first segment the fault equipment area, then classify the electrical equipment in the fault equipment area based on convolutional neural network, and give the types of fault equipment [6]. Reference [7] adopts a fully connected recursive neural network nonlinear predictor, which is also combined with speech coding algorithm. Compared with the previous common predictors, this method can obtain very low transmission coding speed and improve the coding effect.

The research group has carried out relevant research on infrared image intelligent detection technology of substation equipment based on deep learning algorithm and digital image lossless compression technology [8]. On the basis of the research, this paper starts to develop a set of infrared video intelligent diagnosis device for substation equipment based on edge calculation and carry out practical application, so as to realize the local real-time fast processing of infrared video image of patrol inspection of the acquisition device and the timely diagnosis of equipment defects, which greatly improves the safe and stable operation level and power supply reliability of substation equipment.

2. Overall design of device system

Taking FPGA as the hardware platform, the hardware mapping of convolutional neural network software algorithm is carried out to realize the detailed and structured description of the algorithm. The convolutional neural network is divided into three key modules: convolution operation module, sampling operation module and activation function. The architecture of high-performance embedded heterogeneous system based on FPGA is shown in Figure 1.

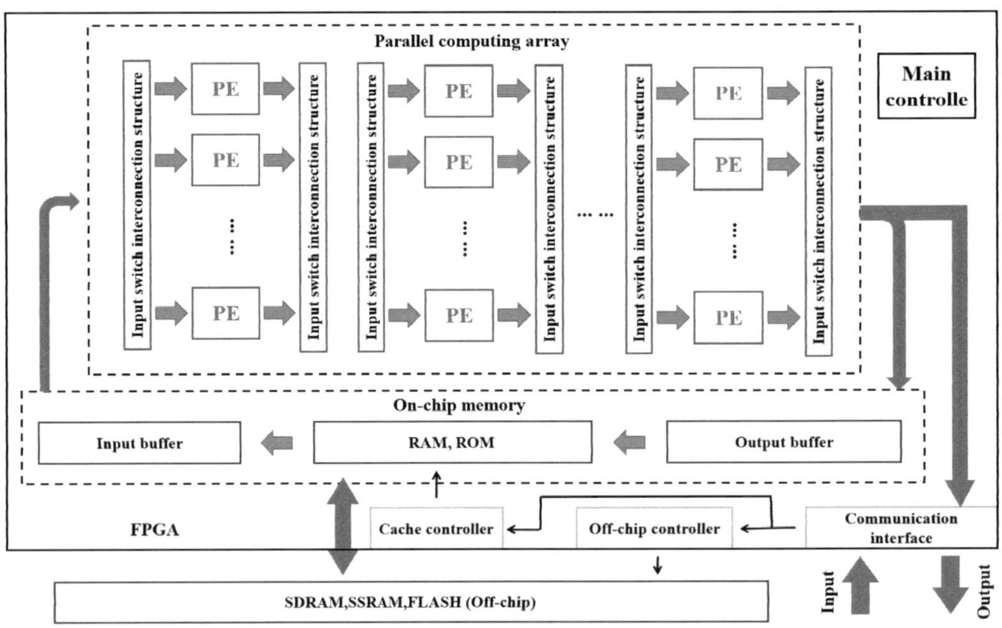

Figure 1. Architecture diagram of high performance embedded heterogeneous system based on FPGA.

The overall architecture is mainly divided into two parts: one is the internal resources on the FPGA chip; The second is the external resources outside the FPGA chip. Internal resources include on-chip memory, parallel computing resources, controller and peripheral standard communication interface, etc; External resources mainly include off chip memory and host, which are respectively used to store a large number of intermediate results generated in the calculation process and user's man-machine interface operation.

The network parameters required for convolution neural network operation, such as convolution kernel parameters and sampling coefficients, are trained by the CPU software of the host computer.

1. The system transmits the network parameters obtained by software training, the input image data to be identified and classified and a series of control instructions to the internal resources of FPGA through the communication interface. The communication interface can be PCI bus, PCI Express bus, USB, serial interface and so on.

2. When FPGA receives the instruction of convolutional neural network to start running, it starts the main program of the system and enters the input image stored in the on-chip memory into the parallel computing array through the input buffer.
3. The data of different computing units are allocated to different computing units in parallel.
1) If the parallel computing array obtains the final output result, the calculation result is fed back to the host CPU through the communication interface;
2) If the output characteristic map of a structure layer is obtained by the parallel computing array, the calculation results are transmitted to the off chip memory for saving, waiting for the calculation of the next structure layer to start, or transmitted to the on-chip memory through the output buffer, and continue to return to the parallel computing array for the calculation of the next structure layer;
3) If the parallel computing array obtains the intermediate result of the output feature map, it will directly return to the parallel computing array through the cache buffer and wait for the next step of computing processing.

On chip cache controller and off chip controller control the operation of on-chip memory and off chip memory respectively. The main controller macroscopically controls the coordination of all resources within the FPGA, such as arm, MCU, and even the IP core designed by itself. In the parallel computing array, input switch interconnection structure and output switch interconnection structure are added to the input and output ends of the computing unit PE respectively to facilitate the driving and distribution of input and output data and ensure the fluency and efficiency of data calculation.

3. Hardware design of infrared image intelligent diagnosis device

Based on the research of embedded computing device, how to integrate infrared acquisition equipment with embedded computing device is the key to the commercialization of this research result.

The prototype design of infrared image intelligent diagnosis device is mainly divided into three units: the first layer is AI processing unit, which solidifies the trained infrared image intelligent recognition and diagnosis model algorithm based on FPGA hardware framework, realizes the intelligent recognition and diagnosis of infrared image, and sends the diagnosis results to the main control and processing unit through serial port; The second layer is the power interface unit, which provides interface format conversion and power management for the whole device; The third layer is the main control unit, which labels the original infrared image according to the feedback information of AI processing unit, and outputs the infrared video data after labeling processing for display. The hardware structure diagram of infrared intelligent diagnosis device is shown in Figure 2.

The main control unit selects CPU chip as the processor, including ARM chip, RAM chip, other supporting management chips and pin interfaces, power interface unit, onboard power management chip, video conversion chip lt8819ex, bt1120 interface, HDMI interface, and DC/DC power module, etc. The AI processing unit of the first layer and the power interface unit of the second layer are connected through 36 pin FFC interface, and the power interface unit of the second layer and the main control unit of the third layer are connected by double row pins.

The infrared intelligent diagnosis device is composed of the device body and the upper computer display software. The device body consists of four parts: power supply module, infrared camera module, AI calculation unit and data exchange module. The specific module block diagram and physical drawing of the device are shown in Figure 3. The appearance diagrams of the device are shown in Figure 4-5.

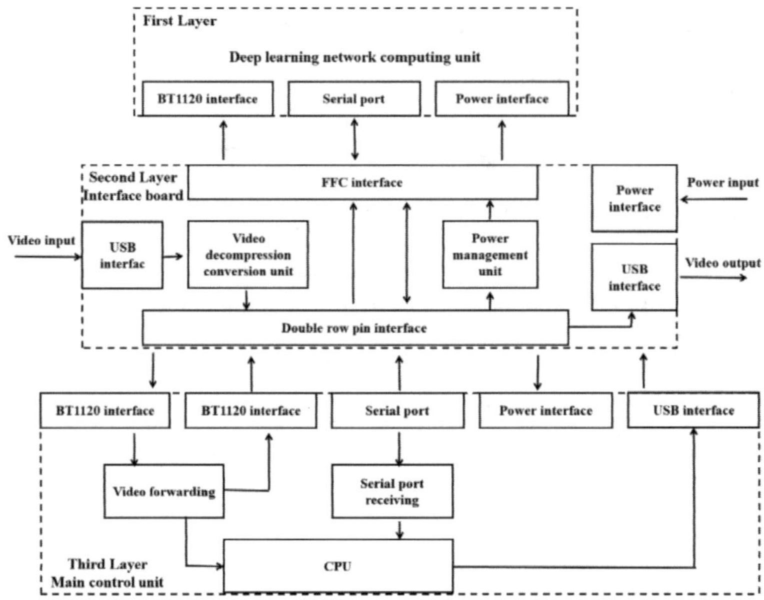

Figure 2. Hardware structure diagram of infrared intelligent diagnosis device.

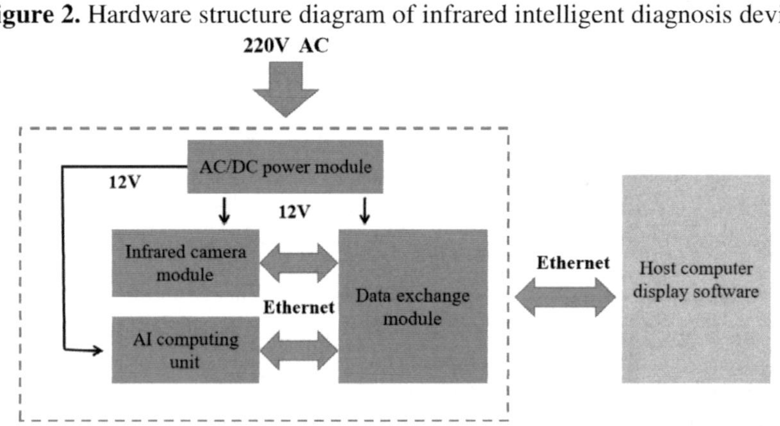

Figure 3. Specific module block diagram and physical drawing of the device.

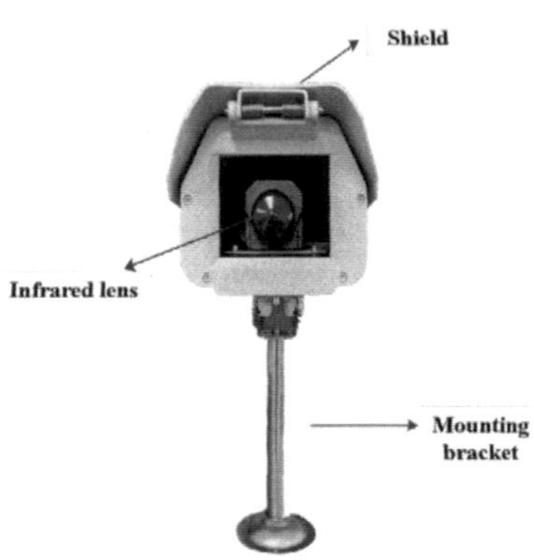

Figure 4. Appearance diagram of the front of the device.

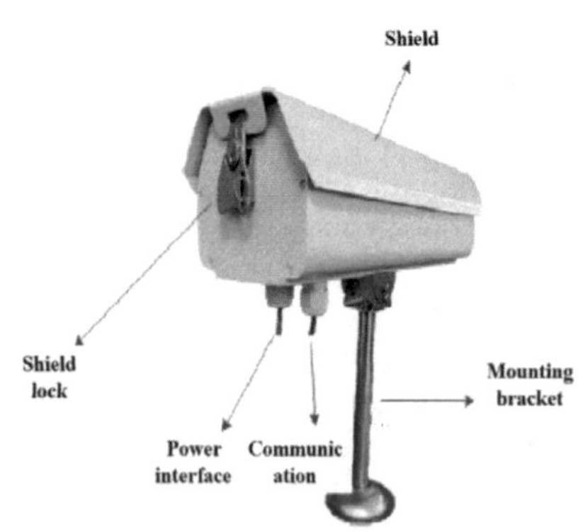

Figure 5. Appearance diagram of the back of the device.

4. Application of infrared video intelligent diagnosis device for substation equipment

The developed infrared video intelligent diagnosis device for substation equipment is used to test 138 known defect data sets. The identification results of current-inductive-heated defect of the typical equipment are shown in Figure 6, and the red box is the defect identification results. The two indexes, defect recognition rate and misjudgment rate, are used to illustrate the performance of intelligent discrimination of current heating defects. The defect recognition rate represents the proportion of the defect pictures correctly recognized by the intelligent diagnosis equipment in the defect pictures, and the misjudgment rate represents the proportion of the misjudgment pictures of the intelligent diagnosis equipment in all images. The calculation formula is as follows:

$$R = \frac{c}{t}$$

Where, R represents the rate of defect recognition; c represents the number of pictures which are Correctly identified; t represents the number of the total pictures.

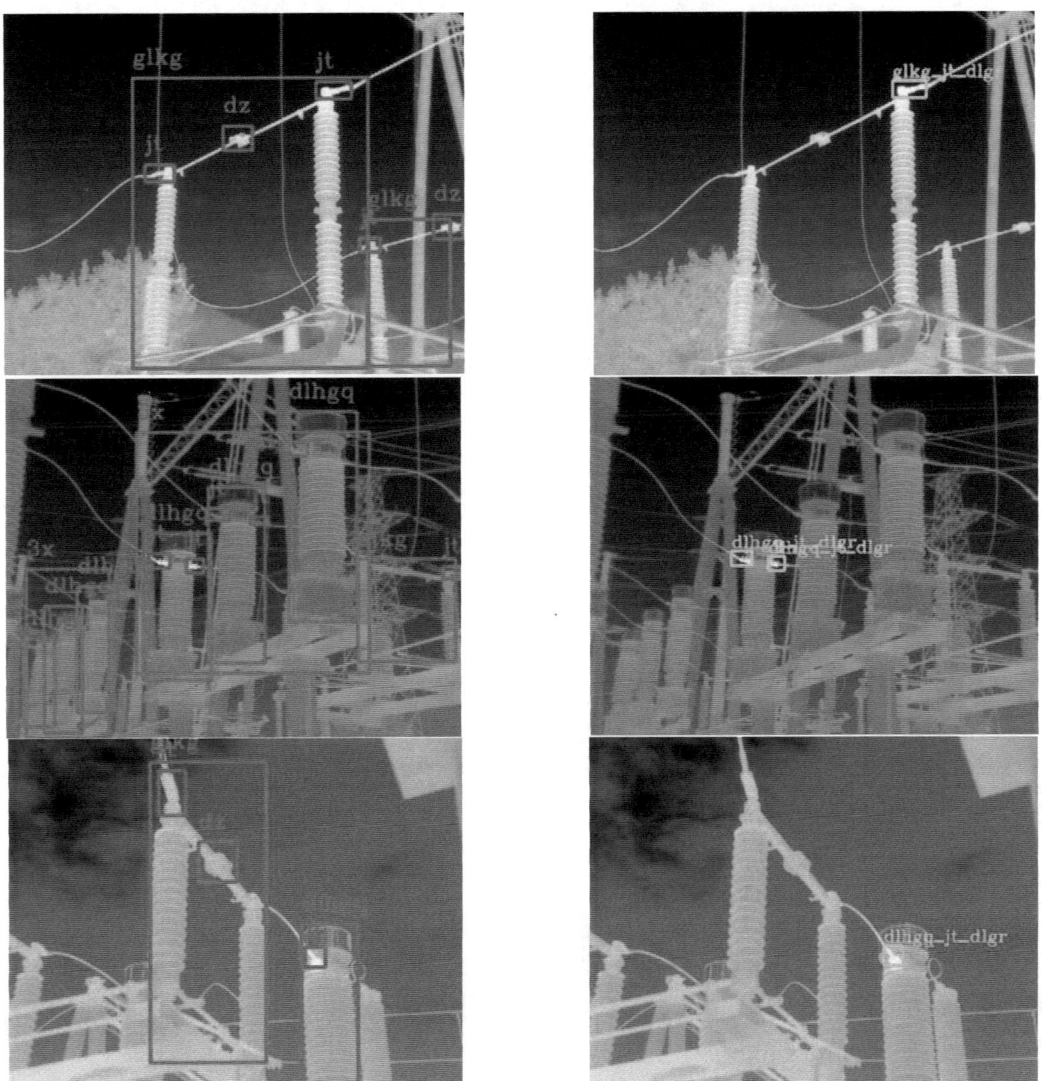

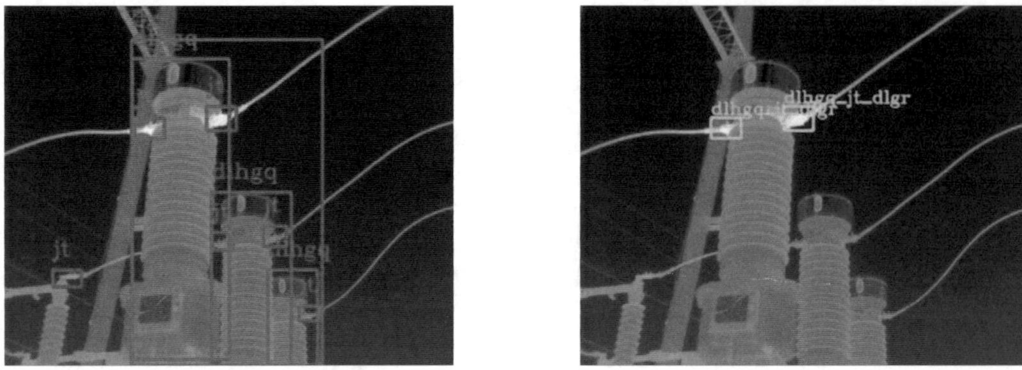

(a) Equipment identification results (b) Diagnosis results

Figure 6. The identification results of current-inductive-heated defect of the typical equipment.

138 defect pictures and 500 normal pictures are used to test the equipment and make statistics on the test results, as shown in Table 1. The defect identification accuracy of the device is 81.16%. The misjudgment rate of the device is 16.45%.

Table 1. Current-inductive-heated defects diagnosis results of the test set.

Type of the picture	Total	Correct diagnosis	Error diagnosis
Normal	500	421	79
Current-inductive-heated defect	138	112	26

18 defect pictures and 500 normal pictures are used for the experiment, and the improved algorithm is used to diagnose the infrared pictures. The diagnosis results of the improved algorithm for voltage-inductive-heated defects in the test set are shown in Table 2. The accuracy of defect identification is 94.44%. The misjudgment rate of the device is 13.90%.

Table 2. Current-inductive-heated defect diagnosis results of the test set of improved algorithm.

Type of the picture	Total	Correct diagnosis	Error diagnosis
Normal	500	429	71
Voltage-inductive-heated defect	18	17	1

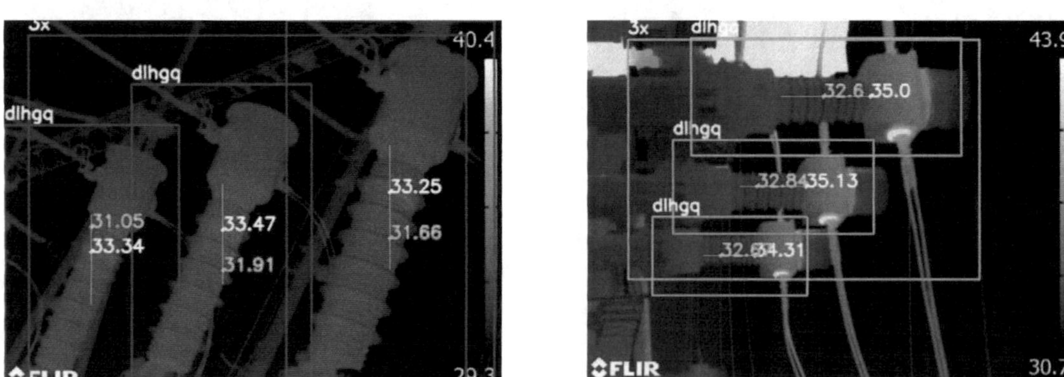

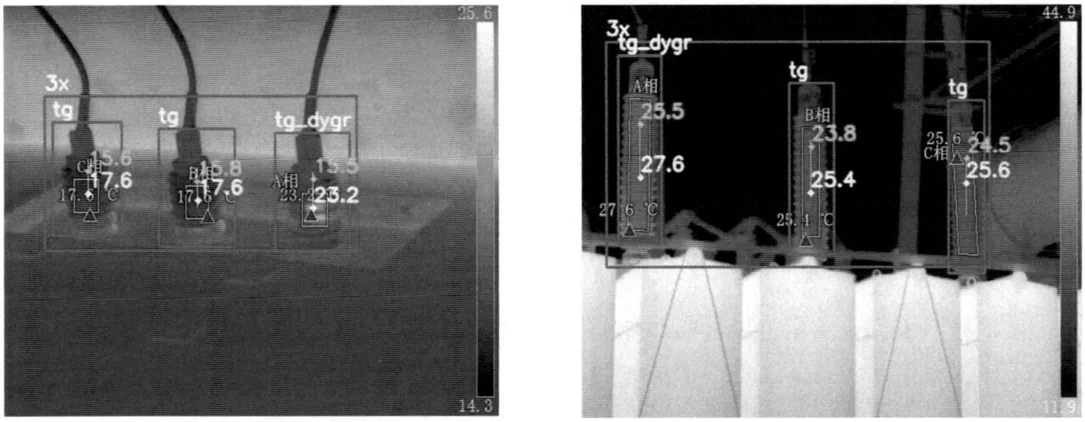

Figure 7. Defect diagnosis results of the voltage-inductive-heated equipment of the test set.

5. Conclusion

Based on the intelligent detection technology of infrared image of substation equipment and the lossless compression technology of digital image, a set of infrared video intelligent diagnosis device of substation equipment based on edge calculation is developed and applied in practice. The device can process typical infrared video images and realize real-time identification and early warning of abnormal temperature of substation equipment. The application of the device can improve the efficiency and accuracy of infrared detection in substation and ensure the safe and stable operation of equipment.

6. References

[1] Zou H, Huang F Z. Multi-target localization for infra- red images of electrical equipment based on improved FAsT-Match algorithm[J]. Proceedings of the CSEE, 2017, 37(2): 591-599.

[2] Chen R G, Feng X Y, Gu C. Application of infrared thermometry technology for the fault diagnosis in GIS [J]. High Voltage Appararus, 2015, 51(9):190-194.

[3] Teow Wee Teo and Mohd Zaid Abdullah, "Solar Cell Micro-Crack Detection Using Localised Texture Analysis," Journal of Image and Graphics, Vol. 6, No. 1, pp. 54-58, June 2018. doi: 10.18178/joig.6.1.54-58

[4] Yefan Zhou, "Research on Image-Based Automatic Wafer Surface Defect Detection Algorithm," Journal of Image and Graphics, Vol. 7, No. 1, pp. 26-31, March 2019. doi: 10.18178/joig.7.1.26-31

[5] Murthy V S, Tarakanath K, Mohanta D K, Insulator Condition Analysis for Overhead Distribution Lines Using Combined Wavelet Support Vector Machine (SVM)[J]. IEEE Transactions on Dielectrics & Electrical Insulation, 2010,17(1):89-99.

[6] Feng Z X, Zhou D G, Jiang Y, Fault region extraction using improved MSER algorithm with application to the electrical system [J]. Power System Protection and Control, 2019, 47(5) : 123-128.

[7] Ma N and WEI G. A code-excited nonlinear predictive speech coding algorithm based on recurrent neural networks[J]. Journal of China Institute of Communications,2000，21(10):31-37.

[8] X. ZHANG, Z R LV, et al., Intelligent Detection Technology of Infrared Image of Substation Equipment Based on Deep Learning Algorithm[C], 2021 IEEE Sustainable Power and Energy Conference, 2021, pp. 3855-3860.

Resilience Assessment of An Expressway Microgrid Considering Network Reconfiguration

Boyuan Cheng, Xinshou Tian and Limin Jia

North China Electric Power University, China Institute of Energy and Transportation Integrated Development, Beijing, China

by.cheng@ncepu.edu.cn

Abstract. With the development of distributed generation, energy supply to the transportation industry through microgrids has become a reality. However, in recent years, affected by environmental changes, extreme weather has occurred frequently. Due to the simple topology, the safe operation of microgrids has been threatened. Therefore, it is necessary to analysis and assess the resilience of microgrids. At the same time, due to the high penetration rate of distributed generation, the microgrid can be reconfigured during extreme disasters to form distributed generation islands to ensure the power supply of local loads. This paper first introduces the resilience curve and resilience index of the expressway microgrid, and then proposes a mathematical model for network reconfiguration to improve the resilience of the microgrid. Finally, the resilience assessment method proposed in this paper is used to assess the resilience of an expressway microgrid under extreme storm weather, then, the reconfigured system is obtained by solving the model, which verifies the effectiveness of this method in improving the resilience.

1. Introduction

In China's transportation industry, nearly 90% of carbon emissions come from road transportation [1]. In the context of the global reduction of carbon emissions, more and more distributed generation (DG), such as photovoltaics (PV), doubly fed induction generator (DFIG), and energy storage systems (ESS), are connected to the distribution network. Now in some expressway areas rich in renewable resources, there have been examples of forming microgrids to supply power to expressway service areas or tunnels [2].

In recent years, due to environmental impact and frequent extreme weather, the safe operation of the power system has been threatened [3]. Resilience is used to assess the ability of the power system to respond to disasters [4]. Compared with the distribution network, the microgrid is more vulnerable in the face of extreme events due to small capacity and simple topology. Therefore, it is necessary to study the resilience of microgrids.

The purpose of this paper is to assess the resilience of expressway microgrids during disasters, and to improve the resilience of expressway microgrids by means of network reconfiguration. The rest of this paper is organized as follows. In Section 2, the resilience index and resilience assessment framework are introduced. The mixed integer linear programming model for grid reconfiguration is introduced in Section 3. In Section 4, an analysis of expressway microgrid verifies the effectiveness of the resilience assessment method and the enhancement strategy. Finally, the conclusions are drawn in Section 5.

2. Resilience of microgrid

2.1. Resilience Characteristics

The resilience curve can be used to represent the resilience level of a power system. Some previous studies proposed a resilience triangle curve [5]. A recent study proposes a resilience trapezoidal curve based on the resilience triangle, which divides the event into three stages: the disturbance stage, the post-disturbance stage, and the restoration stage [6].

The object of this paper is the resilience of the microgrid. Due to the simple topology of the microgrid, the high degree of intelligence, and the high penetration rate of DG, this paper proposes a modified microgrid resilience trapezoidal curve based on the characteristics of the microgrid that is different from the distribution network. The microgrid resilience trapezoidal curve is shown in Figure 1. In the restoration phase, a network reconfiguration method is taken to improve the resilience of the microgrid.

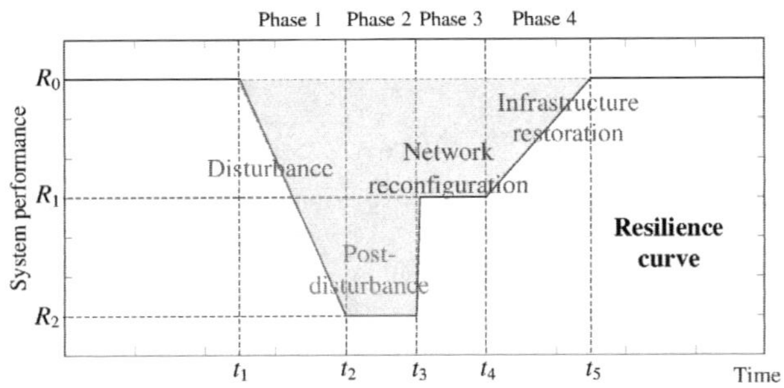

Figure 1. Modified resilience trapezoid curve

The abscissa of the resilience curve is time, and the ordinate is system performance. The system performance adopted in this paper for the resilience assessment of the microgrid is the recovery rate considering the importance level of the load, and the expression is as follows:

$$f(t)=\frac{\alpha_{\mathrm{I}}\sum P_1+\alpha_{\mathrm{II}}\sum P_2+\alpha_{\mathrm{III}}\sum P_3}{\alpha_{\mathrm{I}}\sum P_{\mathrm{I}}+\alpha_{\mathrm{II}}\sum P_{\mathrm{II}}+\alpha_{\mathrm{III}}\sum P_{\mathrm{III}}} \quad (1)$$

where α is the weight of the load, P_1, P_2, P_3 are node load demand, $P_{\mathrm{I}}, P_{\mathrm{II}}, P_{\mathrm{III}}$ are node load supplied.

The resilience index (RI) of the microgrid is the integral sum of the loss system performance in the whole time period.

$$RI=\int_0^T (1-f(t))dt \quad (2)$$

2.2. Resilience Assessment Framework

Combined with the resilience curve proposed above, the steps of microgrid resilience assessment in this paper are shown in Figure 2.

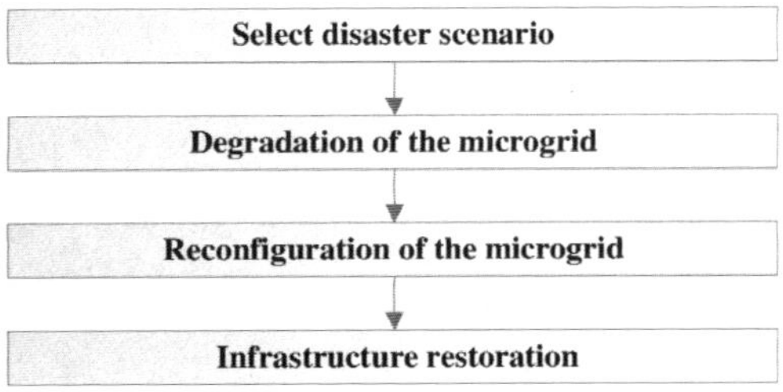

Figure 2. Steps of Resilience Assessment

At the beginning of microgrid resilience assessment, a disaster scenario needs to be selected. After the disaster scenario is selected, the fault of the microgrid can be determined, and the state of the microgrid after the fault can be obtained. After the microgrid is degraded, in order to ensure the power supply of important loads and improve the system performance, the central controller of the microgrid (MGCC) issues instructions to disconnect some lines to form DG islands to supply power to the important loads. At the end of the disaster event, infrastructure restoration is carried out. This paper uses mean time to repair (MTTR) to measure the time of infrastructure restoration [7].

3. Model of network reconfiguration

This section mainly introduces the mathematical model of microgrid network reconfiguration.

The network reconfiguration of the microgrid is a process of finding the optimal topology. The objective function of this model is to ensure that the total power supply of the load is maximized by considering the importance level of the load on the basis of satisfying various constraints:

$$f = \max\{P\} \tag{3}$$

$$P = \alpha_{\mathrm{I}} \sum_{i=1}^{l} k_i P_{1i} + \alpha_{\mathrm{II}} \sum_{i=1}^{m} k_i P_{2i} + \alpha_3 \sum_{i=1}^{n} k_i P_{3i} \tag{4}$$

where α_{I} is the weight of the Class I load, α_{II} is the weight of the Class II load, α_{III} is the weight of the Class III load, k_i is 0-1 variable, which indicates whether the bus is disconnected P_{1i} is the power supply of the Class I load, P_{2i} is the power supply of the Class II load, P_{3i} is the power supply of the Class III load. l is the set of Class I load buses, m is the set of Class II load buses, n is the set of Class III load buses.

The constraints are the following:

(1) DG Island Power Constraints:

$$\sum_{i=1}^{r} P_{1i} + \sum_{i=1}^{r} P_{2i} + \sum_{i=1}^{r} P_{3i} \le \sum_{i=1}^{r} P_{DGi} \tag{5}$$

where r is the set of buses of the DG island, P_{DGi} is DG output.

(2) Active Power Flow Constraints:

$$P_i = U_i \sum_{j=1}^{r} U_j \left(G_{ij} \cos\theta_{ij} + B_{ij} \sin\theta_{ij} \right), \forall i \in r \tag{6}$$

where P_i is the active power flowing into the bus, U_i is voltage of the i bus, U_j is voltage of the j bus, θ_{ij} is the phase angle difference between the i bus and the j bus, G_{ij} is the conductance between the i bus and the j bus, B_{ij} is the reactance between the i bus and the j bus.

(3) Line Connectivity Constraints:

$$k_i = 1, \forall i \in r \tag{7}$$

(4) DG Output Constraints:

$$\underline{P_{DGi}} \le P_{DGi} \le \overline{P_{DGi}}, \forall i \in r \tag{8}$$

where $\overline{P_{DGi}}$ is DG output maximum, $\underline{P_{DGi}}$ is DG output minimum.

(5) Voltage Constraints:

$$\underline{U_i} \le U_i \le \overline{U_i}, \forall i \in r \tag{9}$$

where $\overline{U_i}$ is bus voltage maximum, $\underline{U_i}$ is bus voltage minimum.

(6) Line Capacity Constraints:

$$-\overline{P_{ij}} \le P_{ij} \le \overline{P_{ij}}, \forall i \in r \tag{10}$$

where P_{ij} is the active power flowing through the line, $\overline{P_{ij}}$ is the line capacity maximum.

(7) Radial Network Constraints:

The topology of the DG islands should be radial.

4. Case study

In this section, our goal is to analysis an expressway microgrid using the resilience assessment method, and to verify the effectiveness of the resilience improvement method proposed in Section 3.

Expressway traffic load includes two aspects: infrastructure load and service facility load. Due to the different impacts on social production and life of different loads, the loads are divided into three categories in general. The Class Ⅰ load includes emergency lighting load of tunnel, monitoring load, fire protection load, lighting load of service area, part of living load such as air conditioning, heater, etc. The Class Ⅱ load includes the ventilation load of the tunnel, the remaining lighting load, the monitoring load of the service area, and the communication load. The Class Ⅲ load includes the remaining living load of the service area. In this paper, the microgrid composed of tunnels and nearby service areas is selected as shown in Figure 3. The parameters of each bus are shown in Table 1.

4.1. Base case (without network reconfiguration)

According to the microgrid resilience assessment steps proposed above, the first step is to establish a disaster scenario, and this paper selects extreme storm weather.

Under storm disaster conditions, the output of wind turbines will be greatly affected, and the steady model of the wind turbine can be derived from the following fitting function.

$$P_{wind} = \begin{cases} 0, & V_{wind} \le V_{in} \\ aV^2_{wind} + bV_{wind} + c, & V_{in} < V_{wind} \le V_0 \\ P_0, & V_0 < V_{wind} \le V_{out} \\ 0, & V_{out} < V_{wind} \end{cases} \tag{11}$$

According to (12), it can be known that the working state of the wind turbine is related to the wind speed. In this storm scenario, at 8h, the wind speed reaches V_{out}, and then the DFIG stops working. The unit of PV and ESS can be seen as a power source which provides stable active power output. The MTTR is 10 hours in this scenario. The event time of the microgrid in extreme storm weather is shown in Table 2.

The system performance can be calculated according to the events of the microgrid, and the resilience curve can be drawn as shown in Figure 4.

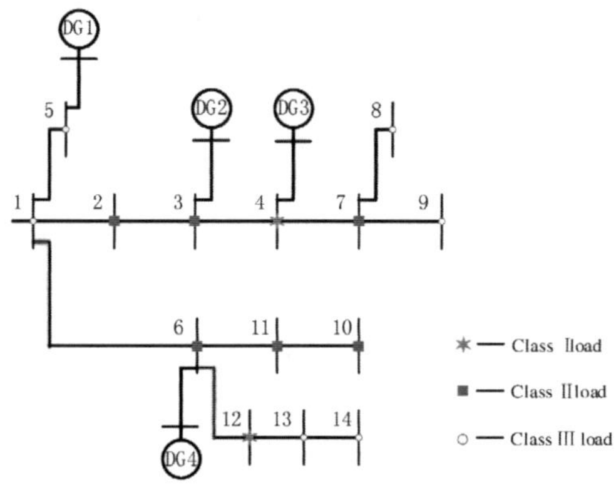

Figure 3. Expressway microgrid

Table 1. Parameters of each bus

Bus	Load (MW)	Class	DG type	Output (MW)
1	2	III	-	-
2	2	II	-	-
3	3	II	DFIG	12
4	1	I	PV + ESS	8
5	2	III	PV + ESS	4
6	3	II	PV + ESS	6
7	2	II	-	-
8	5	III	-	-
9	4	III	-	-
10	1	II	-	-
11	1	II	-	-
12	1	I	-	-
13	1	III	-	-
14	2	III	-	-

Table 2. Event time of the microgrid

Time	Event
5h	Line 1-5 tripped
8h	Wind turbine cut out
10h	Line 1-6 tripped
20h	Infrastructure restoration starts
30h	Restoration complete

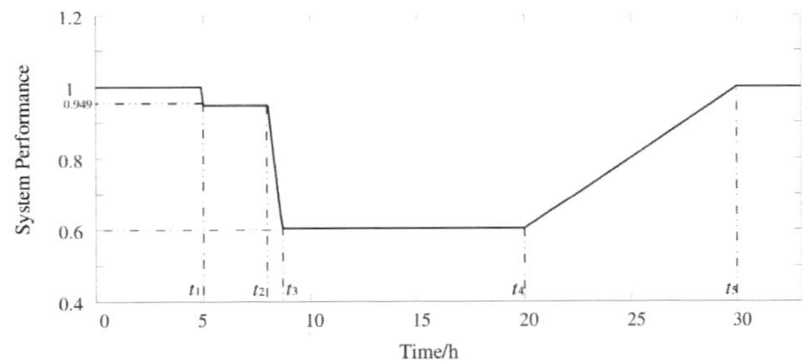

Figure. 4. Resilience curve of base case

It can be seen from the Figure 4 that before t_1=5h, the microgrid topology does not change, and the system performance remains at 1. When t_1=5h, line 1-5 is tripped. The DG on bus5 can only supply power to the load on bus5, and the overall system loses part of active power. In order to maintain the stable operation of the system, the microgrid will reduce a certain amount of load. The load is generally reduced in equal proportions, and the proportion is the ratio of the power provided by the generator to the load demand. According to the load reduction, the system performance can be calculated to drop to 0.949. At t_2=8h, the wind speed exceeds the maximum working wind speed of the wind turbine, the wind turbine starts to stop working, and it generally takes 10 minutes to cut off the wind turbine. At t_3=8.16h, the system function is reduced to 0.6. Then the system function stabilizes at the minimum value until t_4=20h when the infrastructure restoration work starts, the system function starts to rise, and the recovery ends at t_5=30h, and the system function returns to 1.

4.2. Modified case (with network reconfiguration)

The MGCC issues a network reconfiguration command at t_4=14h, other conditions are the same as in case 1. In this paper, this model is programmed in MATLAB2018b and is solved by calling the CPLEX commercial solver through YALMIP toolbox, by solving the model, the results can be obtained as shown in the Table 3. This microgrid finally forms three DG islands, which supply power to the loads in the islands respectively, and the resilience curve is shown in Figure 5.

Table 3. Model results

Index	**Result**
k	[0 1 1 1 1 1 1 0 0 1 1 1 0 0]

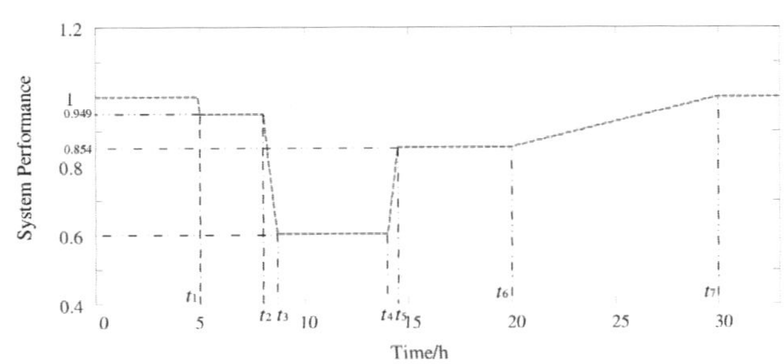

Figure. 5. Resilience curve of case 2

It can be seen from the Figure. 5 that before t_4=14h, the resilience curve is the same as Case 1. At t_4=14h, the microgrid begins to reconfigure the grid, and after 0.5h, the reconfiguration is completed at t_5=14.5h, forming three DG islands, ensuring the power supply of the Class Ⅰ load and the Class Ⅱ load. At this time, the system performance rises to 0.8. Then the system function remains unchanged until t_4=20h when the infrastructure restoration work starts, the system function starts to rise, and the recovery ends at t_5=30h, and the system function returns to 1.

The numerical results are shown in Table 4. It can be seen from Table 4 that after the network reconfiguration, the recovery rate of important loads is greatly improved, and the resilience index of the expressway microgrid is also greatly improved.

Table 4. Numerical results

Index	Case1	Case2
Class Ⅰ Load Supply Rate	62.5%	100%
Class Ⅱ Load Supply Rate	50%	100%
Class Ⅲ Load Supply Rate	50%	0
RI	6.9231	9.6558

5. Conclusion

In order to realize the effective response of expressway microgrid in the face of extreme disasters, this paper constructs the expressway microgrid resilience index and framework, and adopts the method of network reconfiguration to improve the resilience of the system. The main conclusions are as follows:
(1) The integral sum of the system performance loss in the whole time period of the system is proposed as the resilience assessment index, and how to quantify the ability of the expressway microgrid to cope with extreme weather is realized.
(2) A resilience assessment framework for expressway microgrids is built to comprehensively and objectively assess the impact of extreme disasters on expressway microgrids.
(3) A network reconfiguration method is adopted to improve the resilience of the expressway microgrid in the face of extreme disasters. By ensuring the power supply of important loads, the ability of the system to face disaster disturbances is greatly improved.

6. References

[1] L. Jia, J. Ma, P. Cheng and Y. Liu. A perspective on solar energy-powered road and rail transportation in China[J], CSEE Journal of Power and Energy Systems, vol. 6, no. 4, pp. 760-771, Dec. 2020.
[2] Y. Yang, Q. -S. Jia, G. Deconinck, X. Guan, Z. Qiu and Z. Hu. Distributed Coordination of EV Charging With Renewable Energy in a Microgrid of Buildings[J]. IEEE Transactions on Smart Grid, vol. 9, no. 6, pp. 6253-6264, Nov. 2018.
[3] Y. Xu, C. -C. Liu, K. P. Schneider, F. K. Tuffner and D. T. Ton. Microgrids for Service Restoration to Critical Load in a Resilient Distribution System[J]. IEEE Transactions on Smart Grid, vol. 9, no. 1, pp. 426-437, Jan. 2018.
[4] M. Panteli, D. N. Trakas, P. Mancarella and N. D. Hatziargyriou. Power Systems Resilience Assessment: Hardening and Smart Operational Enhancement Strategies[J]. Proceedings of the IEEE, vol. 105, no. 7, pp. 1202-1213, July 2017.
[5] V. Widiputra and J. Jung. Optimal Grid Reconfiguration Algorithm for Improving System Resilience under Extreme Weather Events[C]. 2020 4th International Conference on Green Energy and Applications (ICGEA), 2020, pp. 12-16.
[6] M. Panteli, P. Mancarella, D. N. Trakas, E. Kyriakides and N. D. Hatziargyriou. Metrics and Quantification of Operational and Infrastructure Resilience in Power Systems[J]. IEEE Transactions on Power Systems, vol. 32, no. 6, pp. 4732-4742, Nov. 2017.
[7] M. Nazemi, M. Moeini-Aghtaie, M. Fotuhi-Firuzabad and P. Dehghanian. Energy Storage Planning for Enhanced Resilience of Power Distribution Networks Against Earthquakes[J]. IEEE Transactions on Sustainable Energy, vol. 11, no. 2, pp. 795-806, April 2020

Acknowledgments

This research was supported by the National Natural Science Foundation of China (No. 52007174).

Journal of Physics: Conference Series **2320** (2022) 012015 doi:10.1088/1742-6596/2320/1/012015

Adaptive Restraint Coefficient-Based Differential Protection for Main Transformers in Wind Farms

Zhi Zhang[1], Yang Sun[1] and Zhiyan Li[2]

[1]School of Mechanical Electronic and Information Engineering, China University of Mining and Technology(Beijing), Beijing 100083, China
[2]China Airport Construction Group Co.,Ltd., Beijing 100621, China

zhangzhi3700@163.com

Abstract. Influenced by factors such as fault current frequency deviation, current transformer (CT) saturation and abnormal data, the traditional transformer current differential protection scheme contains the risks of slow operating speed or false operation in the wind power system. To guarantee the safe operation of the power system, by improving the traditional current differential protection scheme, a differential protection scheme based on the adaptive restrain coefficient was proposed for the main transformers of wind farms. The proposed protection adaptively adjusts the restraint coefficient according to the time-domain distribution characteristics of differential current. When the internal fault occurs, the restraint coefficient will be adaptively reduced; the restraint coefficient will adaptively increase when the non-internal fault occurs. The proposed protection scheme significantly surpasses the traditional current different protection in terms of operating performance, because it can accurately identify the internal faults and strongly resist abnormal data, CT saturation and magnetizing inrush current.

1. Introduction

The longitudinal differential protection scheme is one of the main protection schemes of the transformer. Its operating criterion is established according to the current information at both ends of the transformer, so as to rapidly remove internal faults. With strong selectivity and quickness, its reliable operation is critical to the entire power system [1]. However, the topological structure and control strategies of the doubly-fed induction generator (DFIG) differ from those of traditional synchronous power supply. Factors such as fault current frequency deviation, CT saturation, magnetizing inrush and abnormal data will affect the operation performance of transformer current differential protection, which may lead to protection delayed or false operation [2]-[4]. Abnormal data may be caused by faults of measuring devices, interference in information channels, etc. To eliminate the influences of the above factors on transformer current differential protection, experts and researchers have proposed various solutions. By improving the control strategy, the second harmonic is suppressed by adding a notch filter to the rotor side converter to filter out the fundamental frequency component, thereby avoiding protection blocking when internal faults occur [5]. Besides, histogram is used to establish transformer protection criterion, which has the robustness to CT saturation [6]. Because CT saturation has little interference to current phase information on both sides of the transformer, Fréchet distance algorithm is applied to measure the current phase difference on both sides of the transformer and came up with a differential protection criterion insusceptible to CT saturation [7]. A large amount of data is trained by artificial neural network to distinguish fault currents under different operating conditions [8]. Hausdorff distance algorithm is applied to distinguish internal faults by measuring the similarity between waveforms [9]-[10].

Nevertheless, the above-proposed protection can only alleviate the influences of part of the factors, e.g., fault current frequency deviation, CT saturation, magnetizing inrush current and abnormal data, on the operation performance of traditional current differential protection scheme. Besides, these methods are complicated to implement. On the basis of traditional transformer current differential protection, the adaptive restraint coefficient is constructed according to time-domain distribution characteristics of differential current under various conditions, and an adaptive restraint coefficient-based protection scheme is presented for the main transformers of doubly-fed wind farms. This scheme has the characteristics of simple principle and low computation cost, and applies to all fault types because it can adaptively adjust the restraint coefficient under different operating conditions. In addition, it has strong robustness to factors like fault current frequency deviation, CT saturation, magnetizing inrush and abnormal data. The reliability and superiority of the adaptive protection scheme are fully reflected by the simulation results under various operating conditions in PSCAD/EMTDC platform.

2. Performance problems of traditional current different protection

Based on the grid-connected simulation model of the DFIG wind farm as illustrated in Figure 1, the performance problems of the traditional transformer current differential protection scheme in the doubly-fed wind farm are discussed. The power frequency of the system is 50Hz, and the rated power of the wind farm is 100MW. The main transformer adopts Ynd11 type wiring mode, the rated capacity of 120MVA, the rated ratio is 230/37kV, and the short circuit impedance is 12%.

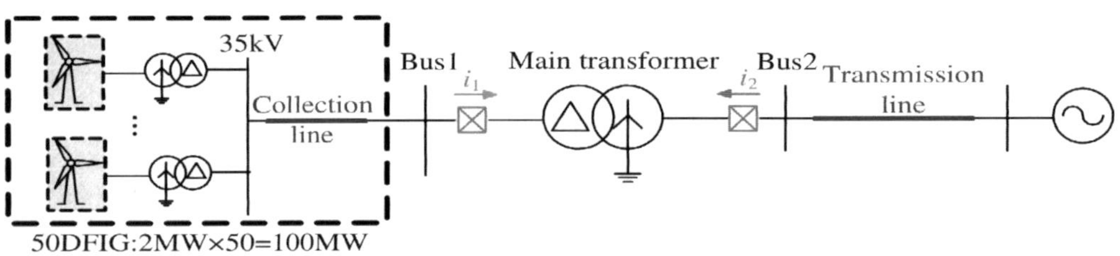

Figure 1. Schematic diagram of doubly-fed wind power system.

The internal three phase fault on wind farm side (WF-side) of the transformer occurs at t=1s. The rotor speed of wind turbines was 0.7p.u. before the fault occurred. Due to the frequency deviation characteristic, the fault current at the WF-side contains the non-fundamental frequency component of 35Hz, as visible in Figure 2(a). In Figure 2(b), the component of second harmonic may surpass the threshold due to incorrect calculation. The ratio restraint differential protection was blocked by the restraint component, which may cause the protection to delay operation.

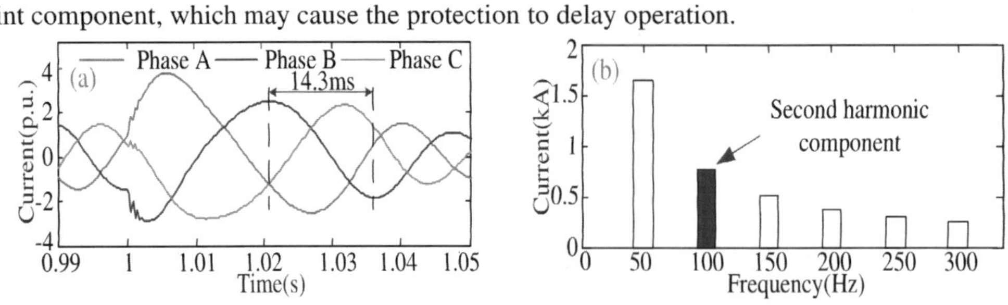

Figure 2. Simulation results of frequency deviation characteristic. (a). Fault currents on WF-side. (b) The component of second harmonic.

Similarly, the superposition of magnetizing inrush and fault current may delay the operation speed of traditional protection. In addition, when the external fault of the transformer is affected by CT saturation or abnormal data, the differential current waveform will be distorted by the influence, resulting in the false operation risk of traditional differential protection.

3. Adaptive restraint coefficient-based protection for main transformers

3.1. Differential current under different operating conditions

In order to directly reflect the restraint principle of traditional transformer differential protection scheme, its vector restraint characteristics are analyzed, as illustrated in Figure 3. I_d and I_r are the differential current and restraint current, respectively. K is the restraint coefficient of the traditional differential protection scheme, which is 0.5, 0.8 and 1.2, respectively. The outside of the circle is the operating region, and the inside of the circle is the restraining region; The dotted circle indicates the minimum operating current. The value of K is positively correlated with the size of the restraining region. The larger the value of K, the larger the restraint region and the less likely the protection will operate. Therefore, when the internal fault occurs, the restraint coefficient of K should be as small as possible and as large as possible in other cases.

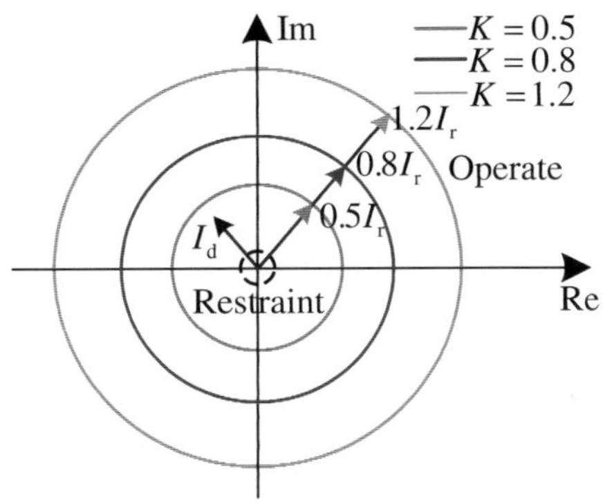

Figure 3. Vector restraint characteristics of traditional current differential protection.

The performance of the traditional transformer differential protection scheme is improved by the time-domain distribution characteristics of differential currents under different operating conditions. Figure 4(a)-(d) are the differential current waveforms under various operating conditions. i_1 and i_2 are the currents on the WF-side and grid-side of the transformer, respectively. As visible in Figure 4(a) and (b), CT saturation will lead to differential current distortion, and their amplitudes are no longer 0. Comparing Figure 4(c) and (d), the amplitudes of differential currents may be similar in magnetizing inrush and internal fault. However, due to the presence of linear transferring zone in CT, the amplitude of the differential current is close to 0 for about 10ms in each sampling period after the transformer is energized without load.

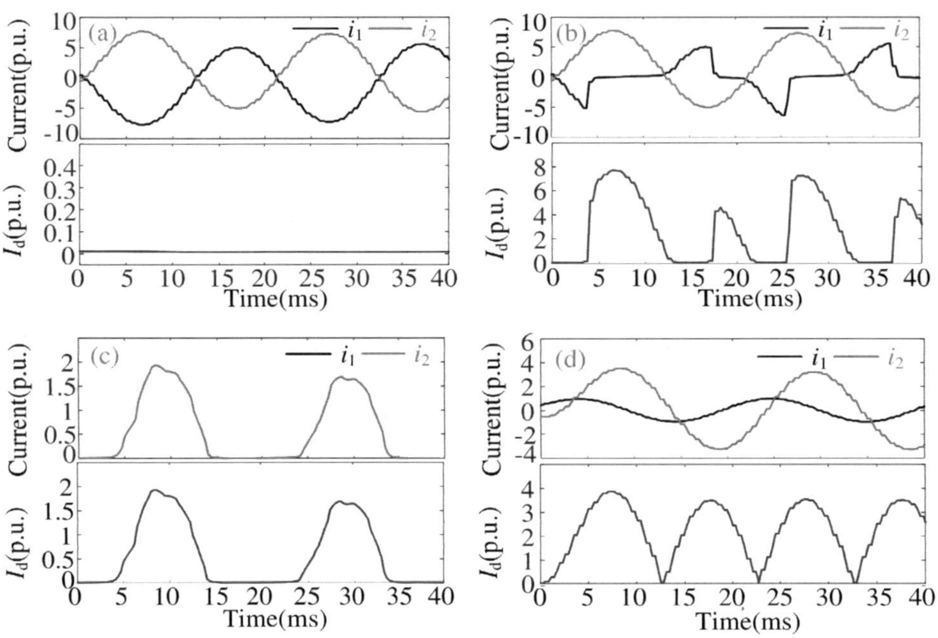

Figure 4. Differential currents under different operating conditions. (a) External fault. (b) External fault accompanied by CT saturation. (c) Magnetizing inrush. (d) Internal fault.

3.2. Basic principle of adaptive differential protection scheme

It can be known from the time-domain distribution characteristics of differential current under different operating conditions. When the non-internal fault occurs, there are regions with amplitude of 0 for differential current in each sampling period, but there is almost no region with amplitude of 0 for differential current after internal faults occurrence. In view of the problems existing in the traditional transformer current differential protection, the differential protection scheme based on adaptive restraint coefficient is put forward. The adaptive restraint coefficient is defined as follows:

$$K_{\text{adp}} = K e^{4x-1} \tag{1}$$

Where K_{adp} is the adaptive restraint coefficient; K is the restraint coefficient of traditional current differential protection, its typical value is set to be 0.8; x is the ratio between the sampling points of differential current less than 0.3p.u. over the total sampling points in the data window, $x \in [0,1]$. The sampling frequency of the system is 4kHz, and each moving data window contains 80 sampling points. The function curve of adaptive restraint coefficient is displayed in Figure 5. K_{adp} is positively correlated with x, as shown in Figure 5. When a non-internal fault occurs, the value of x is large, and the restraint coefficient is adaptively increased to increase the reliability of protection. Besides, due to the influence of internal fault currents, the value of x is small, and the restraint coefficient is reduced adaptively to improve its operating sensitivity.

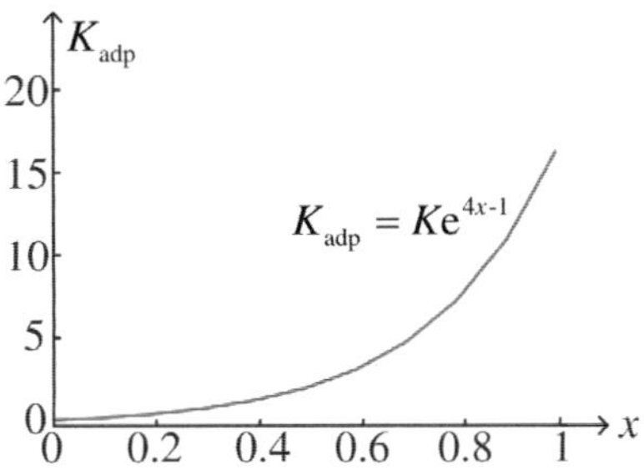

Figure 5. Function curve of adaptive restraint coefficient.

The criterion of adaptive current differential protection is defined as follows:

$$|\boldsymbol{I}_{\mathrm{M}} + \boldsymbol{I}_{\mathrm{N}}| > K_{\mathrm{adp}} |\boldsymbol{I}_{\mathrm{M}} - \boldsymbol{I}_{\mathrm{N}}| \tag{2}$$

Where $\boldsymbol{I}_{\mathrm{M}}$ and $\boldsymbol{I}_{\mathrm{N}}$ are the three-phase currents on both sides of the transformer, respectively.

In order to exhibit the superiority of the adaptive protection scheme, phase A is taken as an example to compare the adaptive restraint coefficient and the traditional restraint coefficient under different operating conditions. As displayed in Figure 6(a), the differential current is almost all greater than the threshold of 0.3p.u. after internal fault occurrence. The adaptive restraint coefficient K_{adp} is smaller than the traditional restraint coefficient K, i.e., $K>K_{\mathrm{adp}}$. Compared with the traditional scheme, the sensitivity of protection is effectively improved. As seen from Figure 6(b), (c) and (d), when the transformer has the non-internal faults, there are some regions where the differential current is less than the threshold of 0.3p.u. in each sampling period, and the adaptive restraint coefficient K_{adp} is larger than the traditional restraint coefficient K, i.e., $K<K_{\mathrm{adp}}$. Compared with the traditional scheme, the protection reliability is effectively improved.

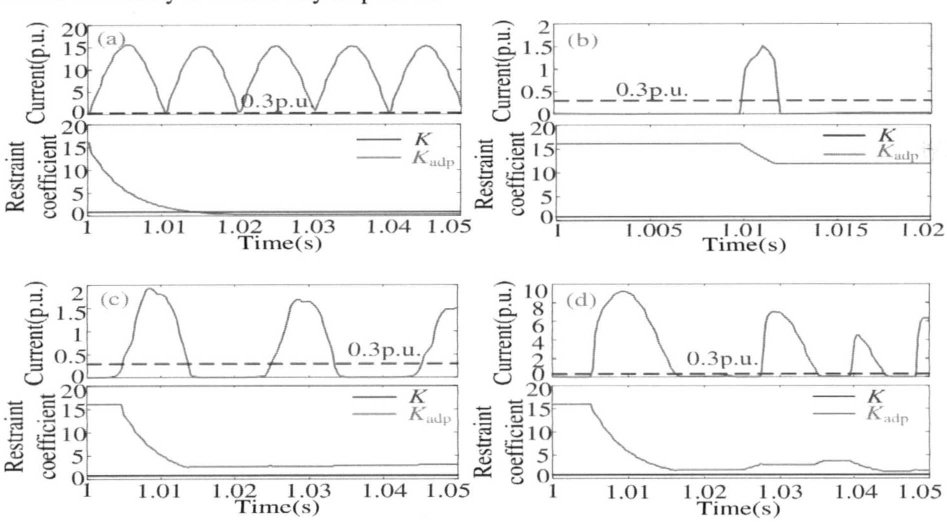

Figure 6. Comparison of restraint coefficient of two protection schemes. (a) Internal fault. (b) Abnormal data. (c) Magnetizing inrush. (d) External fault with CT saturation.

4. Performance evaluation and discussion

Based on the simulation model of doubly-fed wind farm shown in Figure 1, the performance of adaptive restraint coefficient-based protection is verified under different operating conditions. Due to limited space, only some typical simulation results are listed here, and the failure time is set as t=1.0s.

4.1. Internal Faults

The phase A inter-turn faults on grid-side of the transformer. The short-circuited turns are 1%, 5%, and 10%, respectively. The operation performance is shown in Figure 7. As visible in Figure 7(a), when inter-turn faults occur, the severity of faults is positively correlated with the percentage of short-circuit turns, and the differential currents almost all exceed the threshold value of 0.3p.u.. From Figure 7(b), the adaptive restraint coefficients K_{adp} are less than ratios of differential current over restraint current I_d/I_r within 20ms after fault occurrence. Therefore, the adaptive protection scheme can effectively identify inter-turn faults.

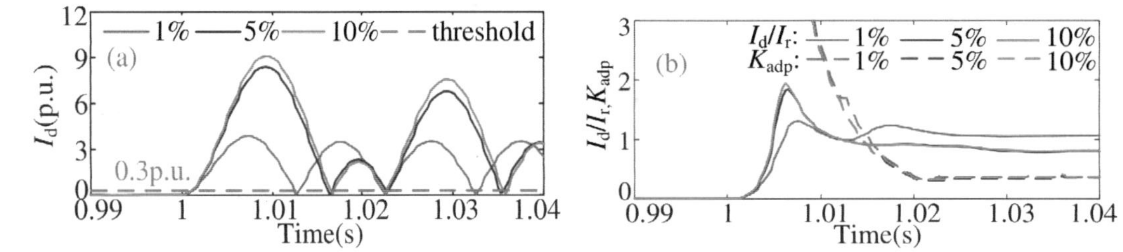

Figure 7. Operation performance under internal faults. (a) Differential current waveforms of the fault phase. (b) Performance of adaptive differential protection.

4.2. External Faults

The external phase-A-to-ground fault on WF-side of the transformer. The operation performance is displayed in Figure 8. From Figure 8(a), the differential currents are less than the threshold of 0.3p.u.. From Figure 8(b), the ratios of differential current over restraint current I_d/I_r are 0, the adaptive restraint coefficients K_{adp} are 16.07. Thus, the protection proposed can identify the external fault has high reliability in operation performance.

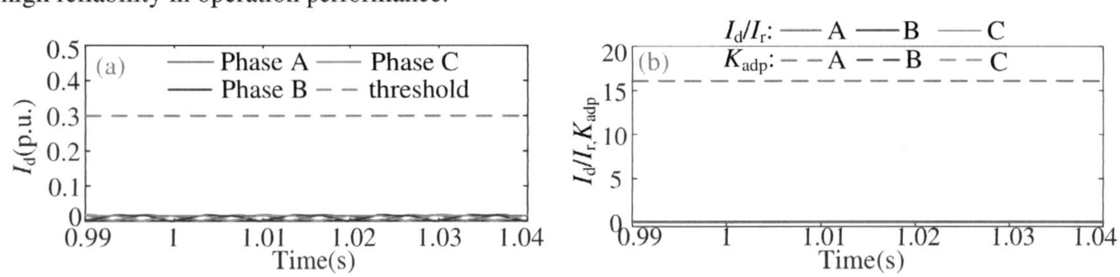

Figure 8. Operation performance under external fault. (a) Differential current waveforms of the fault phases. (b) Performance of adaptive differential protection.

4.3. Magnetizing Inrush

Taking phase A as an example, the performance of adaptive protection scheme under different magnetizing inrush conditions without considering the influence of remanence is assessed. The transformer is energized with no load when the inception phases are 0°, 45° and 90°, respectively. As visible in Figure 9(a), the differential current waveforms under various magnetizing inrush conditions are similar to that under external faults accompanied by CT saturation, the differential currents exceed the threshold of 0.3p.u. at some time. The adaptive restraint coefficients K_{adp} are still greater than the ratios of differential current over restraint current I_d/I_r, which are all 1. The protection proposed does not operate reliably, as seen from Figure 9(b). The proposed protection scheme has good robustness to magnetizing inrush.

 2320 (2022) 012015 doi:10.1088/1742-6596/2320/1/012015

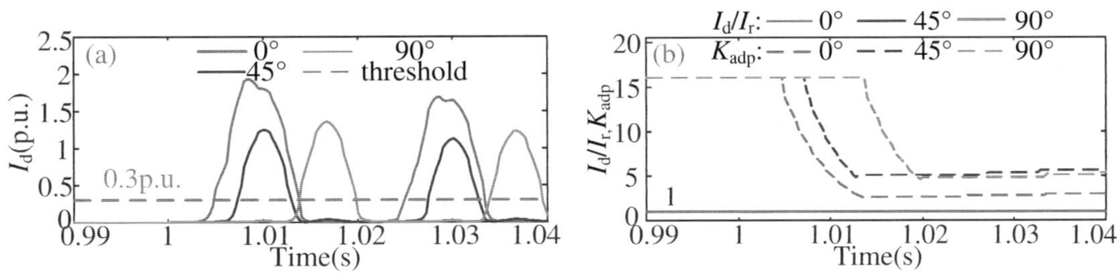

Figure 9. Operation performance under magnetizing inrush. (a) Differential current waveforms under various magnetizing inrush conditions. (b) Performance of adaptive differential protection.

5. Comparison of the proposed scheme and traditional protection method

5.1. Internal fault current superimposed with considerable magnetizing inrush.

The transformer is energized with internal phase A-to-phase C fault on WF-side, and the core remanence of three phases are 0. The comparison results are shown in Figure 10. From Figure 10(a), the differential currents of the fault phases are distorted due to the influence of magnetizing inrush. As seen from Figure 10(b), the ratios of the fault phases exceed the second harmonic restraint coefficient of 15%, and the protection blocking time of phase A and phase C are 180ms and 230ms, respectively. The operating speed of traditional differential protection will be reduced. Within 20ms after the fault occurrence, the adaptive restraint coefficients of the fault phases are less than the ratios of differential current over restraint current I_d/I_r, indicating protection operates reliably, as shown in Figure 10(c). Therefore, the proposed adaptive differential protection scheme has better performance than the traditional current differential protection in this scenario.

5.2. External fault with CT saturation.

The external phase A-to-phase C fault on WF-side of the transformer. The CTs of phases A and C on grid-side are seriously saturated. The simulation results are shown in Figure 11. From Figure 11(a), the current waveforms of WF-side are distorted by CT saturation. The ratio of differential current over restraint current I_d/I_r may exceed the traditional restraint coefficient of 0.8. Besides, I_{d2}/I_{d1} is the ratio of second harmonic component over fundamental frequency component of differential current. The ratios of phases A and C are less than the second harmonic restraint threshold of 15%, indicating the traditional protection false operation, as visible in Figures 11(b) and (c). The performance of the adaptive protection scheme is shown in Figure 11(d). The adaptive restraint coefficients K_{adp} of the fault phases always exceed the ratios of differential current over restraint current I_d/I_r, respectively. Therefore, the protection does not false operation. The protection proposed exhibits superior performance to CT saturation compared with the traditional protection scheme.

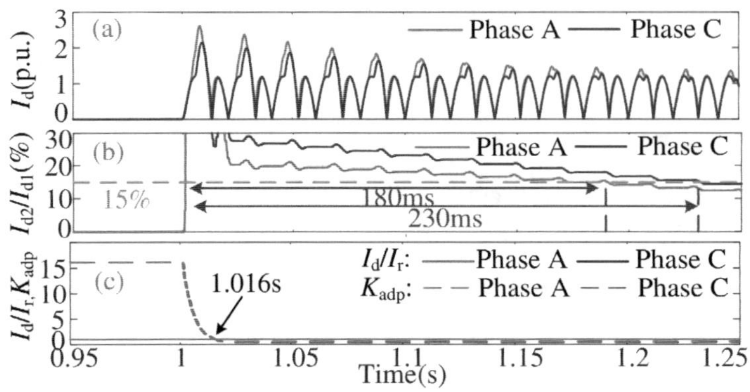

Figure 10. Comparison of performance under internal fault current superimposed with magnetizing inrush. (a) Differential current waveforms of the fault phases. (b) The ratio of second harmonic. (c) Performance of adaptive differential protection.

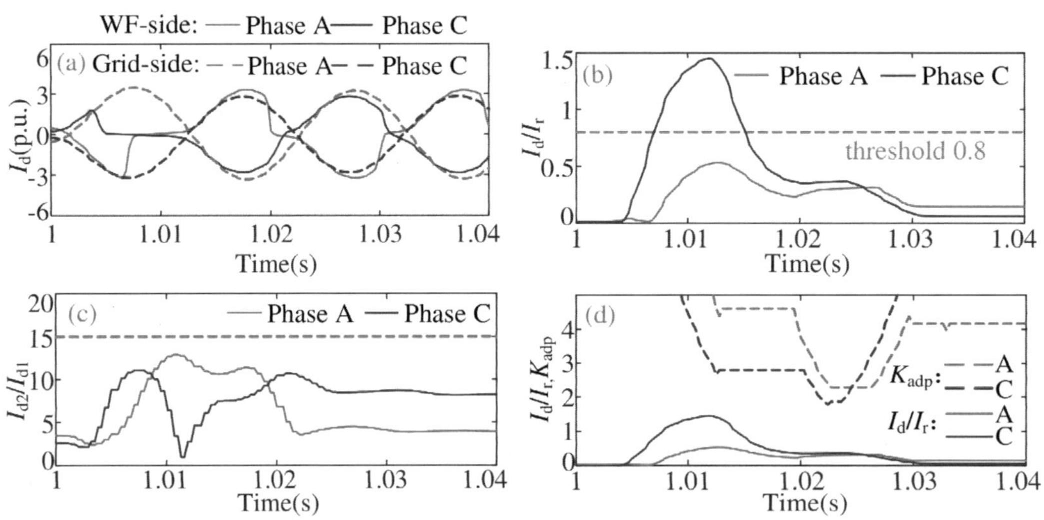

Figure 11. Comparison of performance under external fault with CT saturation. (a) Differential current waveforms of phases A and C. (b) I_d/I_r of phases A and C. (c) I_{d2}/I_{d1} of phases A and C. (d) Performance of adaptive differential protection.

As proved with the above simulation results, the adaptive differential protection proposed can correctly identify internal types, and has obvious advantages over the traditional transformer current differential protection in terms of reliability and sensitivity.

6. Conclusion

To overcome the adverse impact of large-scale wind power integration on traditional transformer current differential protection, an adaptive differential protection scheme applicable to the main transformers of wind farms was proposed based on the limitations of traditional transformer current differential protection in the wind power system. This scheme can achieve adaptive differential protection according to the time-domain distribution characteristics of differential currents. Through simulation, the following conclusions were drawn:

1) With the simple and reliable principle as well as the low computation cost, the proposed scheme does not need to be used in conjunction with the second harmonic restraint criterion. This scheme is obviously superior to traditional transformer current differential protection for its effectively improved protection reliability and sensitivity.

2) The proposed protection can offer superior robustness to frequency deviation, CT saturation, magnetizing inrush current and abnormal data. Moreover, the proposed protection can respond rapidly within 20ms after internal fault occurrence, which guarantees the rapidity of transformer protection.

7. References

[1] Ali Salman Abdul Razzaq, Mohd Fairouz Mohd Yousof, Salem Mgammal Al-Ameri, Mohd Aizam Talib, and Ali Jber Mshkil 2022 Interpretation of inter-turn fault in transformer winding using turns ratio test and frequency response analysis *International Journal of Electrical and Electronic Engineering & Telecommunications* **11** 218-25

[2] M. Stanbury and Z. Djekic 2015 The impact of current-transformer saturation on transformer differential protection *IEEE Transactions on Power Delivery* **30** 1278-87

[3] F. Peng, H. Gao and Y. Liu 2018 Transformer sympathetic inrush characteristics and identification based on substation-area information *IEEE Transactions on Power Delivery* **33** 218-28

[4] K Ramesh and Y Mallikharjuna Rao 2014 Power transformer protection using wavelets *International Journal of Electrical and Electronic Engineering & Telecommunications* **3** 14-21

[5] T. Zheng, J. X. Yan, M. H. Wang, and L. Yang 2021 Impact of inrush current on second harmonic characteristics of doubly fed induction generator *Power System Technology* **45** 3514-23

[6] T. Zheng, T. Huang, Y. Ma, Z. Zhang and L. Liu 2018 Histogram-based method to avoid maloperation of transformer differential protection due to current-transformer saturation under external faults *IEEE Transactions on Power Delivery* **33** 610-9

[7] H. Weng, S. Wang, Y. Wan, X. Lin, Z. Li, J. Huang 2020 Discrete Fréchet distance algorithm based criterion of transformer differential protection with the immunity to saturation of current transformer *International Journal of Electrical Power & Energy Systems* **115** 105449

[8] M. Yazdani-Asramia, M. Taghipour-Gorjikolaiea, S. M. Razavib, and S. A. Gholamian 2015 A novel intelligent protection system for power transformers considering possible electrical faults, inrush current, CT saturation and over-excitation *Int. J. Elect. Power Energy Syst.* **64** 1129-40

[9] J. Liao, X. Zhu, and Q. Wang 2019 Transient current similarity-based protection for interconnecting transformers in wind farms *IEEE Access* **7** 45744–57

[10] H. Weng, L. Hua, X. Lin, Y. Wan, Z. Li, J. Huang, and J. Lu 2018 A novel criterion of the transformer differential protection based on the hausdorff distance algorithm *Proceedings of the CSEE* **38** 475-483+678

Journal of Physics: Conference Series **2320** (2022) 012016 doi:10.1088/1742-6596/2320/1/012016

Study of the IPOS DC/DC Converter for DC Offshore Wind Farm

Huaxing Ding[1], Guibin Zou[1,4], Ning Ding[1,2] and Chuanjie Wang[3]

[1] School of Electrical Engineering Shandong University, Jinan, China
[2] State Grid Shandong Electric Extrahigh Voltage Co., Ltd. Jinan, China
[3] State Grid Henan Electric Power Co., Ltd. Economic Research Institute, Zhengzhou, China

[4] guibinzou@sdu.edu.cn

Abstract. With the development of the offshore wind farms, DC offshore wind farm topology is getting more attention due to its low loss and light weight, where the DC/DC converters with high voltage gain are key equipments. This paper studies the Input-Parallel-Output-Series(IPOS) converter based on Dual-Active-Bridge(DAB) topology, discusses the modulation and control method of the converter, and proposes a fault ride-through scheme. Simulation results validate the proposed topologies and control schemes.

1. Introduction

The development of wind power technology enables offshore wind farms becoming larger and larger. Due to the constraints of land resource and wind energy, large-scale wind farms are often set far away from power nodes. At present, onshore wind farms in some countries have tended to be saturated, while offshore wind energy has not yet been utilized widely, which has huge development potential.
With the increase of distance of the wind farm, the reactive current and overvoltage caused by the capacitance of AC cable to ground become more prominent. The topology of offshore wind power system has developed from AC-collection-AC-transmission to AC-collection-DC-transmission (flexible DC transmission)[1]. In addition, as the wind turbine capacity has reached 10 MW level (such as SG14-222 DD and GE Haliade-X), the offshore wind farm capacity and transmission distance have further increased. Thus the loss of the AC collection network has increased significantly. The DC-collection-DC-transmission topology has received more attention due to its larger transmission capacity, less transmission loss, and free of reactive voltage problem[2]-[4].

Although DC offshore wind farm meets the need of the future development direction of wind power, it also faces technical bottlenecks such as high-voltage DC circuit breakers and large-capacity DC/DC converters. As the key equipment for connecting different voltage levels in offshore DC power grids, DC/DC converters need to have high voltage gain and relatively light weight[5][6]. Commonly used DC/DC converter topologies include boost type, switching capacitor type, LC resonant type, DAB type, etc., but these topologies can't be applied in high-voltage occasions directly due to the voltage and current limitations of power electronic devices, where the Input-Parallel-Output-Series(IPOS) and Modular Multilevel Converter(MMC) topologies meet the requirements. The objective of this paper is to discuss the configuration and application scenario of IPOS topology in offshore wind farm, and to investigate the corresponding steady-state and fault-state control strategies.

2. The principle of IPOS converter

2.1. Circuit topology

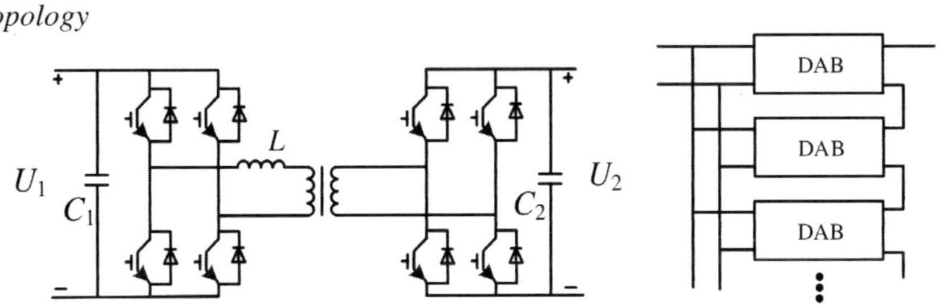

(a)DAB sub-module topology (b)IPOS topology

Figure 1. IPOS converter topology

Figure 1(a) shows the DAB sub-module topology. By using a high-frequency transformer, two AC H-bridge converters are connected at their AC sides. The H-bridge converters use square wave modulation to generate high-frequency square waves on the AC side. Figure 1(b) shows the topology of IPOS converter. The low-voltage sides are connected in parallel to implement current sharing, while the high-voltage sides get connected in series to withstand high voltage. The IPOS topology avoids the direct series connection of switching devices, and realizes electrical isolation.

The traditional single-phase-shift(SPS) modulation method of DAB adjusts the magnitude and direction of power flow by adjusting the phase-shifting angle between two H-bridges. Since there is only one control degree, the soft switching condition is difficult to realize. Therefore, extended-phase-shifting(EPS), double-phase-shifting(DPS), and triple-phase-shifting(TPS) methods are successively proposed, in which control degrees of phase-shifting are added to the power electronic devices inside the primary and secondary side for optimization[7].

2.2. Modulation principle

The EPS method is adopted in DAB sub-modules. Its principle is as follows. In Figure 1(a), U_1 and U_2 are the DC voltages at each side of the sub-module, respectively; the transformation ratio is n; L is the sum of leakage inductance converted to the primary side and the external series inductance; C_1 and C_2 is supporting capacitor for both sides, respectively. The voltage transfer ratio $K=U_1/nU_2$. Figure 2 shows the working waveform of EPS. U_{h1} is the output voltage of the primary AC-side; U_{h2} is the secondary AC-side output voltage converted to the primary side; U_L , I_L are voltage and current of the inductor, respectively; T_{hs} is the half switching cycle; f is the system working frequency; D_1, D_2 are the phase-shifting proportions of the internal and external phase-shifting, respectively. According to the relative size of D_1 and D_2, EPS can be divided into two operation modes: Figure 2(a) is the operation mode 1 ($0<D_1<D_2<1$), and Figure 2(b) is the operation mode 2 ($0<D_2<D_1<1$).

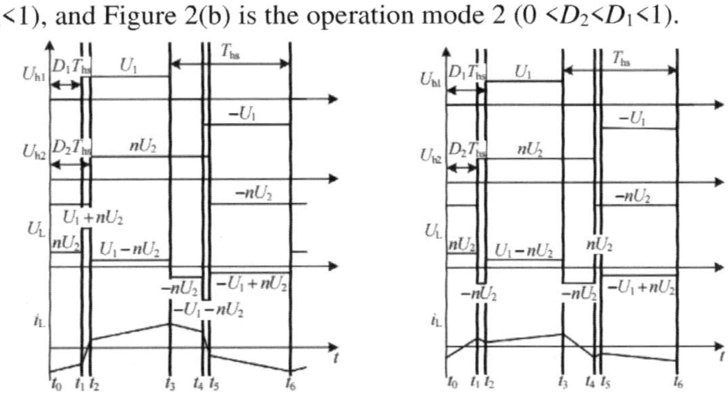

(a)Mode 1 (b)Mode 2

Figure 2. EPS working waves

According to the half-wave odd symmetry of the inductor waveform and the volt-second balance characteristic of the inductor,
In mode 1,

$$\begin{cases} i_{\mathrm{L}}(t_0) = -\dfrac{nU_2}{4fL}[K(1-D_1)+(2D_2-1)] \\ i_{\mathrm{L}}(t_1) = -\dfrac{nU_2}{4fL}[K(1-D_1)+(2D_2-2D_1-1)] \\ i_{\mathrm{L}}(t_2) = \dfrac{nU_2}{4fL}[K(2D_2-1-D_1)+1] \\ i_{\mathrm{L}}(t_3) = -\dfrac{nU_2}{4fL}[K(1-D_1)+(2D_2-1)] \end{cases} \tag{1}$$

In mode 2,

$$\begin{cases} i_{\mathrm{L}}(t_0) = -\dfrac{nU_2}{4fL}[K(1-D_1)+(2D_2-1)] \\ i_{\mathrm{L}}(t_1) = \dfrac{nU_2}{4fL}(KD_1+1-K) \\ i_{\mathrm{L}}(t_2) = \dfrac{nU_2}{4fL}[K(D_1-1)+(2D_2-2D_1+1)] \\ i_{\mathrm{L}}(t_3) = -\dfrac{nU_2}{4fL}[K(1-D_1)+(2D_2-1)] \end{cases} \tag{2}$$

According to the average power transmission formula and reference power formula (maximum power in SPS modulation)

$$P = (\int_0^{T_{hs}} U_{h1} i_L(t)dt)/T_{hs} \tag{3}$$

$$P_{\mathrm{N}} = \frac{nU_1U_2}{8fL} \tag{4}$$

The per-unitized transmission power can be deduced as

$$p = \begin{cases} 2(-D_1^2+2D_1D_2-D_1-2D_2^2+2D_2), 0 \le D_1 \le D_2 \le 1 \\ 2(2D_2-2D_1D_2-D_1+D_1^2), 0 \le D_2 \le D_1 \le 1 \end{cases} \tag{5}$$

2.3. Control method

The switching loss has a great influence on the transmission efficiency. To ensure efficient transmission, it is necessary to achieve soft switching[8]. Take mode 1 for example, to ensure the primary side power electronic switches in Zero-Voltage-Switching(ZVS) state needs $I_{\mathrm{L}}(t_1) \le 0$; to ensure the secondary side power electronic switches in ZVS state needs $I_{\mathrm{L}}(t_2) \ge 0$. Take the above two conditions into equation (1) as

$$\begin{cases} D_2 \ge [(K+2)D_1+1-K]/2 \\ D_2 \ge (KD_1-1+K)/2K \end{cases} \tag{6}$$

Under the condition of ZVS, the current stress is optimized using the method proposed in [9][10], and the optimized internal and external phase-shifting proportion D1 and D2 are obtained.
The closed-loop control strategy is shown in Figure 3. The voltage transfer ratio K is obtained by sampled input voltage V_{i} and output voltage V_{o}. The difference between the reference voltage V_{oref} and

V_o is imported to a PI regulator, and the output side capacitor current I_c is obtained. The sum of I_c and I_o gets multiplied by V_{oref} to get the output power P. Thus the per-unitized p is obtained by dividing P with P_N. The optimal D_1 and D_2 are determined according to the size of p and K, *respectively.*

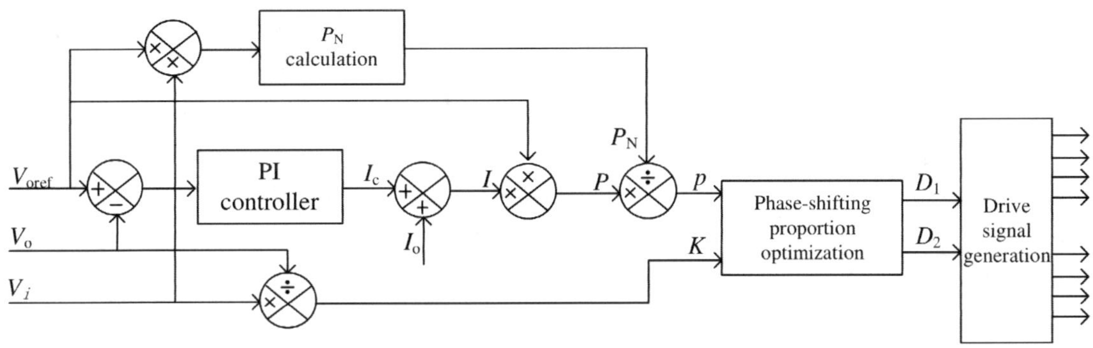

Figure 3. DAB control loop

3. Fault ride-through scheme

The current withstanding capability of power electronic switches such as insulated gate bipolar transistor (IGBT) is generally 1.5 to 2 times the rated current. However, the overcurrent after the DAB output port is short-circuited will quickly lock the IGBT, reducing the transient control capability of the DAB, which will affect the protection and fault recovery. Reference [11] analyzes the causes and suppression methods of DAB fault current DC bias, but it still cannot effectively limit the magnitude of the fault current. Reference [12] achieved the purpose of limiting the short-circuit current by improving the control strategy during the fault. However, the control method makes the secondary side H-bridge of the DAB blocked. In this paper, the fault ride-through is achieved by improving DAB topology and control strategy.

3.1. The improved DAB sub-module topology

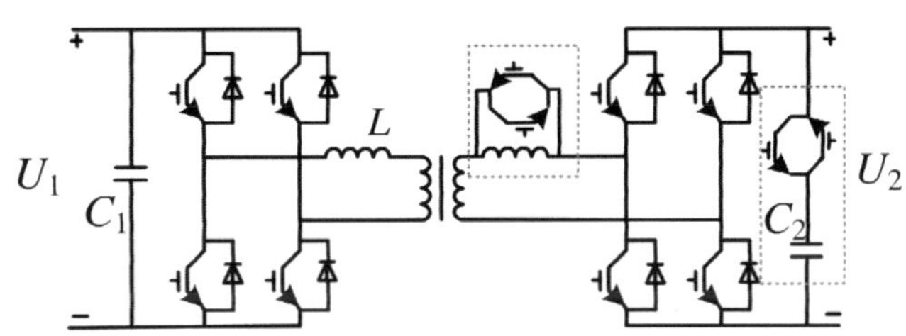

Figure 4. The improved DAB sub-module topology

Figure 4 shows the improved DAB sub-module topology. To prevent the rapid discharge of the secondary side capacitor caused by the short circuit, a power electronic switch is connected to the secondary side capacitor in series. The switch is turned off after detecting overcurrent to prevent the discharge of the capacitor. In addition, adding a current-limiting inductor to the secondary AC side can generate a large equivalent impedance, which has a strong inhibitory effect on overcurrent. During the fault, the switching devices connected in parallel with the inductor are turned off to enable the series inductor.

3.2. Control strategy

It can be known that the DAB switching loss in each working cycle is fixed, resulting in the proportional relation between the switching frequency and the total switching loss. Therefore, increasing the frequency can be used to increase loss, thus limiting current growth. According to [13] , the peak value of the primary side inductor current during fault is

$$I_{peak} = \frac{U_1(1-D_1)}{Lf} \tag{7}$$

According to the above formula, the maximum current is related to the frequency and the internal phase-shifting proportion. Since the secondary side voltage drops to 0 during the fault, the control mentioned in previous section is no longer effective, so the fixed phase-shifting proportion control can be used to increase D_1 to suppress the fault overcurrent. The fault ride-through control strategy is shown in Figure 5.

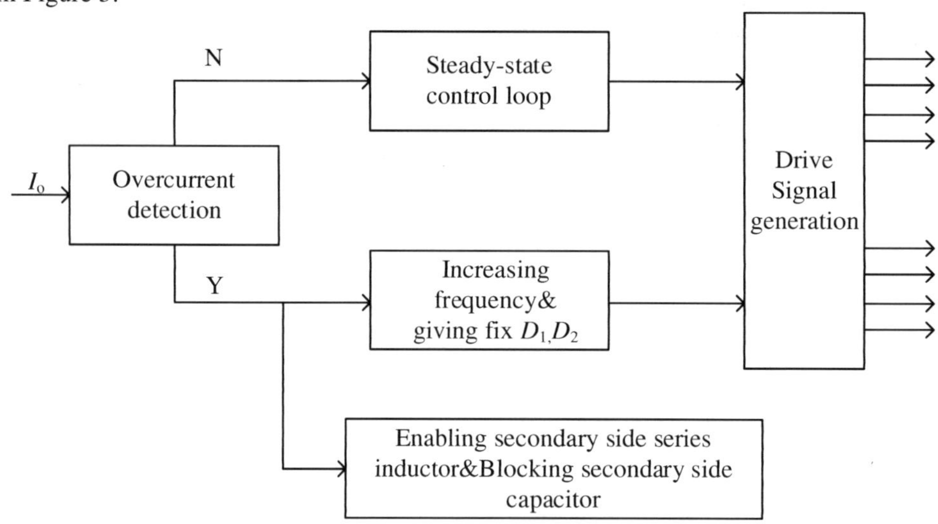

Figure 5. The fault ride-through control strategy

4. Simulation analysis

Due to the limited scale of DAB sub-modules, when the IPOS converter is applied to the high-voltage transmission of offshore wind power, a large number of sub-modules are required, which brings inconvenience to the construction of the offshore platform. Therefore, the IPOS converter is more suitable to applied as the medium-voltage level DC/DC converter, connecting wind turbine terminal and the medium-voltage collecting bus, as shown in Figure 6.

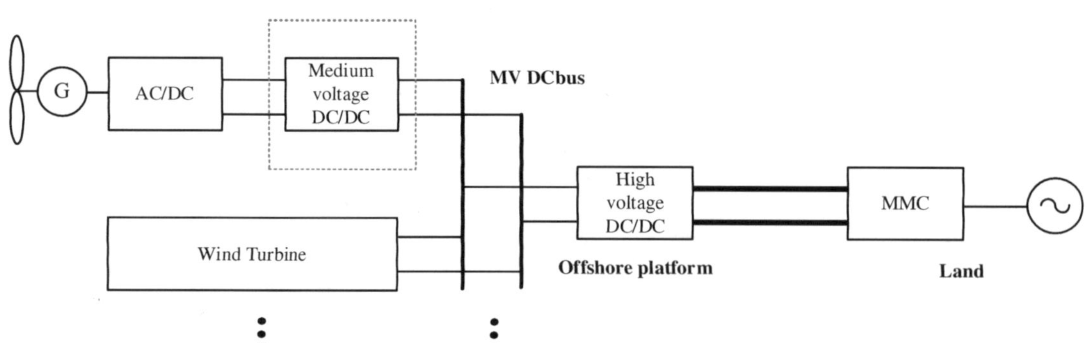

Figure 6. DC offshore wind farm topology

Permanent-magnet synchronous generators (PMSG) are gradually becoming the mainstream in offshore wind turbines, and their terminal voltages range from 0.69 kV to 10 kV level depending on the converter type and capacity. In this paper, the terminal voltage of 6.6 kV is used, and the voltage of the medium-voltage collecting bus is set to 66 kV. The simulation is completed using PSCAD/ EMTDC. The specific parameters are shown in Table 1.

Table 1. Simulation parameters

Parameters	Values
DC voltage U_1	6.6 kV
DC voltage U_2	66 kV
SM number N	10
System frequency f	10 kHz
transformer ratio n	1
SM capacitance C_1	100 μF
SM capacitance C_2	470 μF
SM inductance L	50 μH
SM inductance L_t	20 μH

4.1. Steady-state analysis

The steady-state simulation results are shown in Figure 7. Figures 7(a) and 7(b) show that the DAB sub-module and IPOS converter stabilize the output voltage at 6.6 kV and 66 kV, respectively; Figures 7(c) and 7(d) are the current and voltage waveforms of the primary and secondary switches of the DAB sub-module, respectively. It can be seen that the primary side switches and the secondary side switches both achieve ZVS turn, verifying the effectiveness of the steady-state closed-loop control strategy.

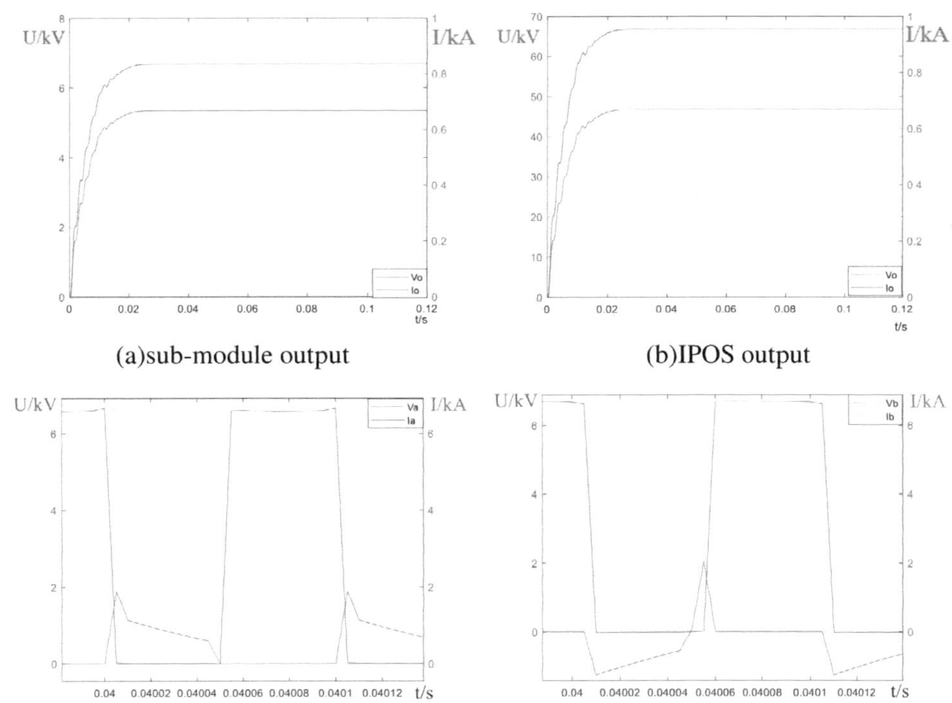

(a)sub-module output (b)IPOS output

(c) primary side switches (d)secondary side switches

Figure 7. Steady-state waveforms

4.2. Fault state analysis

To simulate the fault state of the converter, the short-circuit fault is set at the converter output port at 0.05 s, and the duration is set as 0.03 s. Figures 8(a) and 8(b) are the output voltage and current of the IPOS converter without fault state control, respectively. When the fault is cleared, the output voltage has a large fluctuation due to the secondary side capacitor's charging and discharging, which will affect the stability. The peak value of the fault current is close to 7 kA (almost 10 times the steady-state current) at the beginning of the fault, which will cause the forced blocking of the converter; Figures 9(a) and 9(b) show the output voltage and current of the IPOS converter with fault state control. It can be seen that due to the additional control, the overvoltage is decreased, and there is no need to re-establish the voltage when the fault is cleared. The peak value of the fault current is suppressed to 1.4 kA (about twice the steady-state), which meets the current withstand capability of the IGBT and can realize the fault ride-through. With the low-voltage ride-through control of wind turbines, the wind farm can operate without disconnecting from the grid during the main grid voltage drops, supporting the grid to recover normal voltage.

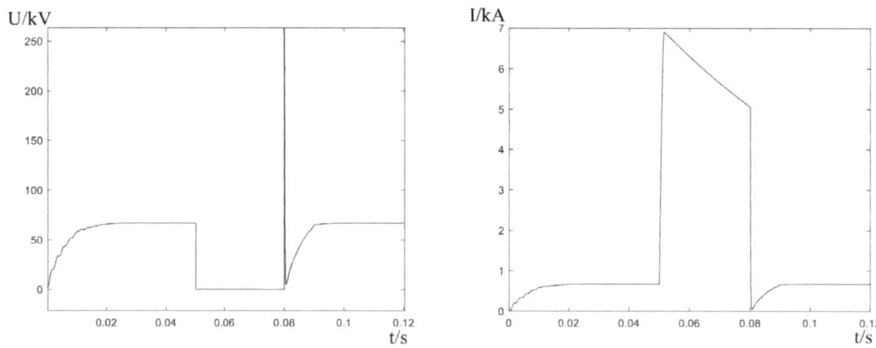

(a)Output voltage (b)Output current

Figure 8. IPOS converter output waveforms without fault control

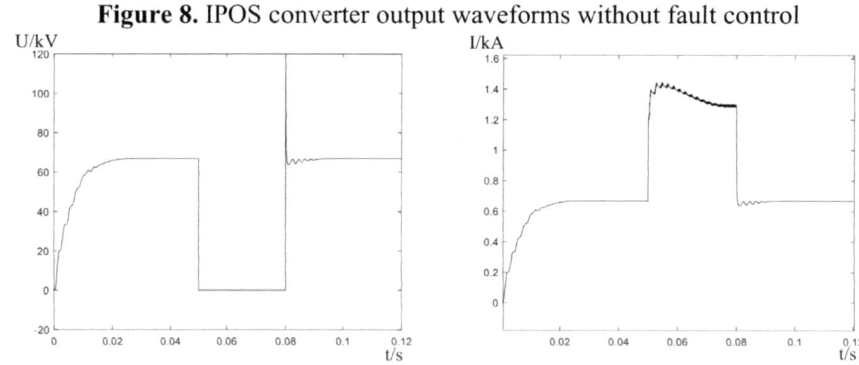

(a)Output voltage (b)Output current

Figure 9. IPOS converter output waveforms with fault control

5. Conclusion

With the development of the DC offshore wind power system, the DC/DC converter with high voltage gain has become indispensable. This paper studies the IPOS converter based on the DAB topology, discusses the EPS modulation method of the DAB converter, designs the closed-loop control strategy, and proposes a scheme of fault current limiting. The effectiveness of the above strategies has been confirmed by simulation results.

6. References

[1] T. Kawaguchi, T. Sakazaki, T. Isobe and R. Shimada, "Offshore-Wind-Farm Configuration Using Diode Rectifier With MERS in Current Link Topology," in IEEE Transactions on Industrial Electronics, vol. 60, no. 7, pp. 2930-2937, July 2013.

[2] D. Elliott et al., "A Comparison of AC and HVDC Options for the Connection of Offshore Wind Generation in Great Britain," in IEEE Transactions on Power Delivery, vol. 31, no. 2, pp. 798-809, April 2016.

[3] J. Esch, "High-power wind energy conversion systems: State-of-the-art and emerging technologies," in Proceedings of the IEEE, vol. 103, no. 5, pp. 736-739, May 2015.

[4] F. Deng and Z. Chen, "Control of Improved Full-Bridge Three-Level DC/DC Converter for Wind Turbines in a DC Grid," in IEEE Transactions on Power Electronics, vol. 28, no. 1, pp. 314-324, Jan. 2013.

[5] X. Cai et al., "Present Status and Future Development of Offshore All-DC Wind Farm," in Proceedings of the CSEE, vol. 36, no. 8, pp. 2036-2048, April 2016.

[6] Akash Javali and Pius Pinto A J, "A NEW STEP-UP DC-DC CONVERTER FOR ISOLATED POWER SYSTEM," in International Journal of Electrical and Electronic Engineering & Telecommunications, Vol. 1, No. 1, pp. 183-189, March 2015.

[7] B. Zhao, Q. Song, W. Liu and Y. Sun, "Overview of Dual-Active-Bridge Isolated Bidirectional DC–DC Converter for High-Frequency-Link Power-Conversion System," in IEEE Transactions on Power Electronics, vol. 29, no. 8, pp. 4091-4106, Aug. 2014.

[8] V. Khyalap and S. Colaco, "SOFT SWITCHING CONTROL OF SYNCHRONOUS BUCK CONVERTER WITH ZCD CIRCUIT," in International Journal of Electrical and Electronic Engineering & Telecommunications, Vol. 1, No. 1, pp. 292-297, March 2015.

[9] J. Huang, Y. Wang, Z. Li and W. Lei, "Unified Triple-Phase-Shift Control to Minimize Current Stress and Achieve Full Soft-Switching of Isolated Bidirectional DC–DC Converter," in IEEE Transactions on Industrial Electronics, vol. 63, no. 7, pp. 4169-4179, July 2016.

[10] H. Guo et al., "Optimal Control Strategy of Dual Active Bridge DC-DC Converters With Extended-phase-shift Control," in Proceedings of the CSEE, vol. 39, no. 13, pp. 3889-3898, July 2019.

[11] B. Zhao, Q. Song, W. Liu and Y. Zhao, "Transient DC Bias and Current Impact Effects of High-Frequency-Isolated Bidirectional DC–DC Converter in Practice," in IEEE Transactions on Power Electronics, vol. 31, no. 4, pp. 3203-3216, April 2016.

[12] C. Zhuo et al., "Research on fault current limitation and active control for power electronic transformer in direct current grid," in IET Generation, Transmission and Distribution, vol. 15, no. 1, pp. 121-134, January 2021.

[13] Y. Chen, Y. Zhang, "Transient Overcurrent Analysis of Dual Active Bridges Under Short Circuits of Loads," in Proceedings of the CSEE, vol. 41, no. 2, pp. 667-680, January 2021.

Journal of Physics: Conference Series **2320** (2022) 012017 doi:10.1088/1742-6596/2320/1/012017

Research of the Loss of Power Transformer Structure

Liu Lei, Li Longfei, Chu Houguang, Zhang Fei, Jin Ming, Zhang Weining, Ge Zhijie and Du Zhaoguang

State Grid Xinjiang Electric Power Co., Ltd. Electric Power Science Research Institute, Wulumuqi ,830011

liulei_buaa@163.com

Abstract. This paper first introduces the research status of magnetic flux leakage and eddy current loss of power transformer at home and abroad, and discusses the significance and purpose of magnetic flux leakage and eddy current loss of power transformer metal components. At the same time, for the problems encountered in the calculation process of this paper, the corresponding solutions and the corresponding practical assumptions to simplify the calculation are given. In this paper, COMSOL software is used to establish the mathematical model of large power transformer and the actual three-dimensional finite element model of transformer, and the three-dimensional leakage magnetic field of large capa city transformer is accurately calculated. A surface impedance method based on double scalar magnetic potential is proposed. Using this method, the leakage magnetic field of structural parts of power transformer, such as pull plate, clamp, oil tank and so on, is analyzed and calculated The size and distribution of magnetic and eddy current losses. The simulation results are compared with the theoretical values to verify the validity and accuracy of the surface impedance method based on double scalar magnetic potential in analyzing the eddy current loss of large power transformer. Then this method is further applied to the analysis and calculation of magnetic leakage field and eddy current loss of a 6300kVA large power transformer to analyze the magnetic leakage field and eddy current loss of its metal components.

1. Introduction

In recent years, with the rapid development of China's electrical technology industry, power transformer as an energy conversion device in the power transmission and transformation system, its single capacity is very high, the voltage level also increases, the loss and heating caused by large capacity transformer will waste a lot of electric energy. Local overheating caused by large losses in its structural parts[1], Will make transformer part of the device deformation and even damage, resulting in many transformers producing unusually serious high temperature, may eventually cause a large area of power system collapse and then cause huge economic losses to the country, and then affect the safe operation of the system[2]. Therefore, it is of great significance to study the size and distribution of transformer leakage field, eddy current loss, and local overheating caused by it.

In the 1980s, many methods to solve the eddy current field was put forward by scholars at home and abroad. Such as the first proposed $A-V-A$ method. This method has the advantage of not needing special treatment for the multi-continuous region but needs to use vector magnetic potential A to solve the whole region. Because the models of transformers and structural components are complex, there are too many unknowns when using this method to solve eddy current problems, and the computer needs large memory[3]. In order to solve the above problem of method A , experts and scholars put forward A method named $T-\Omega$ to calculate eddy current field. The magnetic scalar position Ω is used in the abnormal eddy current region, and the electric vector position T is used in the eddy current region and the excitation source region. The advantages of T method include less unknown quantity, easy implementation, and good stability of the solution. However, in the case that the multi-connected region cannot be solved directly, numerical discretization errors and deletion errors exist in this method to fill the holes surrounded by conductors through low permeability and zero conductivity air materials[4].

In 2003, Schmidt and other foreign experts and scholars used three-dimensional finite element analysis to calculate the eddy current loss of transformer tanks with different capacities and optimized and analyzed the structural parts such as box body to study the eddy current loss of new nonlinear materials[5]. In 2004, some experts put forward a method to calculate the three-dimensional eddy current field generated by magnetic flux leakage of transformers, mainly analyzing the eddy current distribution of clamps and fuel tanks to find out the possible location of local overheating[6]. In recent years, the use of the finite element method continues to mature and computer technology develops rapidly. People have gradually changed from the calculation method of numerical analysis of electromagnetic field to the application of finite element software for commercial purposes, including the calculation of leakage magnetic field and stray loss of multidimensional transformers by Comsol, Ansys, and MagNet. Great progress has been made in two - dimensional and three-dimensional magnetic leakage field analysis[7].

2. Mathematical model and calculation method of eddy current loss

FIG. 1 is a typical nonlinear three-dimensional multi-connected domain eddy current problem. V_1 is the steel plate, Γ_c is its boundary, the steel plate may have holes in it. Artificially selected air zone V_3 surrounds the source current zone V_2, V_4 is the residual air region, These may include ferromagnetic materials without regard to eddy current losses. Make $\Omega_K = V_4$, and $\Omega_J = V_2 + V_3$. $\Gamma_{K.J}$ is the interface of Ω_K, and Ω_J. Γ_1 is called the boundary of the first type and Γ_2 is the boundary of the second type.

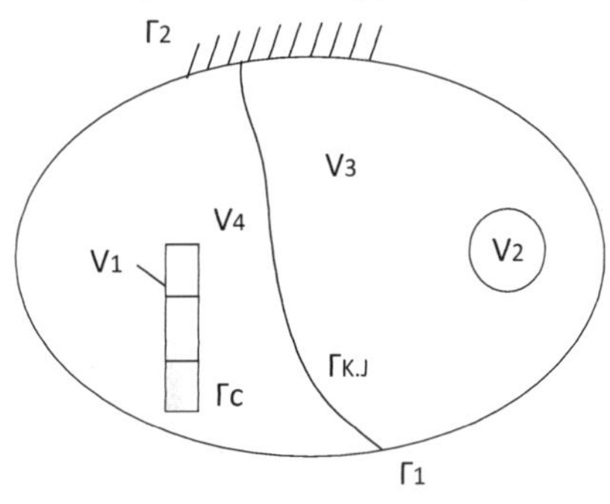

Figure 1 Three-dimensional nonlinear eddy current problem

When the surface impedance method is calculated based on the simplified scalar magnetic potential [8], all regions except the steel plate can be treated as state variables with simplified scalar magnetic potential φ, and the outer surface of the steel plate is treated as impedance boundary. The external magnetic field of the steel plate can be expressed as:

$$\boldsymbol{H} = \boldsymbol{H}_s - \nabla\varphi \tag{1}$$

Where, H_s is the magnetic field generated by the source current.

Impedance boundary conditions are as follows:

$$Z_0\nabla_t^2\varphi - Z_0\nabla_t\boldsymbol{H}_s = j\omega\mu_0\left(\frac{\partial\varphi}{\partial n_c} - \boldsymbol{H}_{sn_c}\right) \tag{2}$$

Where, Z_0 is the surface impedance, t is the tangential unit vector of the steel plate surface, n_c is the normal unit vector of the steel plate surface, and ∇_t is the Laplacian operator used for calculation on the surface.

To solve the nonlinearity of the magnetic properties of steel plates, a surface impedance method based on double scalar magnetic potential can be used. When calculating the air region around the steel plate, the full scalar magnetic potential ψ is taken as the state variable, and when calculating the air region where the source current coil is located, the simplified scalar magnetic potential φ is taken as the state variable. The magnetic field of the full scalar magnetic potential region can be expressed as:

$$\boldsymbol{H} = -\nabla \psi \tag{3}$$

When the simplified scalar magnetic potential region is used, the magnetic field can still be represented by Formula (1), and the impedance boundary conditions obtained are:

$$Z_0\left(\frac{\partial^2\psi}{\partial t_1^2}+\frac{\partial^2\psi}{\partial t_2^2}\right) = j\omega\mu\frac{\partial\psi}{\partial n_c} \tag{4}$$

The equations of the two scalar magnetic potentials as weighted allowances are as follows:

$$\int_{\Omega_k}\mu\nabla N_i \square \nabla\psi d\Omega+\int_{\Omega_J}\mu_0\nabla N_i \square \nabla\varphi d\Omega+\frac{Z_0}{j\omega}\int_{\Gamma_c}\nabla_t N_i \square \nabla\psi d\Gamma = -\int_{\Gamma_{K.J}} N_i\mu_0 H_{sn} d\Gamma \tag{5}$$

With the aid of the above equation and surface impedance Z_0, and using the Poynting theorem, the eddy current losses of each structural part of the transformer can be calculated.

3. Study on the loss of transformer structural parts

In this paper, Comsol simulation software is used to model a three-phase large transformer with a capacity of 6300MVA. The external dimension of the transformer is 5600mm×2000mm×2700mm, and its structure is complex, including a pull plate, clamp, oil tank, lung lobe shield, and other structural parts. In addition, the calculation time will be very long and occupy in the overall simulation calculation of the THREE-DIMENSIONAL transformer. Therefore, to facilitate the simulation calculation, In this paper, some components which have little influence on the distribution of transformer leakage field are omitted in the construction of the transformer model. Finally, part of the 3D power transformer model built in this paper is shown in Figure 2.

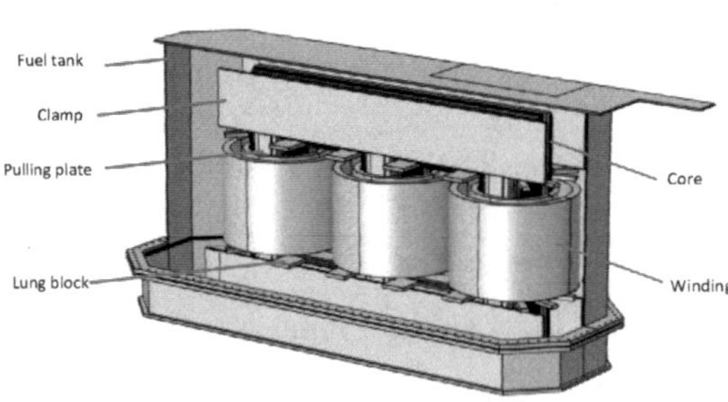

Figure 2. A part of transformer model

3.1. Analysis of magnetic leakage and eddy current loss of clamp

The magnetic density distribution of clamps with or without shielding obtained through simulation calculation is shown in Figures 3 and 4. The eddy current density distribution of the clip is shown in FIG. 5 and FIG. 6. The figure shows that clip of flux density and eddy current loss distribution is very uneven, clip pieces of flux density and eddy current is mainly concentrated in the clip a corresponding winding at the top of the edge and bottom edge position, and the edges of the flux density and eddy current loss than on the edge of the larger, and the up and down in the clip part away from the side of the winding, the flux density and eddy current loss are much smaller, so, Most eddy current losses are concentrated in a very small area. If no measures are taken, the temperature in this area may rise too high and local overheating may occur. The eddy current loss of the clip without shielding is 22.34kW obtained by the frequency domain solver.

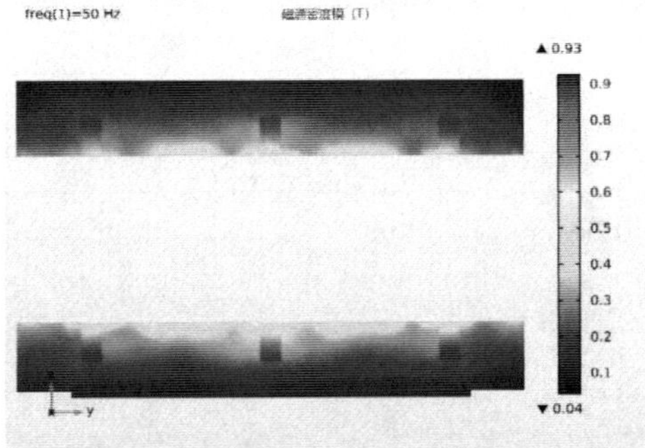

Figure 3. Magnetic density of clip without shielding

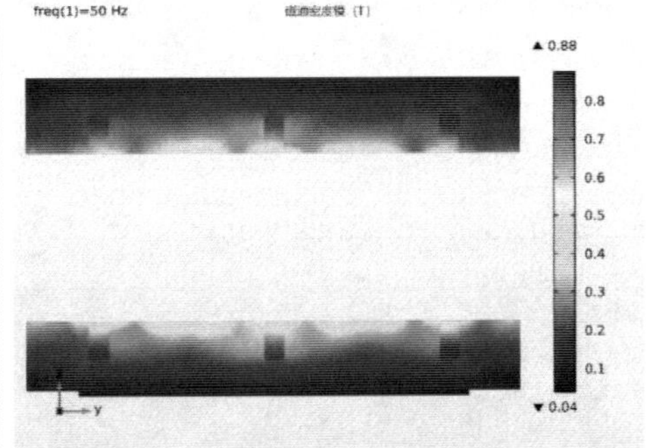

Figure 4. The magnetic density of clamp after shielding

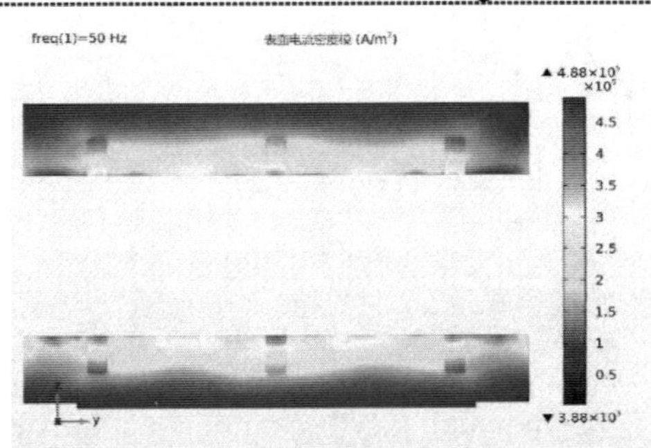

Figure 5. Eddy current density of clamp

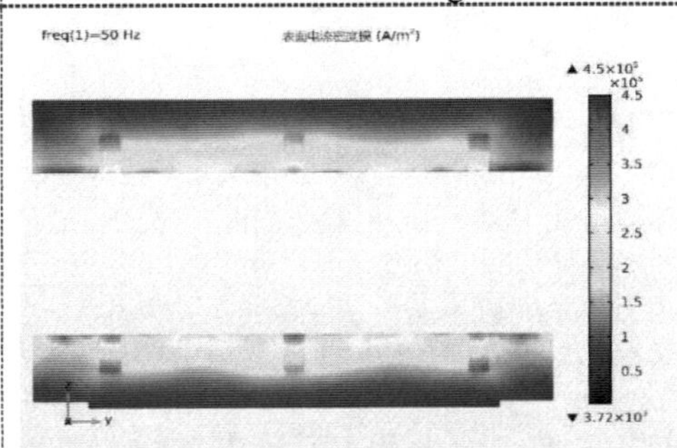

Figure 6. Eddy current density of clamp after shielding

3.2. Analysis of magnetic leakage and eddy current loss of drawplate

The drawing plate is between the core and the winding, which can stabilize the core and bear the tension. Because the pull-plate is located in the region of high magnetic leakage, high eddy current loss will occur, resulting in high-temperature rise. Therefore, designers need to calculate transformer stray loss accurately when the size of the core-drawn plate changes. Accurate evaluation of local overheating is of great significance to the research of coil insulation and mechanical characteristics of transformers. The magnetic flux leakage density of the drawn plate obtained through simulation calculation is shown in Figure 7 and Figure 8, and the eddy current density of the drawn plate is shown in Figure 9 and Figure 10.

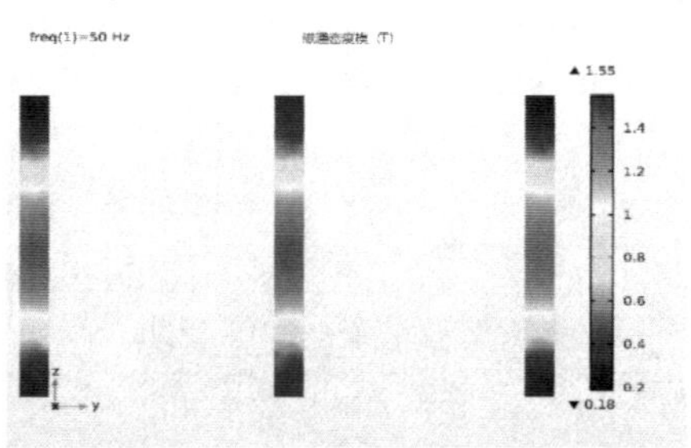

Figure 7 Magnetic density of pull plateFig

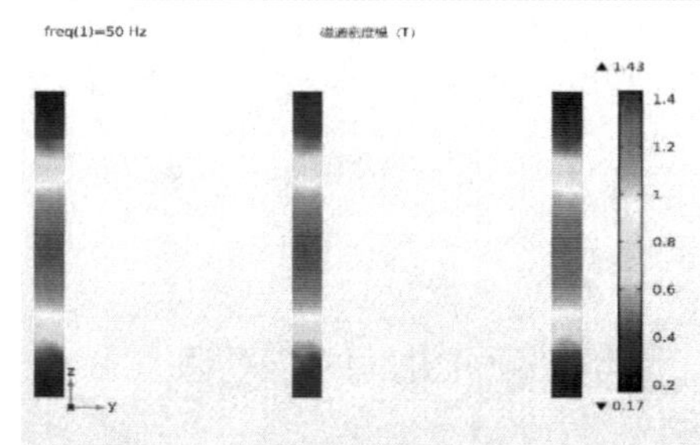

Figure 8. The magnetic density of pulling plate after shielding

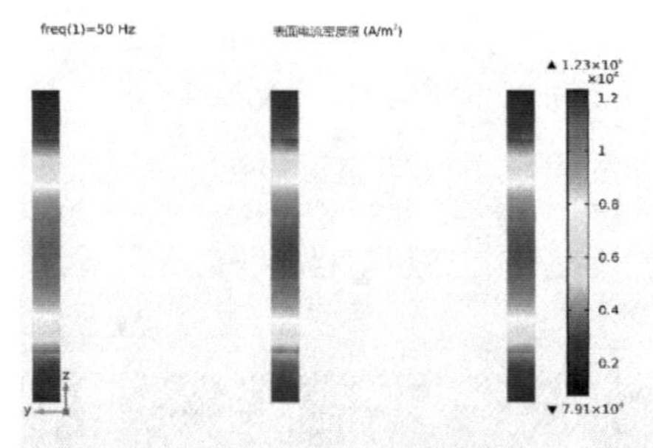

Figure 9. Eddy current density of pull plate

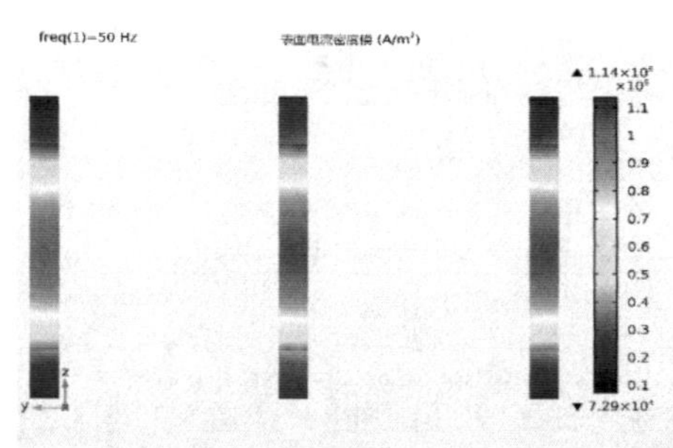

Figure 10. Eddy current density of pull plate after shielding

3.3. Analysis of leakage magnetic field and eddy current loss of oil tank

The size and distribution of tank wall magnetic flux leakage obtained by finite element simulation calculation are shown in Figure 11 and Figure 12, and the size and distribution of tank wall eddy current density are shown in Figure 13 and Figure 14.

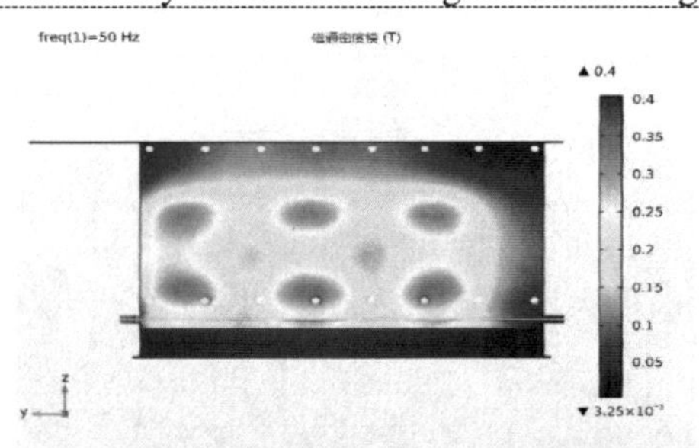

Figure 11. Magnetic density of inner rear wall of the oil tank

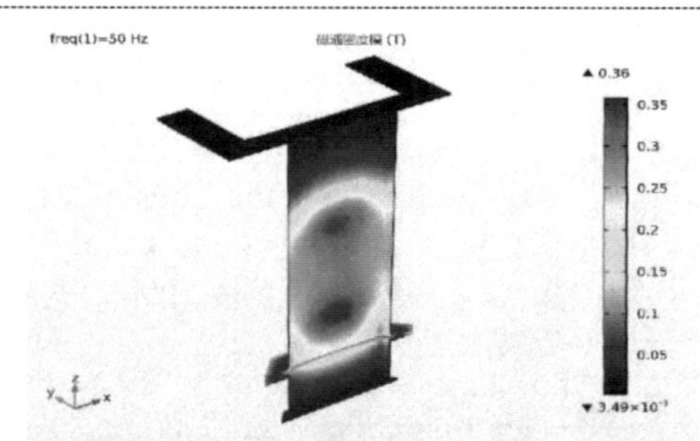

Figure 12. The magnetic density of the left wall inside the oil tank

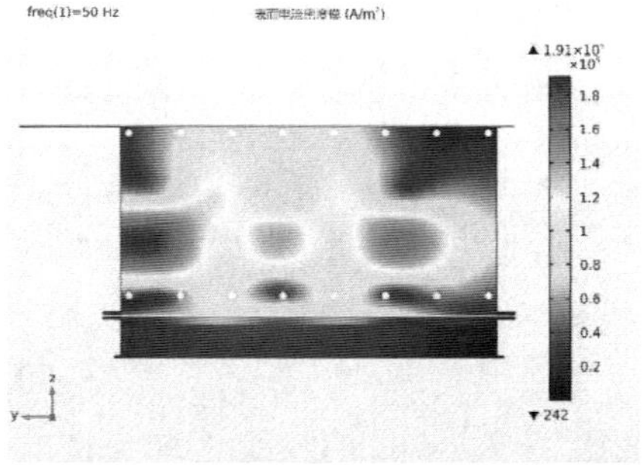

Figure 13. Eddy current density of inner rear wall of the oil tank

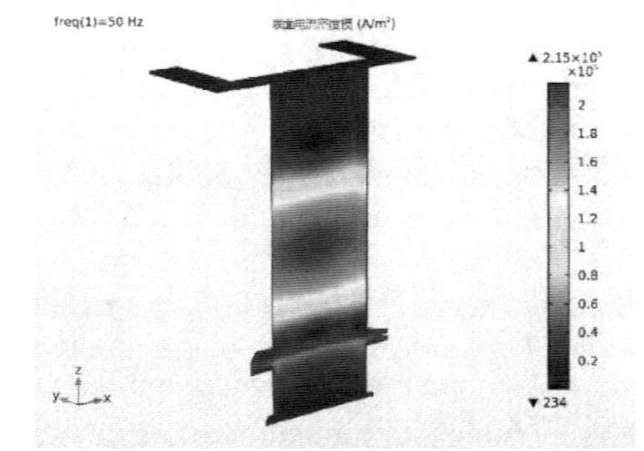

Figure 14. Eddy current density of the inner left wall of the oil tank

As can be seen from the graph, the size of the transformer oil tank wall leakage and eddy current distribution is very uneven, and the tank wall of magnetic flux leakage and eddy current have focused on and corresponds to the location of the on both ends of the winding, and when the construction of the model, the transformer core and winding inside the fuel tank when close to the tank of the left wall, lead to the left wall tank be winding magnetic flux leakage in the path of the main circulation on the fuel tank, The magnetic leakage and eddy current loss of the inner left wall are very high. In this way, most of the eddy current losses are concentrated in the left and rear walls of the very small area, it is easy to appear that the temperature in this area is too high, a local overheating phenomenon.

4. Conclusion

This paper takes a three-phase three-column transformer SFSZ11-6300kVA/110 as the main research object. Based on its actual size, a 3D model is established, and Comsol finite element simulation analysis software is used to simulate and analyze the magnetic leakage field and eddy current loss generated by the power transformer. By comparing and analyzing the size and distribution of magnetic leakage field and eddy current loss on each metal structure, the following conclusions are drawn:

(1) Internal leakage magnetic field distribution of large power transformer is very uneven, and the leakage magnetic field is mainly distributed in the operation of the transformer winding, and in the middle of the winding of leakage magnetic field in the area of distribution, mainly axial magnetic field lines bend degree is small, but at the end of winding near the various structures of suffered serious bend, make the parts of the axial magnetic flux leakage increase greatly. That is, the strong magnetic flux leakage and eddy current density of the clamp and the drawplate appear at the upper and lower end corresponding to the transformer winding, and the strong magnetic flux leakage of the transformer oil tank also appears at the end of the winding and close to the left wall of the winding.

(2) The metal structure on the transformer leakage magnetic field distribution and the eddy current loss calculation results after the study, to reduce the eddy current loss of all structures take reasonable measures of local overheating and curb, sheets and adopts a lung to shape magnetic shielding, by comparing before and after the block the size of the magnetic flux leakage and the calculation of eddy current loss, It is concluded that eddy current loss can be effectively reduced by shielding, while the oil tank is cut off by opening the flow path of magnetic leakage field to reduce the eddy current loss.

5. References

[1] Liu Dongsheng, Zhang Junjie, Zhao Feng, et al. Research on Three-dimensional Eddy Current Field Calculation and Local Overheating Prevention Technology of Ultra-High Voltage And Extra-Large Capacity Transformer [J]. *Transformer*, 2009,46 (9) : 1-5.

[2] Zhanhai Song, Yifang Wang, Shuai Mou, et al. Tank losses and magnetic shunts in a three phase power transformer[C]. International Conference on Electrical Machines and Systems (ICEMS), 2011, 1-4.

[3] Biro O, Preis K, Reinhart W et al. Performance of different vector potential formulations in

solving multiply connected 3-D eddy current problem[J]. *IEEE Transactions on Magnetics*, 1990,26(2): 438-441.

[4] Kameari A. Three-dimensional eddy current calculation using finite element method with A-V in the conductor and vacuum[J]. IEEE Transactions on Magnetics, 1988, 24(1): 118-121.

[5] Schmidt E, Hamberger P, Seitlinger W. Finite element calculation of eddy current losses in the tank wall of power transformers[C]. Electric Machines and Drives Conference, 2003: 1167-1173.

[6] Schmidt E, Hamberger P. Design optimization of power transformers, part 2-eddy current analyses for tank wall and core clamping parts[C]. International Conference on Power System Technology, 2004: 1375-1380.

[7] Kang Yahua. Calculation and analysis of eddy current Loss and temperature rise of power transformer [D]. Shenyang University of Technology,2007.

[8] Holland S.,O'Connell G.P.,Haydock L. Calculating stray losses in power transformers using surface impedance with finite elements[J]. *IEEE Trans.Magn*,1992,28(2)： 1355-1358.

[9] Zhang Liangxian, Chen Musheng, Yu Jian, et al. Eddy Current Loss Calculation and Shielding Analysis of UHV Converter Transformer [J]. *Transformer*, 2013,50 (3) : 15-20.

Acknowledge

This research was supported by the science and technology project of State Grid Corporation (SGXJDK00PJJS2100096).

Journal of Physics: Conference Series **2320** (2022) 012018 doi:10.1088/1742-6596/2320/1/012018

Simulation Modeling of Switching over Current of Circuit Breaker in Intercity Train

Weinan Cao, Ning Sun, Yong Wang and Meng Li
Crrc Qingdao Sifang Co. Ltd., No. 88 Jinhong East Road Chengyang District, Qingdao 266111, China

894981734@qq.com

Abstract. The main interference source of axle counter is the over-current caused by the operation of circuit breaker by intercity train. This paper focuses on the simulation modeling technology of switching over-current of intercity train circuit breaker. Firstly, the mechanism of the interference when the circuit breaker is closed is studied by theoretical analysis; Then, the interference source model is established when the circuit breaker is closed, and the equivalent circuit model of interference path including vehicle body, vehicle body grounding system, traction network and high-voltage cable is obtained through calculation and analysis. The current of transformer primary grounding wire is measured on the actual vehicle, and the accuracy of the model is verified by comparing with the simulation model. Finally, the spectrum characteristics of the interference current in the rail are obtained by Fourier transform. In this paper, a simulation modeling method for analyzing the operating overcurrent of intercity vehicle circuit breaker is proposed, which can effectively evaluate the magnitude of the interference current generated by closing the circuit breaker into the rail, and provide a research basis for the follow-up study of the interference characteristics of the radiation field of rail current on the trackside axle counter.

1. Introduction

With the technological breakthrough in the field of transportation in China. Urban rail transit is more and more widely used. The traditional urban rail transit is the same as the subway, which uses DC power supply. In order to improve the running speed of the train and improve the power supply efficiency, some new rail lines currently use 25kV AC power supply. The new power supply mode is inevitably accompanied by the electromagnetic compatibility problem between urban rail systems. At present, many lines have axle counters disturbed when the train circuit breaker is closed, an error has been reported.

Axle counter is a widely used track section idle and occupation inspection equipment, which is the most important aspect in railway transportation safety and plays a decisive role in railway operation safety. Therefore, the fault of axle counter seriously affects the normal operation of the train and the safety of personnel[1].

At present, the research of axle counter is just beginning at home and abroad. Zamani and Mirabadi calculated the electromagnetic flux and induced voltage of the receiving coil, the direction of the axle counter sensor is optimized by using response surface method (RSM), and the optimal installation angle of the excitation coil and the induction coil is determined[2]. Meng Lei of Harbin Institute of technology studied the structure and working principle of one side axle counter sensor[3]. The three-dimensional simulation model of the sensor is established, and the distribution and variation law of the magnetic field with and without wheels are analyzed by using the finite element simulation software, and the influence law of the coil space position, wheel size, excitation signal frequency and wheel dynamic process on the induced electromotive force on the induction coil is studied. In the study of WANG Zi-cheng et al, Maxwell's three-dimensional model of axle counter and SIMULNK simulation model of traction power supply system are established, and co-simulation is carried out based on Simplorer platform to analyze the influence of traction Backflow on two different types of axle counter[4]; Fu Li et al. tested the interference of axle counter of metro vehicles, and finds that when the traction system of metro vehicles is put into operation, there is a high amplitude current harmonic content on the ground wire, which may interfere with the normal operation of axle counter. Finally, by reducing the capacitance and impedance of EMI filter circuit and optimizing SHEPWM modulation algorithm, the torque control accuracy of the system is improved, and the harmonic content of 38kHz and 42khz current in the main circuit of the system is reduced from 74.68dBA,73.97dBA to 55.94dBA,56.08dBA[5]; in some studies, the single side axle counter model was established by different methods, and its working principle and electromagnetic characteristics were analyzed[6][7].

To sum up, most of the EMC simulation researches on the axle counter at home and abroad focus on the electromagnetic characteristics and rail radiation characteristics of the axle counter head. At the same time, because the axle counter was used together with the DC Metro in the past, the research on the vehicle system also focused on the harmonic interference of IGBT in the traction power supply system, and did not study the interference caused by the breaking of the vehicle circuit breaker; the new type of urban rail transit trains uses AC power supply, and the operating over-current of on-board vacuum circuit breaker is the main interference source of AC power supply trains. At the same time, in the actual operation lines, such as Wenzhou S1 line, Chengdu 18 line, Beijing new airport line, the operation of circuit breaker causes the interference of axle counter. Therefore, it is necessary to study the operation over-current of circuit breaker in intercity train.

In this paper, a simulation modeling method for analyzing the operating overcurrent of intercity vehicle circuit breaker is proposed, which can effectively evaluate the magnitude of the interference current generated by closing the circuit breaker into the rail, and provide a research basis for the follow-up study of the interference characteristics of the radiation field of rail current on the trackside axle counter.

2. Interference mechanism of closed circuit breaker

When the high-speed EMU disconnects the circuit breaker, parallel oscillation is formed between the excitation inductance of the main transformer and the equivalent capacitance to the ground, and the energy stored by them is converted to each other, thus generating switching overvoltage. The schematic diagram is shown in Figure 1.

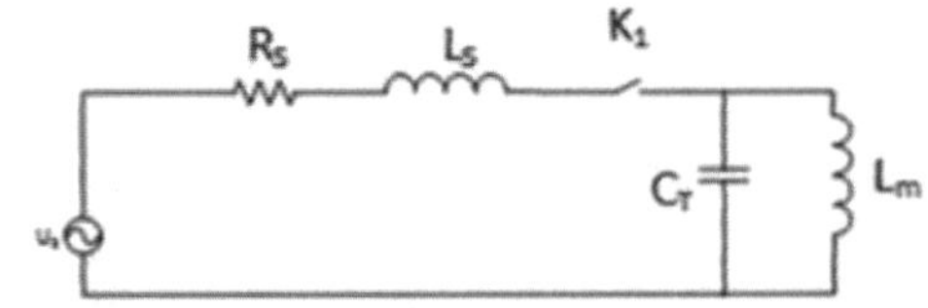

Figure 1. Over current equivalent circuit diagram of operating circuit breaker.

When the circuit breaker is closed, the traction main transformer is also in no-load state, but the magnetic field energy stored in the inductance and capacitance of the main transformer is 0. When the circuit breaker is closed, the traction substation power Us charges the equivalent capacitance of the transformer and cable through the inductance of the catenary circuit. The value of is small, and the high-frequency oscillation process will appear in the circuit, resulting in switching overvoltage, at the same time, because the circuit impedance is small, it will also produce large operating over-current. When the circuit breaker is closed, the operating over-voltage current is related to the inductive and capacitive components in the circuit. When , there may be an operating over-current close to twice the

Journal of Physics: Conference Series **2320** (2022) 012018 doi:10.1088/1742-6596/2320/1/012018

steady-state current, and then the over-current slowly decays with the increase of time, and the decay speed is related to the parameter values of the resistance and inductive components in the circuit [8].

3. Interference source and interference path model

Based on the overcurrent generation mechanism and actual vehicle results, this part models the source and interference path of the vehicle circuit breaker operating overcurrent, including the vehicle body and high-voltage cable model, the traction network model, the vehicle high-voltage system model, and the vehicle grounding system model. The specific modeling methods are as follows.

3.1. Body model and high voltage system model

Vehicle high voltage system models are required to include voltage transformers, traction transformers, circuit breakers and high voltage cables. The voltage transformer is equivalent to an excitation winding inductor and is connected to the body connection. Because the secondary side of the transformer is open when the circuit breaker is closed, the connection between the traction transformer model and the circuit breaker is equivalent to its excitation inductor and excitation resistor in parallel and then series the original edge leakage and the original edge resistance, the circuit breaker is equivalent to the ideal switch. The overall equivalent circuit diagram of its high-voltage system is shown in Figure 2.

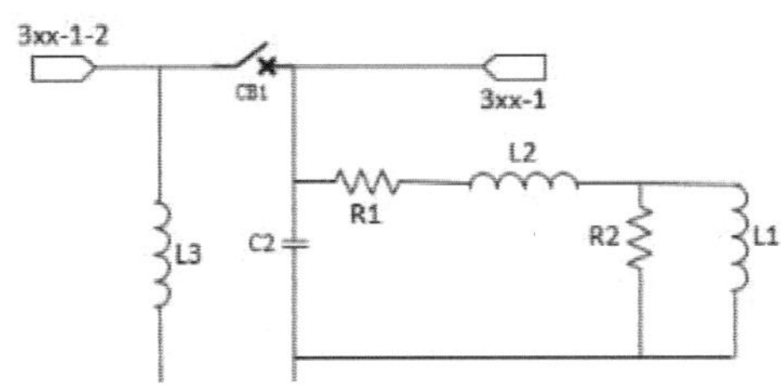

Figure 2. High-voltage system equivalent circuit.

The relevant parameters in the model are shown in Table 1.

Table 1. Relevant Parameters.

Description	Value	Parameter
Transformer original edge resistance/ Ω	1.79	R1
Transformer excitation resistance / Ω	6337.00	R2
Transformer excitation inducer/H	200.00	L1
Transformer original side leakage /mH	2.90	L2
Voltage transformer excitation inductor/H	100.90	L3
Transformer primary distribution capacitors /nF	0.5	C2

The intercity car has 4 car bodies and its high-voltage cable layout is shown below.

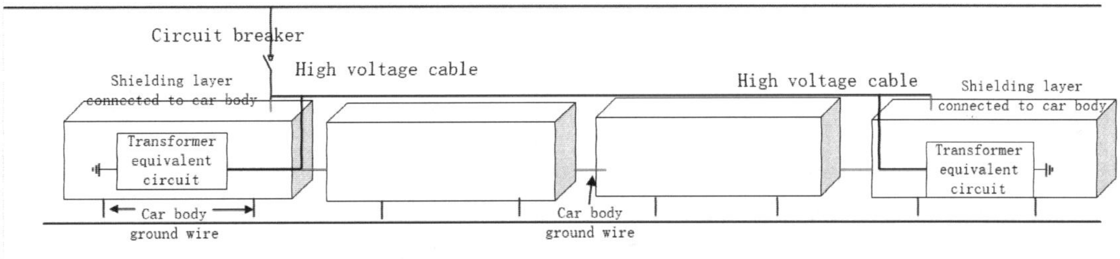

Figure 3. Body structure and high-voltage cable layout.

Because overcurrent has many high frequency components, it is necessary to consider the distribution parameters of car body and high-voltage cable, build the model of car body and high-voltage cable in the simulation software according to the actual parameters and layout, and embed the software calculated distribution parameter model into the circuit model.

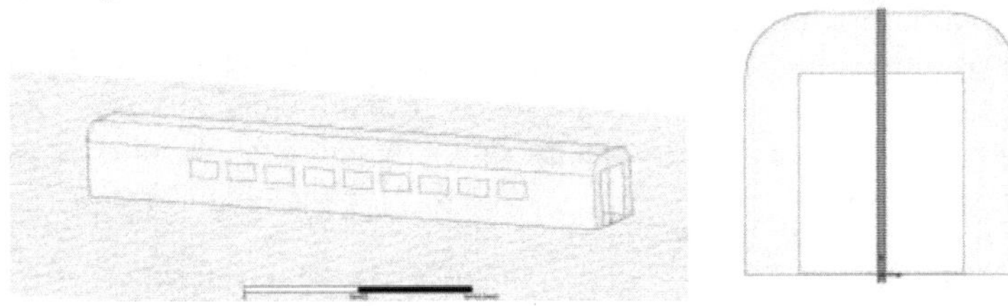

Figure 4. Car body and high-voltage cable 3D model.

3.2. Traction network model

Most of China's current intercity vehicle operation lines use the direct power supply mode of bringing back the flow line, its traction power supply system is composed of traction network and traction substation, by contact network, rail and reflux lines and other facilities together constitute the traction network transmission network. The traction network can be seen as a power supply circuit consisting of multiple conductors, the basic structure of which is shown in the following image:

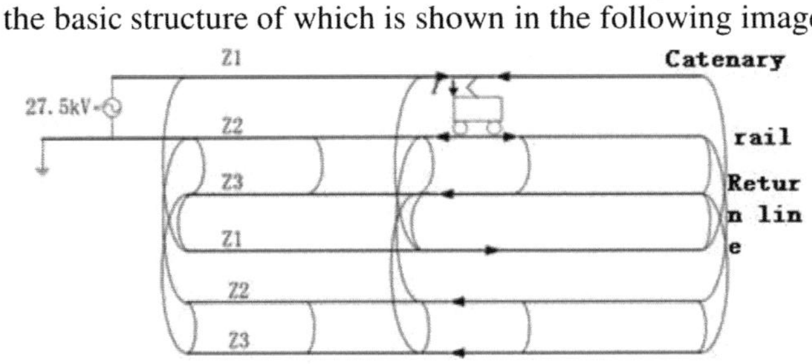

Figure 5. Traction network multi-conductor structure diagram.

A traction power supply system can be thought of as a multi-conductor transmission system consisting of a total of 6 conductors up and down. This paper only studies the uplink line, and since the axle is only mounted on one side of the rail, the rail needs to be divided into two, and finally according to the diagram above, the traction net 4 conductor equivalent circuit shown in Figure 4 is obtained: contact wire, return line, rail (R1, R2). The unit length impedance parameters of the parallel multiconductor unit of the traction network are mainly divided into: capacitive coupling parameters (mutual capacitance, ground-to-ground capacitance), impedance coupling parameters (mutual resistance, ground resistance), self-inductive, self-resisting calculation, parameters calculated using the Carson formula, and then the construction of a traction network equivalent circuit model[9], Figure 6 is a 200m traction network equivalent circuit, which can then be cascaded as needed.

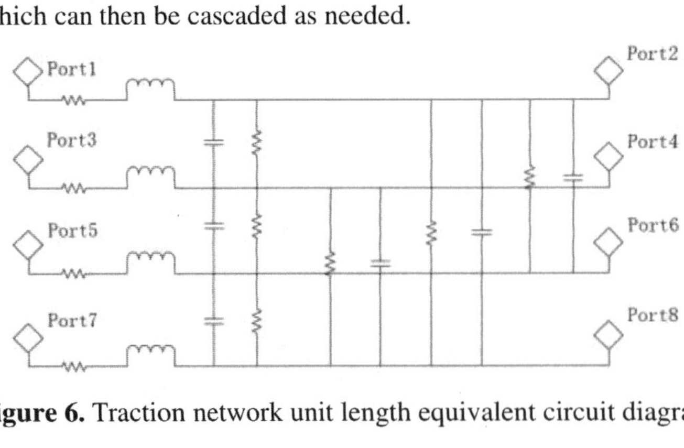

Figure 6. Traction network unit length equivalent circuit diagram.

The main parameters in the traction network equivalent circuit are shown in the table below.

Table 2. Main parameters.

Parameter	Value
The rail self-inducing	0.002588Ω
Rail self-resistance	0.0001356H
Contact the network self-inducting	0.0002102H
Contact the net self-resistor	0.03135Ω
Rail - contact network mutual capacitance	3.5e-11F

3.3. Vehicle grounding system model

Vehicle grounding system model, need to include ground resistor, working ground and protection ground impedance, rail impedance. The equivalent circuit model is constructed according to the ground structure of the actual body as shown in Figure 7.

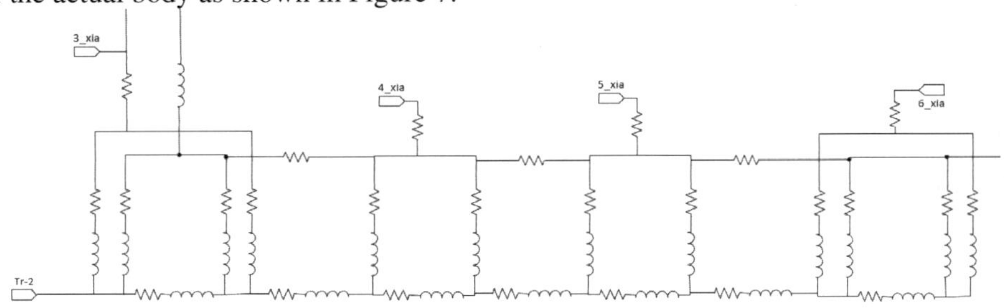

Figure 7. Vehicle grounding system equivalent circuit diagram.

3.4. Overall model of interference sources and interference paths

The traction network model, the vehicle high-voltage system model, and the vehicle grounding system model are integrated with the body model and the high-voltage cable model, as shown in Figure 8, where the traction network length is 5km, with a length of 1km each.

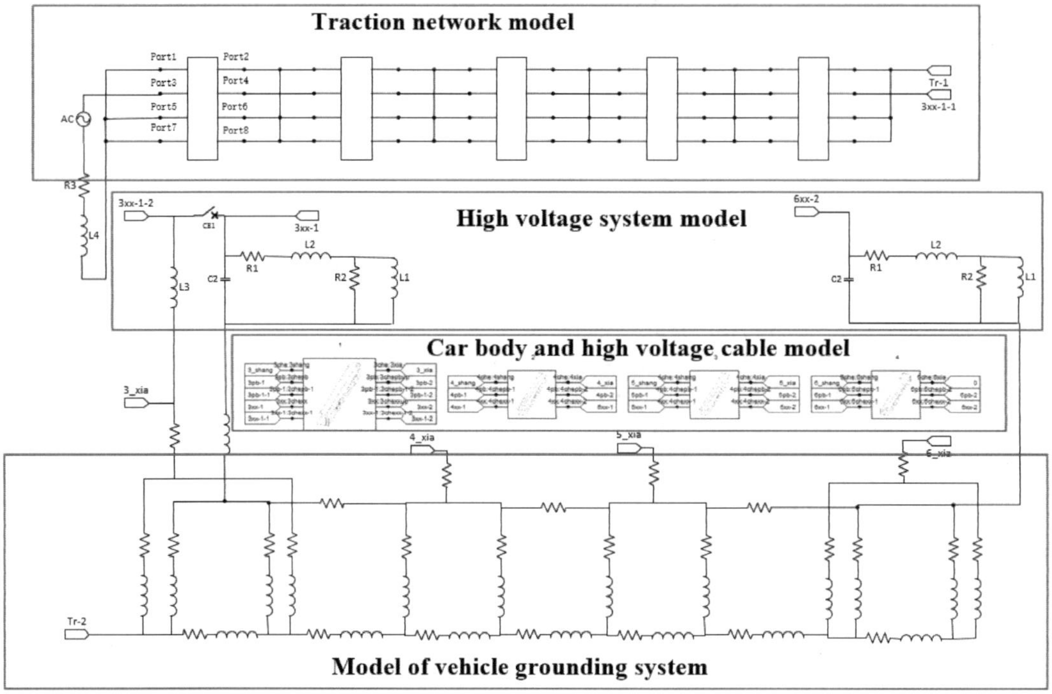

Figure 8. Complete circuit diagram of the vehicle traction system.

3.5. Simulation model accuracy analysis

Select an intercity model to test the transformer ground current when the circuit breaker is closed, as shown in the test diagram, the frequency band range of the test probe is 10kHz to 100MHz.

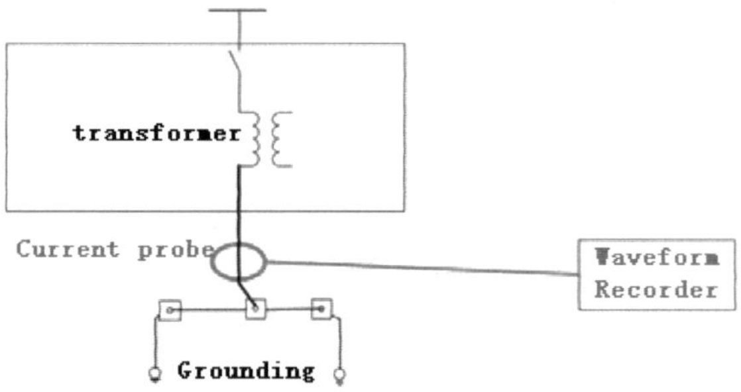

Figure 9. Test diagram.

The test instrument is the recorder and the sensitive current probe, the test condition is static, the circuit breaker is closed, multiple measurements, to ensure that the measured overcurrent at the time of voltage peak closure, the measured current waveform diagram is shown in Figure 10.

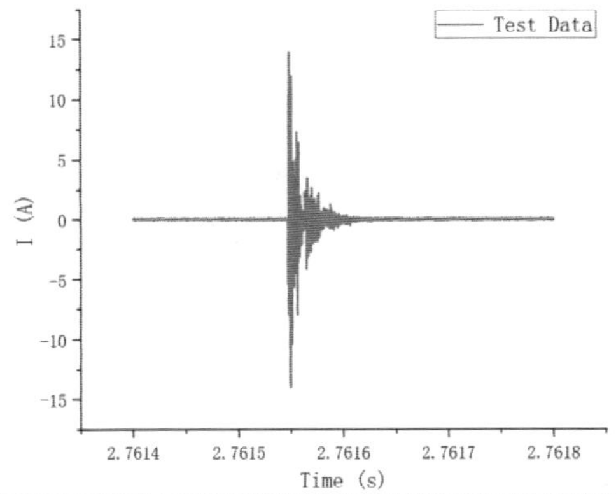

Figure 10. Transformer primary ground current simulation waveform.

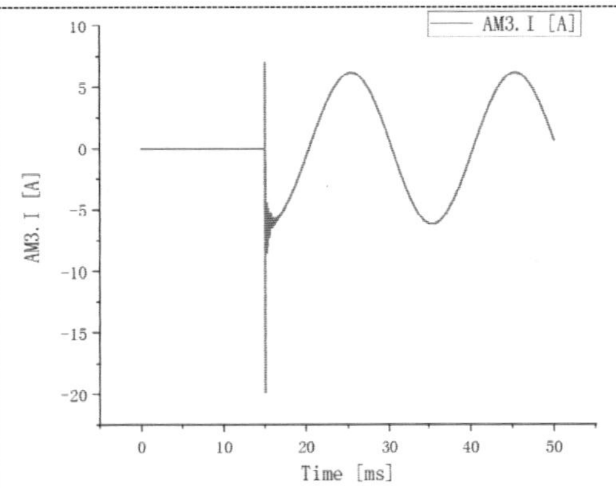

Figure 11. Transformer primary ground current measurement waveform.

The primary ground current waveforms for transformers at the same location in the simulation model are as follows.

The data at the intermediate fluctuation is used to obtain a spectrum comparison chart of the two, because the actual train uses the shaft operating frequency band is 38kHz and 42kHz, so compare the spectrum from 0 to 100kHz.

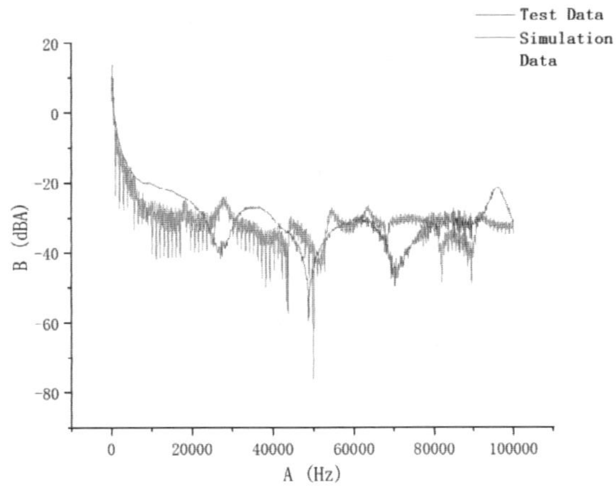

Figure 12. Simulation vs. Test Current Spectrum.

We can see that the simulation data in the range of 0 to 100kHz is roughly consistent with the test data, so the built model can be considered to be more accurate for the operating frequency band of the axle.

3.6. Spectrum analysis of rail over current

The current probe is added to the rail position 3km from the vehicle body of the simulation model to measure the current in the rail where the circuit breaker is closed instantaneously, and its time domain and spectrum waveform are shown in Figures 13.and 14.

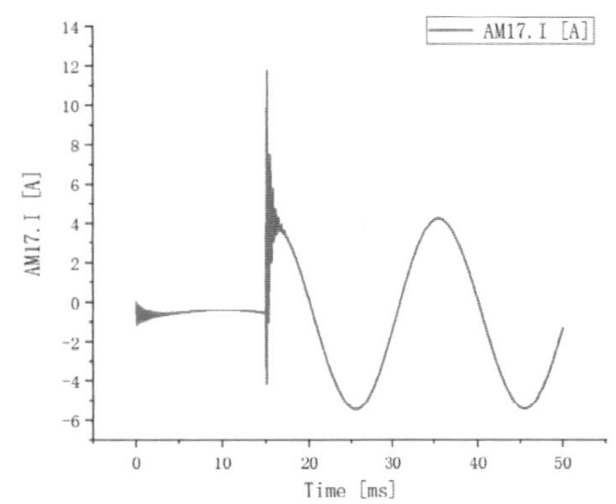

Figure 13. The time domain waveform of the rail current.

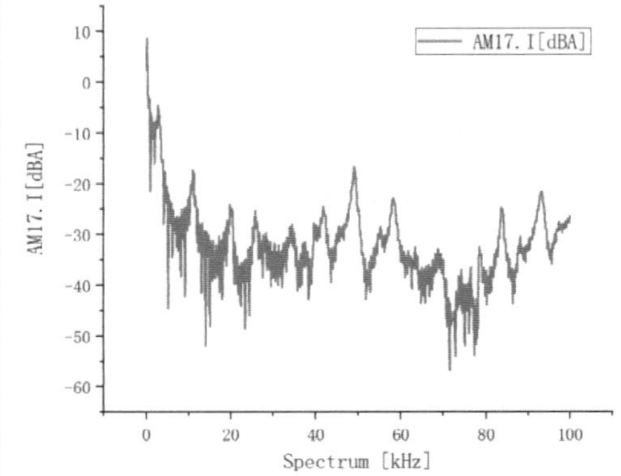

Figure 14. Rail overcurrent spectrum.

We can find that in the frequency band range of 38kHz to 42kHz, the overall interference signal is about -25.4dBA, and the main interference bands include 0-4kHz, 42kHz, 48kHz, 58kHz, 85kHz and 95kHz.

4. Conclusions

This paper focuses on the simulation modeling technology of switching over-current of intercity train circuit breaker. Finally, the spectrum characteristics of interference current in rail are obtained by Fourier transform. It is found that there are some interference components in axle counter working frequency band. In this paper, a simulation modeling method for analyzing the operating overcurrent of intercity vehicle circuit breaker is proposed, which can effectively evaluate the magnitude of the interference current generated by closing the circuit breaker into the rail, and provide a research basis for the follow-up study of the interference characteristics of the radiation field of rail current on the trackside axle counter.

5. References

[1] Ngo Minh Khoa, Mai Vu Cuong, Huynh Quoc Cuong, and Nguyen Truong Tan Hieu, "Performance Comparison of Impedance-Based Fault Location Methods for Transmission Line," International Journal of Electrical and Electronic Engineering & Telecommunications

[2] Zamani, A. , and Mirabadi, A. . (2011) "Analysis of sensor orientation in Railway axle counters using Response Surface Methodology. " *SASTech.*

[3] Meng Lei. (2014)."Analysis and research of unilateral axle counting sensor electromagnetic system." M.A. Thesis. Harbin: Harbin Institute of Technology.

[4] WANG Zi-cheng, ZHANG Ya-dong, GUO Jin, and LUO Rong.(2018) "Impact Analysis of Traction Return Current on Axle Counter." *Railway standard design* 062.002(2018):166-172..

[5] FU Li, CONG Peicheng, and MA Lan. (2020) "Analysis of the Trouble of Traction System Interfering the Axle Counter on Metro Vehicles and Scheme of Solution. " *Rolling Stock* 58.08 (2020):5.

[6] Zhou Changyi, Liu Yanlong Jin Xianji, and Tong Weiming.(2016) "Magnetic field analysis of one side axle counter sensor based on COMSOL. " *China Railway* 000.002 (2016):51-55.

[7] Han Guangyao. (2015) "Research of unilateral axle counting sensor based on electromagnetic induction principle. " M.A. Thesis. Harbin: Harbin Institute of Technology.

[8] WAN Yusu. (2017) "Study On The Characteristics And Influencing Mechanism Of Switching Overvoltage For High-speed EMU. " M.A. Thesis. Chengdu: Southwest Jiaotong University.

[9] Duan Xudong. (2014) "Research on transient circuit model for integrated grounding system of high-speed railway. " M.A. Thesis. Chengdu: Southwest Jiaotong University.

Scenario Generation for Wind Power Using Generative Adversarial Networks

Shi Tiancheng[1], De Gejirifu[2], Cong Hao[1], Guo Wenzhang[1], Zhong Yalin[1] and Qian Long[1]

[1] State Grid Anhui Electric Power Co., Ltd. Economic and Technological Research Institute, Hefei Anhui, 230022, China.
[2] State Grid Electric Power Co., Ltd. Economic and Technological Research Institute, Beijing, 102209, China

18591761290@163.com

Abstract. Scenarios generation is a critical part in planning and operation in high renewable energy penetratied power systems. However, the statistical assumptions of traditional parametric methods may not hold for all types of wind farms. In this paper, a data-driven artificial intelligence approach is presented to generate wind power output scenarios based on generative adversarial networks (GANs). Unlike traditional probabilistic model-based techniques which are typically difficult to scale or sample, the proposed method is data-driven and captures patterns of wind power generation. First, the GAN network structure is constructed, and the Wasserstein distance is employed as the discriminator's loss function. The GAN training then enables the generator to learn random noise and actual history data. Finally, the scenario generation approach based on Monte Carlo simulation and GANs are compared. It shows that the scenarios generated by proposed method can accurately describe the uncertainty of wind power output.

1. Introduction

Because environmental issues, such as climate change, are getting more and more serious, renewable energy sources (RESs) have emerged as key alternatives to existing power generating methods. Wind power, one of the most promising renewable energy sources, has drawn dramatically attention in recent years[1]. Its enormous potential for pollution reduction has provided major advantages to power systems. In contrast, the rising penetration of wind power has increased the power system's unpredictability. Wind power's intermittent and unpredictable nature posed several obstacles to power system planning and scheduling, such as peak shaving and frequency control issues.

The stochastic formulation assists decision makers in mitigating the risk of planning and scheduling in the face of uncertainty. The effectiveness of such strategies will be heavily influenced by the choice of uncertainty representation. Constructing an appropriate uncertainty set, for example, is critical in assessing the conservativeness of robust optimization. Conventional wind power forecasting, on the other hand, generates a wind power output value for each time slot with no uncertainty information. Probabilistic predictions, such as probabilistic intervals, quantiles, or sample paths, match the stochastic formulation better in terms of decision making and risk assessment. The process of generating probabilistic forecasts is referred to as scenario generation[2].

The methods which is commonly adopted in literature for generating wind power output scenarios can be classified into two categories: parametric methods and non-parametric methods[3].

Parametric approaches are based on the probability distribution assumption. Scenarios of parametric approaches are commonly generated by sampling techniques such as Monte Carlo sampling (MCS) or Latin Hypercube sampling (LHS). In addition, Copula techniques are commonly utilized in such systems to capture dependence hierarchies. For example, a Monte Carlo simulation method based on Copula function and first order Markov process is adopted in [4] to generate multivariate wind speed and PV power spatio-temporal series. In [5], LHS is used to generate scenarios that depict wind and PV power output forecasting errors. Whereas, the statistical assumptions may not applicable to all types of wind farms[6].

To tackle this issue, non-parametric models have been extensively studied. The advantage of non-parametric approaches are distribution-free characteristics. For example, in [7], a non-parametric neural network-based prediction intervals (PIs) are used to quantify forecast uncertainty of wind power. Another important type of nonparametric method is generative adversarial networks. Reference [8]combined clustering algorithm and improved generative adversarial network model to obtain more abundant electric vehicle charging current scenarios. Reference [9] applied the gradient penalty term to improve the fitting effect of generative adversarial networks and improve the generation quality of wind power scenarios. Reference [10] proposed a conditional generative adversarial network model for random load modeling of distributed energy resources, using date, temperature and historical data as conditions. GANs are trained semi-supervised and do not require time-consuming manual labelling. Furthermore, GANs is a data-driven technique that can model probability distributions using historical data without making any explicit assumptions. In this paper, we propose a wind power scenario generation method based on Deep Convolutions Generative Adversarial Network (DCGAN) to effectively generate wind power scenarios because of their benefits. DCGANs an improvement of GAN. It introduces convolutional network into the structure of GAN, and improves the quality of data generated by GAN based on the strong feature extraction ability of convolutional network.

The rest of this paper is organized as follows. Section 2 introduces the theoretical foundation of GAN. The wind power scenario generation method is proposed in Section 3. Section 4 illustrates case study and results analysis. The paper is concluded in Section 5.

2. Methodology

GAN was first proposed by Ian goodflow[11]. The central concept is Nash equilibrium, in which generators and discriminators compete with each other. During game training process, the generator's generating ability and the discriminator's discrimination ability get stronger over time until an equilibrium point is reached. At this point, the discriminator is unable to discriminate between the generator's true samples and false samples. The generator and discriminator are trained to increase their generating and discriminating abilities, that is, the process of Nash equilibrium.

The loss functions of the generator and the discriminator can be defined as:

$$L_G = -E_Z\{D[G(Z,\theta_G),\theta_G]\} \tag{1}$$

$$L_D = -E_X\{D(S_{train},\theta_D) + E_Z\{D[G(Z,\theta_G),\theta_D]\}\}\} \tag{2}$$

where L_G and L_D are the loss functions of generator and discriminator respectively; θ_G represents the network parameters of the generator; θ_G represents the network parameters of the discriminator; $G(Z,\theta_G)$ indicates that the generator receives random noise Z ,

A series of sampling operations are carried out through neurons with different functions to output the generated sample data. $D(S_{train},\theta_D)$ and $D[G(Z,\theta_G),\theta_G]$ represents the output of the discriminator when the training sample and the generated sample are used as inputs respectively; $E_Z(\square)$ and $E_X(\square)$ represent the expected value of noise distribution and the expected value of training sample distribution respectively. When the input comes from real sample data, the expected output of the loss function L_D is as close to 1 as possible; When inputting samples generated from the generator, the loss function L_G is expected. The output should be as close to 0 as possible.

According to the training rules of GAN, the generator and discriminator can be fixed respectively to form a minimax game problem. Let the objective function of GAN be $V(G,D)$, then the game problem can be expressed as:

$$\min_{G}\max_{D} V(G,D) = E[D(S_{train},\theta_D)] - E\{D[G(Z,\theta_G),\theta_D]\} \tag{3}$$

3. Scenarios Generation methods for wind power

3.1 Data preprocessing

Since the activation function of neural network has good nonlinearity when input in [0,1] interval, it is necessary to normalize the data. In order to ensure the unity of data magnitude, the maximum and minimum normalization is used to standardize the data and realize the equal scaling of the original data. The value of each data is scaled to a value between 0 and 1, as follows:

$$x_{w,t}^{*} = \frac{x_{w,t} - x_{w,\min}}{x_{w,\max} - x_{w,\min}} \tag{4}$$

where $x_{w,t}^{*}$ is the normalized value of sample data. $x_{w,t}$ is multivariate wind power output at time t. $x_{w,\min}$ and $x_{w,\max}$ represent the minimum and maximum power output, respectively

3.2 Generation procedure

The wind power output scenario generation model is shown in Fig 1.

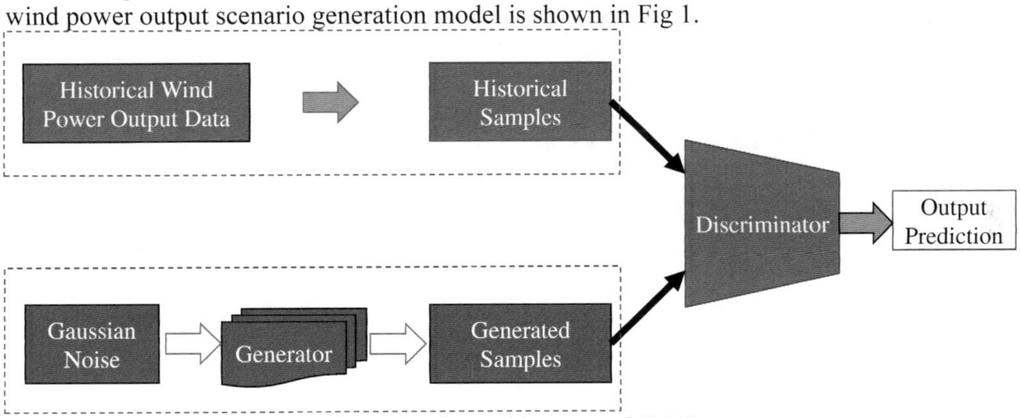

Figure1. The architecture of GANs

The model architecture adopted in this paper is the GAN structure based on the DCGAN structure. DCGAN employs two convolutional neural networks to enhance the structure of standard GAN, increase network stability, improve data sample generation quality, and accelerate convergence. In this study, we use the Wasserstein distance rather than the original Jensen-Shannon divergence. This is due to the fact that training using Jensen Shannon divergence causes the generator to create a single power profile pattern with the maximum likelihood. Because we aim to build scenarios that mimic the diversity of renewable energy generating processes, we can capture all modes in training samples by training with Wasserstein distance. The Wasserstein distance directly computes the distance between two distributions.

The generator's produced wind power scenarios and the preprocessed real load scenarios are mixed together and sent to the discriminator for identification. The discriminator computes the loss function using the initial network parameters and updates them via back propagation. Through the alternating training of two antagonistic networks, GAN updates the loss function and network weight. In this model, the learning rate of the generator is adjusted to 5 times that of the discriminator to guarantee that the discriminator has enough steps to advance to the optimal position every time the generator is updated. In other words, the discriminator is trained five times whereas the generator is trained just once. Repeat the training method described above until the two network loss functions converge. At the moment, the generator is capable of producing realistic samples.

4. Case study

4.1 Data description
Wind power generation data from a wind farm in Anhui Province is used to create training and validation datasets. The original data has a 5 minute resolution. We rearranged the samples, and utilized 80% of them as training data, the rest 20% samples are seen as test data sets. With minimal changes to the algorithm, our technique can simply manage scenario development across numerous sites by leveraging past data from these places as inputs. As a result, the correlations of wind power output are automatically learnt.

4.2 Scenario generation
First, we trained the model to ensure that GANs can generate scenarios. Figure 2 depicts the training evolution for this training dataset. The output of the discriminator exhibits a considerable difference between generated and actual historical data during the first 10,000 iterations, indicating that the discriminator can readily discriminate between the sources of the samples. The Wasserstein distance had already converged to near zero after 15,000 training rounds. We continue training until 21,000 iterations to demonstrate that the procedure is stable, and that once converged, the empirical distribution of a batch of generated scenarios is quite close to the empirical distribution of a batch of training cases. As a result, we are approaching the best answer.

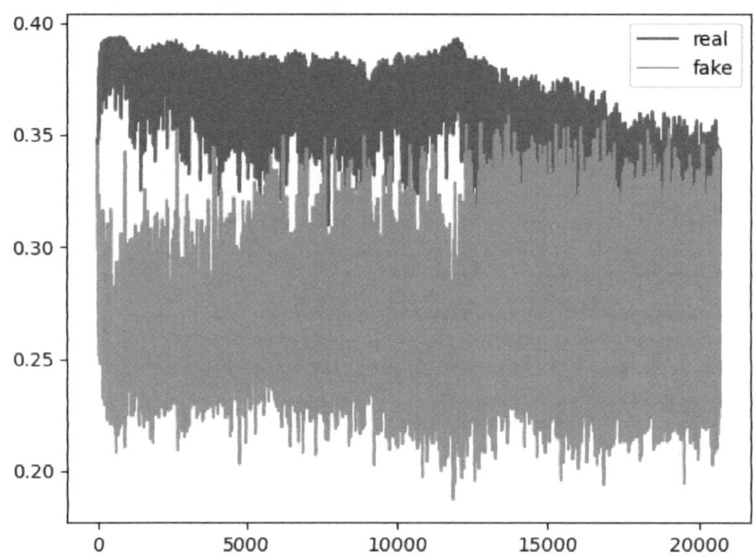

Figure 2. Training evolution for GANs

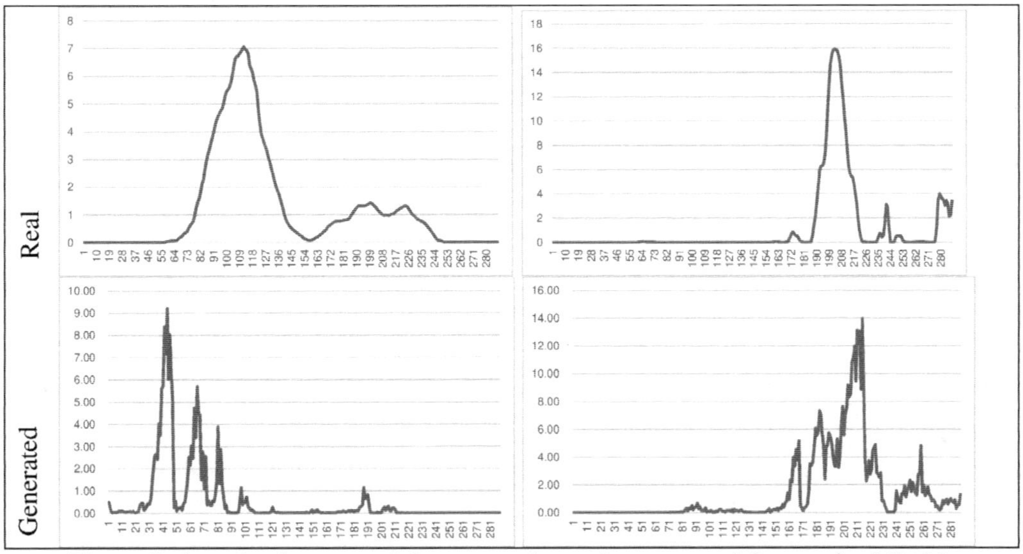

Figure 3. Samples from actual datasets and samples generated from trained GANs

In Figure3, some produced samples are compared to selected samples from the validation set. We can see that the generated scenarios closely match situations from the real wind power outputs. The generator can progressively learn the properties included in the wind power output data. The created scenarios depict the diversity of mode variation effectively. For example, the scenarios show typical characteristics of renewable generating profiles, such as high peaks, diurnal fluctuations, quick power ramps, and so on. such as periodicity and peak, but it will create some low-quality data in the process. The quality of the generated data will increase with more training.

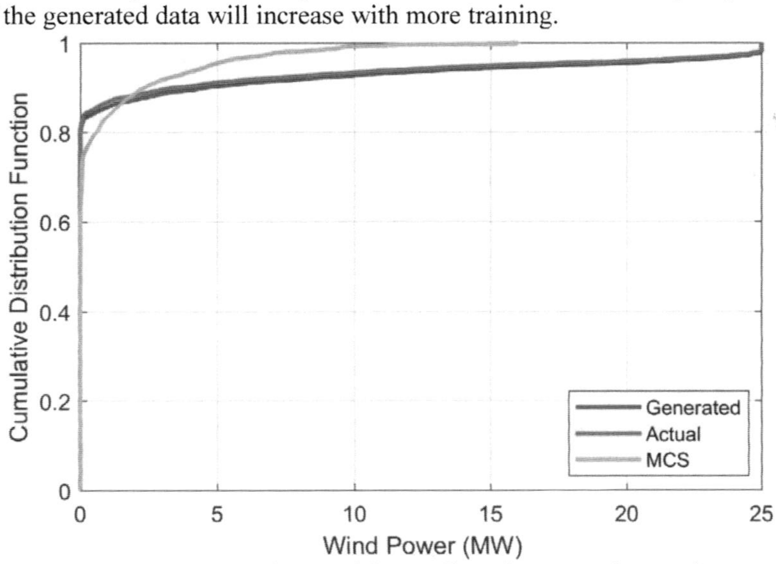

Figure 4. Comparisons of the quality of generated scenarios

In Fig.4, we exhibit the cumulative distribution function (CDF) of historical wind power output samples and GANs-generated samples to compare the stochastic behaviors. We discover that the two CDFs are almost on top of one other. This demonstrates GANs' ability to produce samples with the right distributions. Furthermore, we compared the results of the scenario generation methods based on MCS and GAN. It can be seen that both methods can generate scenarios with close distribution to the real scene, but the distribution deviation between the scenario generated by the MCS method and the real scenario is greater.

In order to intuitively compare the effectiveness of two scenario generation methods, the fitness index F_n is introduced. The expression of F_n is shown as:

$$F_n = 1 - \sqrt{\frac{\sum_{t=1}^{T}(x_t - x_t^*)^2}{\sum_{t=1}^{T}(x_t^*)^2}} \tag{5}$$

where x_t is the generated wind power output scenarios, x_t^* is actual history data. T is number of all time slot. The fitness index of the proposed method is 0.91, and the fitness index of MCS is 0.86. Therefore, the proposed method show superiority over MCS.

5. Conclusion

This work provides a technique for generating wind power output scenarios based on GAN that uses DCGAN architecture to avoid the tough challenge of original GAN training. The experimental results reveal that the generated data is diverse and of excellent quality. Results show that the proposed method has superiority over MCS. The method proposed in this paper avoids simplification and the setup of random variables to suit a certain probability distribution. Infinite scene data may be created by feeding the generator samples with a normal distribution, which is extremely useful for random scene analysis. The distribution is simulated through the mining of historical data and the training of the model, so that the generated data is more in accordance with the actual situation. In the future, we can consider studying the random scene generation method for spatially and temporarily adjacent wind power output scenarios.

References

[1] J. Deng, H. Li, J. Hu, and Z. Liu, 2021 A new wind speed scenario generation method based on spatiotemporal dependency structure. *Renewable Energy*. **163** 1951-62.

[2] J. Liang, and W. Tang 2020 Sequence Generative Adversarial Networks for Wind Power Scenario Generation *IEEE J. Sel. Areas Commun.* **38** 110-8.

[3] Y. Zhang, Q. Ai, F. Xiao, R. Hao, and T. Lu 2020 Typical wind power scenario generation for multiple wind farms using conditional improved Wasserstein generative adversarial network *Int. J. Electr. Power Energy Syst.* **114** 105388.

[4] C. Pan, C. Wang, Z. Zhao, J. Wang and Z. Bie 2019 Copula Function Based Monte Carlo Simulation Method of Multivariate Wind Speed and PV Power Spatio-Temporal Series *Energy Procedia* **159** 213-8.

[5] L. Lu, W. Yuan, C. Su, P. Wang, C. Cheng, D. Yan, and Z. Wu 2021 Optimization model for the short-term joint operation of a grid-connected wind-photovoltaic-hydro hybrid energy system with cascade hydropower plants *Energy Convers. Manage.* **236** 114055.

[6] J. Liang and W. Tang 2019 Wind Power Scenario Generation for Microgrid Day-Ahead Scheduling Using Sequential Generative Adversarial Networks *2019 IEEE International Conference on Communications, Control, and Computing Technologies for Smart Grids (SmartGridComm)* 1-6.

[7] H. Quan, D. Srinivasan and A. Khosravi 2015 Incorporating Wind Power Forecast Uncertainties Into Stochastic Unit Commitment Using Neural Network-Based Prediction Intervals *IEEE Trans. Neural Networks Learn. Syst.* **26**, 2123-35.

[8] F. Yang, S. Yin, S. Zhou, D. Li, C. Fang and S. Lin 2021 Electric vehicle charging current scenario generation based on generative adversarial network combined with clustering algorithm *Int. Trans. Electr. Energy Syst.* **31** 12971.

[9] C. Jiang, Y. Mao, Y. Chai, M. Yu and S. Tao 2018 Scenario Generation for Wind Power Using Improved Generative Adversarial Networks *IEEE Access* **6** 62193-203.

[10] Yi. Wang, G. Hug, Z. Liu, N. Zhang 2020 Modeling load forecast uncertainty using generative adversarial networks *Electr. Power Syst. Res.* **189** 106732.
[11] I. Goodfellow, P. Jean, M. Mehdi, X. Bing, W. David, O. Sherjil 2014 Generative adversarial nets Advances in neural information processing systems pp. 2672-80,

Acknowledgment
This work was supported by the science and technology project of State Grid Anhui Electric Power Co., Ltd. Economic and Technological Research Institute.

Lightning Transient Analysis Comparison and Wind Turbine Models Using PSCAD/EMTDC Circuit Simulator

Tserensambuu Chinges[1, a], Qingmin Li[1, b] and Jiyao Zhao[1, c]

[1]North China Electric Power University, School of Electrical &Electronic Engineering, Beijing, China

[a]chingee@must.edu.mn, [b]lqmeee@ncepu.edu.cn, [c]zhaojiyao1996@126.com

Abstract. This paper aims to compute the lightning electromagnetic transient overvoltage and current on the wind turbine using the PSCAD/EMTDC simulation program. The double-peaked lightning current, tower, grounding system models, and dangerous situation of the grounding system are investigated by the PSCAD simple circuit branch segment codes and defining components. A comparison analysis has been performed to see the influence of lightning current wave shapes and grounding characteristics. The simulation results show that 43 kA and 31 kA lightning currents produce 200 kV overvoltage on the turbine grounding in case of 1000 Ω·m soil resistivity. However, current amplitudes on the equipotential grounding system become 28 kA and 20 kA. The influencing factors of the surge protection device are determined using good accuracy computer simulation.

1. Introduction

Lightning transient behavior of the wind turbine electrical devices depends on the characteristics of the lightning current waveform and the configuration of the grounding system. Most of the failures have occurred on low voltage electrical circuits inside the wind turbine [1]. Such damages are possible coming from the overvoltage rise on the grounding system due to the direct lightning strike, induced overvoltage on the cable lines, and lightning surges coming from the nearby lightning-stricken turbines [2,3]. Proper analysis or modeling is required to predict possible lightning transient characteristics on the wind turbine to keep the reliable operation and protect the electrical equipment.
Previously wind turbine comprehensive models based on the electromagnetic field and circuit methods have been discussed in the literature [4] that analyzed induced voltage and currents of the cables inside the tower. In [5-7] based on the electro-magnetic transient type programs, grounding potential rise generated on the nacelle/tower base and surge protective devices were emphasized when considering the single crest first and subsequent lightning strokes.

In this paper, we attempt to model wind turbine lightning transient physical process using the PSCAD/EMTDC circuit simulator. Comprehensive models have been constructed and analyzed focused on the overvoltage and current distributions in the turbine and transformer grounding system, The influences of the double crest lightning waveforms, different grounding soil resistivity, and low voltage surge arrester installed in the transformer substation were discussed.

2. Wind turbine grounding system arrangement and analysis under consideration

The configuration of a wind turbine grounding system with a transformer substation is shown in figure 1. It has a 15m equipotential grounding cable connecting the turbine and transformer grounding. It is assumed that lightning discharge current travels through cable, thus producing overvoltage and current in the transformer substation. The investigation is focused on the following conditions by changing lightning current and grounding characteristics.

- Lightning current distributions in the considered arrangement
- Transient overvoltage on the turbine/transformer substation grounding system
- Current to the phase line flowing through low voltage surge protective device

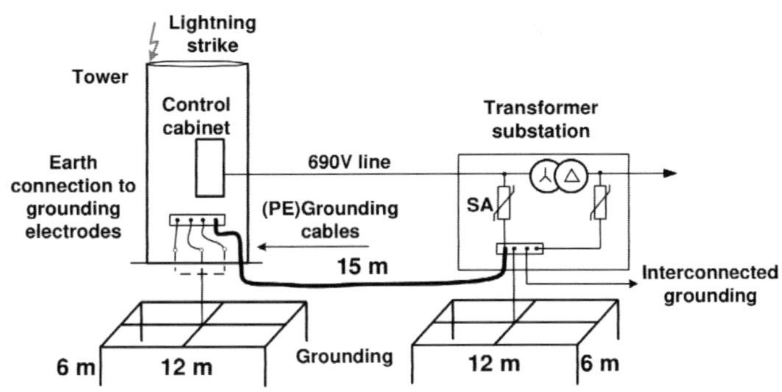

Figure 1. Configuration of the wind turbine lightning protection system.

3. Description of the models and validations

Lightning strike, tower, grounding system, and other electrical device models of the wind turbine configuration have been implemented in PSCAD/EMTDC software.

3.1. Lightning strike model

The lightning current strokes were modeled by an injected ideal current source parallel with 1000 Ω lightning channel impedance. Most of the research adopted standard lightning waveform with single peaks without a concave front. However, some measured median lightning current waveforms showed double peaks with concavity [8]. The first and subsequent lightning current waveforms in figure 2 were represented in PSCAD by implementing the Heidler functions.

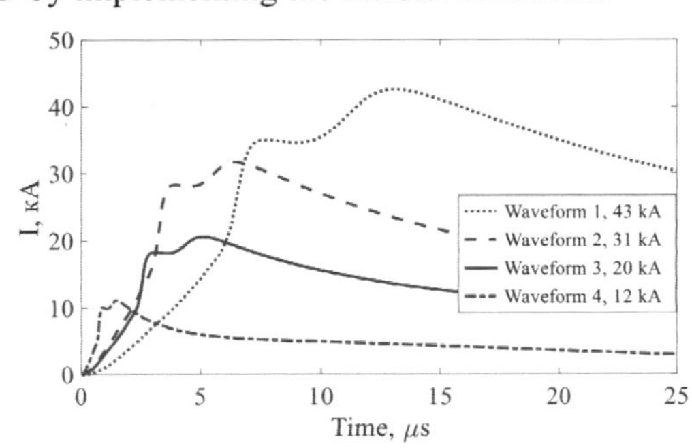

Figure 2. Lightning current wave shapes with different crests

Table 1. The values of coefficients

Lightning Crests	(α)	(β)	(δ)	(γ)
45 kA	0.87	0.94	1.03	1.1
31 kA	0.65	0.47	0.7	1.1
20 kA	0.44	0.37	0.7	1.1
12 kA	0.25	0.44	0.7	1.1

The measured patterns of waveforms were obtained by fitting multiplied coefficients, given in table 1. We name the waveforms 1, 2, 3, 4 by 43 kA, 31 kA, 20 kA, and 12 kA lightning current, respectively.

3.2. Tower model

To obtain good accuracy of the transient response of voltage on the grounding and current distributions in the tower base, the tower surface needs to be divided into suitable segments. Each of the segments is equal to 4 m which is much shorter than injected current source wavelength. The tower is then represented by a multi-conductor network consisting of vertical and horizontal wires, as shown in figure 3.

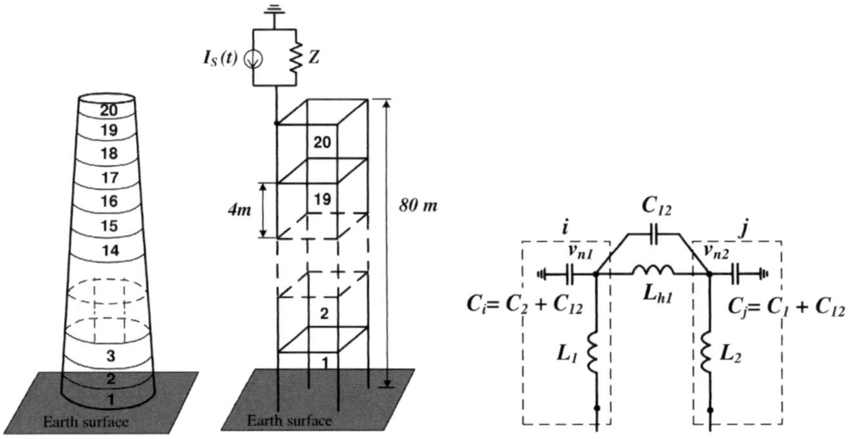

(a) (b) (c)

Figure 3. Configuration of typical tower model and equivalent circuit. (a) 80 m tower with a radius of 2.3 m, (b) Represented tower model in this study (c) Section of the multi-conductor equivalent circuit

The electric network parameters are evaluated based on the electromagnetic field theory. The self and coupling potential coefficients P_{ii}, and P_{ij} were calculated based on the average potential equations which depend on the absolute values of x,y, and z coordinates. The self and mutual inductance of the wires are deduced by Neumann's integral formulas.

Obtained capacitance and inductance values are input parameters to the branch segment manager of the PSCAD. Under the script tab of the branch segment, the following codes are required to be written at every node and branch of the network conductors. The example of branch segment code is **BRN = $n2 $n1 0.0 $L₃ $C₁₂**. The tower parameter's defining components are shown in figure 4.

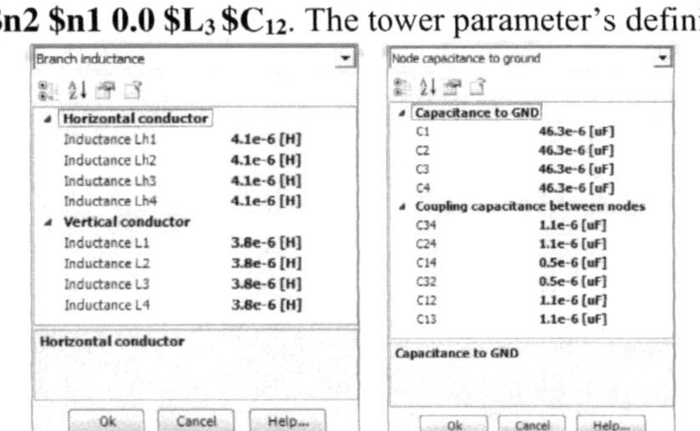

(a) (b)

Figure 4. Component configuration lists for electric parameters. (a) Inductance segment, (b) Capacitance segment

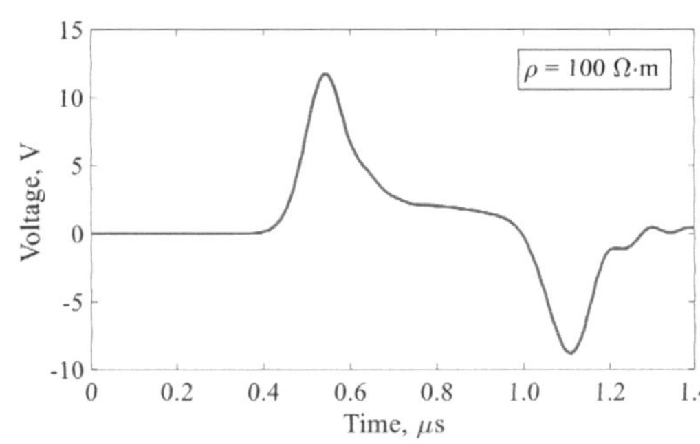

Figure 5. The voltage waveform at the tower base

The branch codes are represented the multi-conductor network and the conductance matrix is constructed and interfaced with the EMTDC. The EMTDC automatically computes node voltages and branch currents using the circuit theory and well-known the trapezoidal algorithm. The formulation of computing single inductive branch current and node voltage are given in equation (1-3):

$$i_{L3}(t) = \frac{v_{n1}(t-\Delta t)-v_{n2}(t-\Delta t)}{2L/_{\Delta t}} + I_{hist} \tag{1}$$

$$I_{hist} = 2\frac{v_{n1}(t-\Delta t)-v_{n2}(t-\Delta t)}{2L/_{\Delta t}} + i_{L3}(t-2\Delta t) \tag{2}$$

$$v_{n1}(t) - v_{n2}(t) = [Y]^{-1}[I_{hist} + i_s(t)] \tag{3}$$

where, Δt is the time constant 0.01 μs, Y^{-1} is the inverse conductance matrix, I_{hist} is the Dommel's current source.

The simulation results were compared with electromagnetic analysis based on the FDTD method [9] and tower bottom voltage gave good agreement. Figure 5 shows the voltage waveform when 0.25 μs rise time with 1 A amplitude ramp wave current is injected at the top of the tower.

3.3. Grounding model

The typical mesh grounding system of 12x12 m dimensional horizontal electrodes buried in 2m with 6 m vertical electrodes are chosen in this study, as shown in figure 1. To compute grounding impulse current distribution, electrodes are discretized into 3m segments. Each segment is represented by the lumped circuit of series resistance, inductance, shunt conductance, and capacitance elements, as shown in figure 6. The electrical parameters are calculated based on Sunde's formulas.

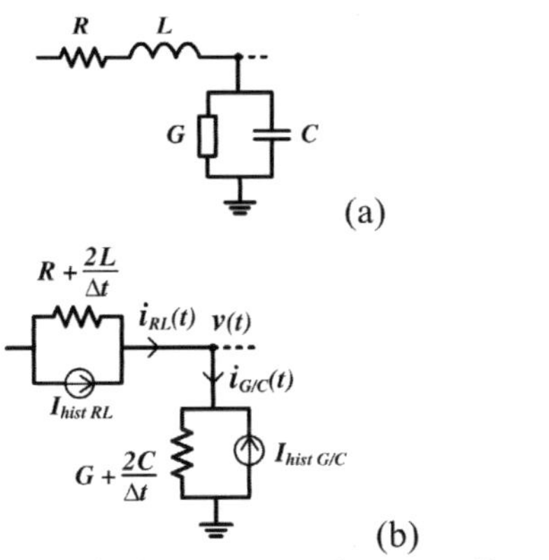

Figure 6. A single segment of a grounding electrode. (a) equivalent circuit, (b) EMTDC computation time-domain circuit

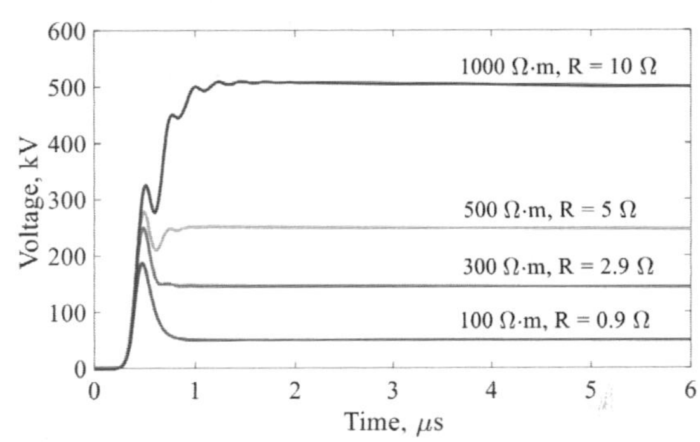

Figure 7. Voltage waveforms of the current injected point with different soil resistivity. (Current rise time is 0.25 μs with a 50 kA ramp wave)

The example of defining branch currents in the electrodes is given in equations (4-5).

$$i_{RL}(t) = (R + 2L/_{\Delta t})v(t) + I_{hist\ RL} \tag{4}$$

$$i_{G/C}(t) = (G + 2C/_{\Delta t})v(t) + I_{histG/C} \tag{5}$$

where, G is the shunt conductance of the electrodes.

Most of the grounding potential rise analysis has emphasized transient behavior at low and high soil resistance. Figure 7 shows the voltage rise of the grounding system as a function of soil. The steady state component R is determined by the ratio of the constant voltage to the current peak. Soil ionization phenomena were neglected in this model. The simulation result agrees well with the electromagnetic model based on the NEC-4 [10]. Earth impedances 3.8Ω (100Ω·m), 5.8Ω (300Ω·m), 7.3Ω (500Ω·m), 9.8Ω (1000Ω·m) corresponding to the subsequent stroke waveform-4 are obtained from the computation.

3.4. Electrical components

3.4.1. Low voltage surge arrester. In this paper HY5WZ – 51/134 and YH10W – 0.8/3.0 typical widely used zinc oxide surge arresters are modeled in the PSCAD based on Pincepti's model. The nominal discharge current for the YH10W arrester is 2 kA.

3.4.2. Equipotential grounding cable. YJV 1x120 mm^2 equipotential grounding cable consisting of core and insulation is considered. The core diameter is 0.061 m, insulation diameter 0.084 m, $\varepsilon = 2.3$ polyvinyl chloride. Three different models are implemented into PSCAD which are lumped parameter circuit, frequency dependent, and Bergeron line model with surge impedance of 30 Ω. In this paper lumped parameter cable model is adopted. The propagation of lightning surge is 40 m/μs.

3.4.3. Power transformer and cable line. 1.5 MW 0.69/35 kV transformer is modeled by 3 phase 2 winding transformer component in PSCAD master library to see the mechanism of surge arrester current to phase line. 690V cable was modeled by frequency dependent model adjusting its geometrical sizes.

4. Simulation results and discussion

After validations of the tower, grounding system models, following comparison results between lightning current and soil resistivity are presented. An example of current distributions in the wind turbine grounding system is shown in figure 8.

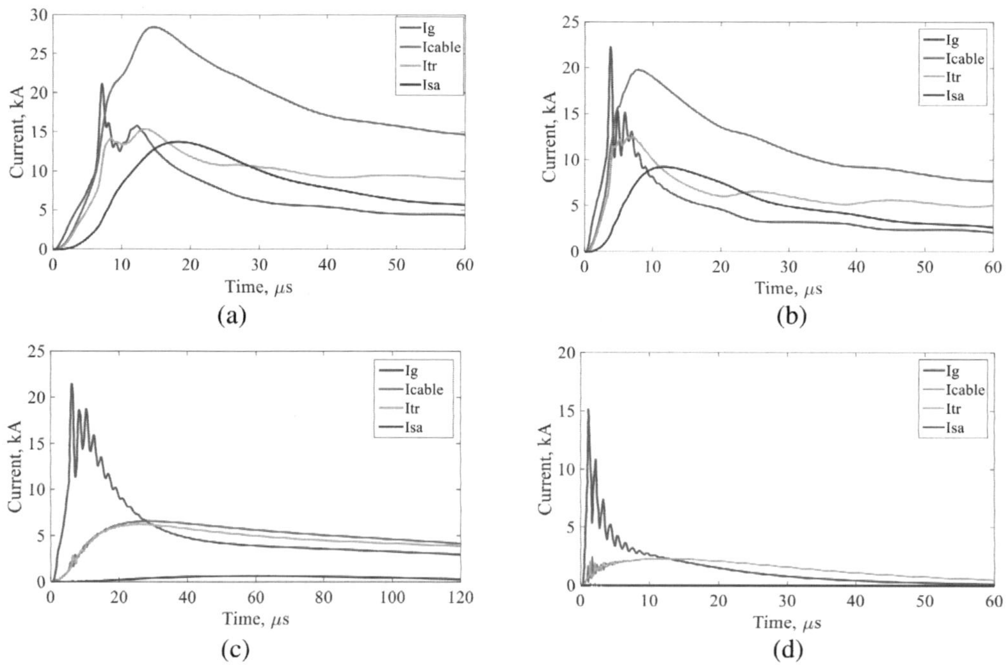

Figure 8. The current distribution of the wind turbine. (a) 43 kA with ρ=1000 Ω·m, (b) 31 kA with ρ=1000 Ω·m, (c) 20 kA with ρ=100 Ω·m, (d) 12 kA with ρ=1000 Ω·m, (I_g - current to the turbine grounding, I_{CABLE} - current to the grounding cable, I_{tr} - current to transformer grounding, I_{SA} - current to the phase conductor)

Figure 9 shows an example of discharge current I_{sa} through a low voltage surge arrester installed inside the transformer substation.

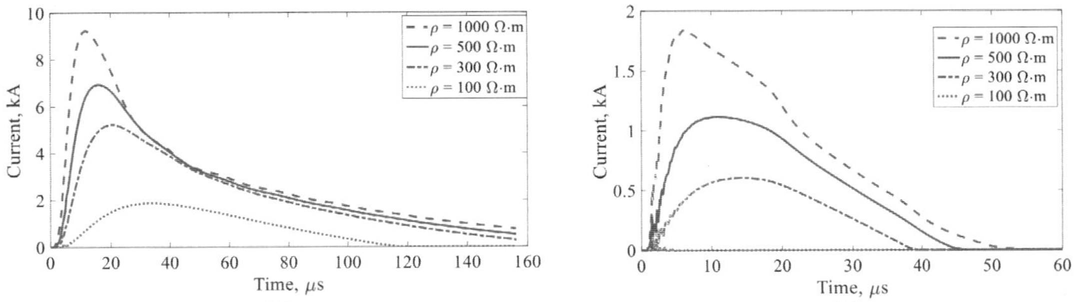

(a) (b)

Figure 9. Current waveforms for two lightning crests as a function of soil resistance. (a) 31 kA, (b) 12 kA

It is clear from figure 9, that the steepness of the current decreases as earthing impedance varies. Observation of current waveform can be important information for the selection of surge arresters in which wind turbines built in high soil resistivity and high probable lightning currents.

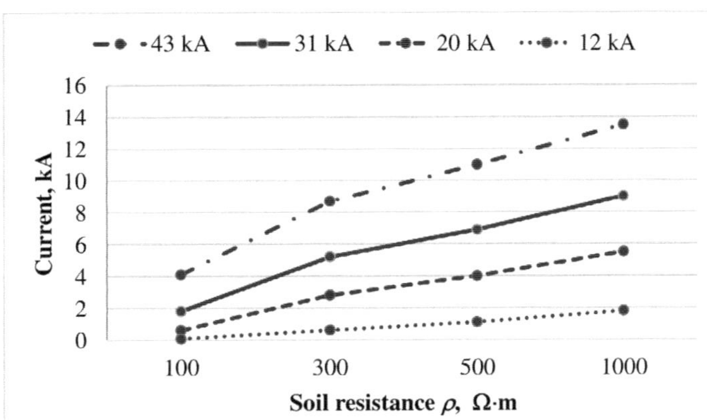

Figure 10. Maximum discharge current as a function of soil resistance and lightning currents

The highest current occurred at soil resistivity of 10 Ω impedance, as shown in figure 10 graphs. The impedance value of grounding decreases, therefore the current I_{sa} is mitigated to an acceptable percentage. The present results agree well as mentioned in [11].

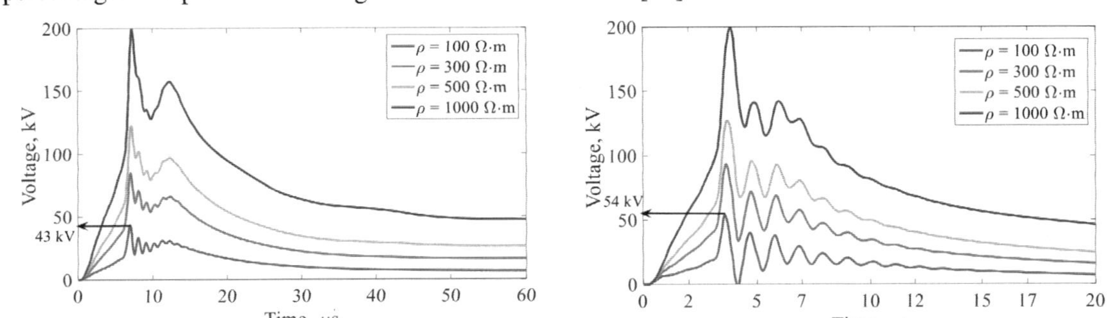

(a) (b)

Figure 11. Transient overvoltage produced on the turbine grounding when 43 kA (a) and 31 kA (b) amplitude lightning current strokes

It can be seen that the 31 kA lightning waveform produces an overvoltage of 54 kV which is more than the overvoltage of 43 kV generated by the 43 kA lightning waveform, as shown in figure 11 and figure 12. This phenomenon is explained by decreasing the lightning current front time, a more steep curve produced due to the inductive behavior of grounding electrodes with 100 Ω·m low soil resistivity. This problem is already described in [12], the present result agrees well for 1000 Ω·m soil resistivity.

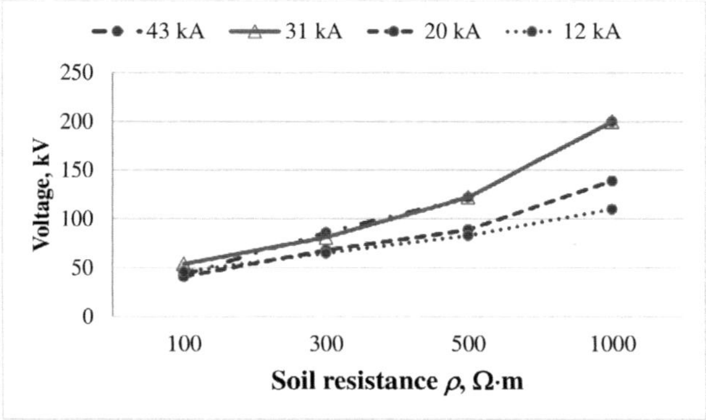

Figure 12. Maximum generated overvoltage on the turbine grounding as a function of soil resistance and lightning current.

An example of overvoltage on the transformer grounding system due to the surge traveling from the equipotential grounding cable is shown in figure 13 and figure 14.

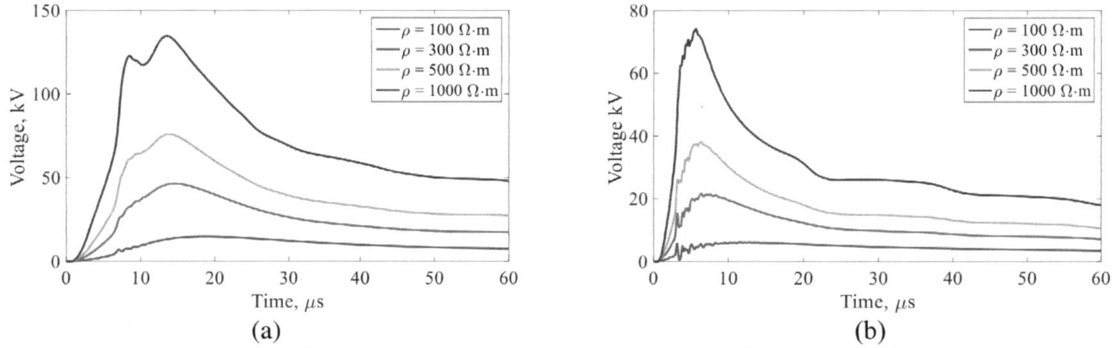

Figure 13. Transient overvoltage produced on the transformer grounding system when 43 kA (a) and 20 kA (b) currents

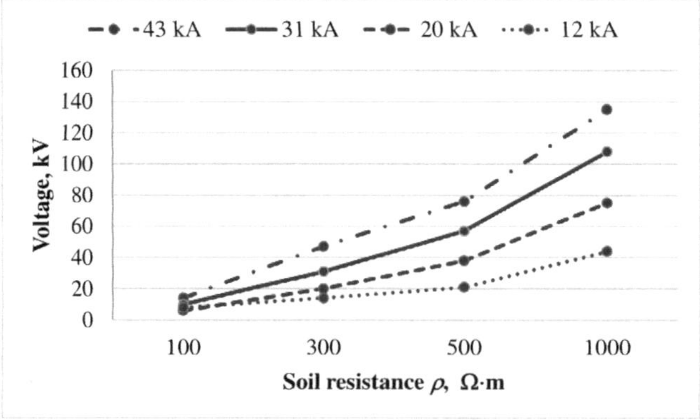

Figure 14. Maximum overvoltage on the transformer grounding as a function of soil resistivity and lightning currents.

The overvoltage characteristics depend on the 15m grounding cable current I_{cable} rise time and amplitude, and cable modeling methods. It can be seen that voltage decreases linearly as decreasing the lighting current and soil resistivity. The value of voltage depends on the length grounding cable, earth impedance, and surge speed.

5. Conclusion

The lightning transient simulations have been performed using PSCAD/EMTDC. The models computed lightning current distribution of the considered wind turbine arrangement and grounding system overvoltage under different soil resistivity. Following rules are established. The models can be used on real wind farms to define grounding overvoltage produced on the high soil resistivity sites. Even in low soil sites lightning strokes with short rise time produce higher transient overvoltages. The current characteristics of surge arresters depend on the rise time and lightning peak currents. Models have limitations thus they can be improved by assumptions of computing mutual couplings between conductors through modeling codes.

6. References

[1] Sarajcev, P. (2016). Lightning Protection of Wind Turbines and Associated Phenomena. In Alternative Energy and Shale Gas Encyclopedia (eds J.H. Lehr, J. Keeley and T.B. Kingery).

[2] K. Yamamoto and S. Sumi, "Validations of lightning protections for accidents at a wind farm," *2013 International Symposium on Lightning Protection (XII SIPDA)*, 2013, pp. 329-334

[3] Zhang, Shi et al. "Study on Fault Mechanism of Lightning Flashover Based on EMTP - RV in Wind Farm." *Journal of Physics: Conference Series* (2019).

[4] Chen, Hongcai & Zhang, Yang & Du, Ya Ping & Cheng, Qingsha. (2020). Comprehensive transient analysis for low-voltage system in a wind turbine under direct lightning. International Journal of Electrical Power & Energy Systems.

[5] Sarajcev, P.; Meglic, A.; Goic, R. Lightning Overvoltage Protection of Step-Up Transformer Inside a Nacelle of Onshore New-Generation Wind Turbines. *Energies* 2021, *14*, 322

[6] Zhang, J., et al.: The impact of lightning strike to multi-blade on the lightning overvoltage and stresses of arresters in offshore wind farm. *IET Renew. Power Gener.* 15, 2814– 2825 (2021)

[7] S. Sekioka, "Simulation of Lightning Overvoltages in Substation for Lightning Strike to Wind Turbine," *2021 35th International Conference on Lightning Protection (ICLP) and XVI International Symposium on Lightning Protection (SIPDA)*, 2021, pp. 01-06

[8] Oliveira A, J, Schroeder M.A.O, R. A. R. Moura, M. T. Correia de Barros and A. C. S. Lima, "Adjustment of current waveform parameters for first lightning strokes: Representation by Heidler functions," *2017 International Symposium on Lightning Protection (XIV SIPDA)*, 2017, pp. 121-126

[9] Y. Ikeda, N. Nagaoka, Y. Baba and A. Ametani, "A frequency dependent circuit model of a wind turbine tower using transient response calculeted by FDTD," *2012 International Conference on Lightning Protection (ICLP)*, 2012, pp. 1-6

[10] M. R. Ahmed and M. Ishii, "Electromagnetic analysis of lightning surge response of interconnected wind turbine grounding system," *2011 International Symposium on Lightning Protection*, 2011, pp. 226-231

[11] T. Kisielewicz, F. Fiamingo, Z. Flisowski, B. Kuca, G. B. Lo Piparo and C. Mazzetti, "Factors influencing the selection and installation of surge protective devices for low voltage systems," *2012 International Conference on Lightning Protection (ICLP)*, 2012, pp. 1-6

[12] W. L. M. de Azevedo, W. C. da Silva, A. R. J. de Araújo and J. P. Filho, "Computation of Transient Voltages on Wind Turbines on Frequency-Dependent Soil Caused by Lightning Strikes," *2021 Workshop on Communication Networks and Power Systems (WCNPS)*, 2021, pp. 1-6

Road Extraction using High Resolution Satellite Images based on Receptive Field and Improved Deeplabv3+

Yun Wang, Suye Wang and XueBin Hong

Department of Computer, Shantou Polytechnic, Shantou,515078, China

984145687@qq.com

Abstract. Road extraction from high resolution remote sensing images is an important and challenging computer vision task. This paper presents a road segmentation based on Receptive Field and Improved Deeplabv3+, which obtains the best training image set by calculating the edged energy function after simply clipping. To solve the problem of data homogeneity across as well as convergence, we innovatively use the initialization method Leaky-He to extract the layer backbone network in the network structure. Using the DeepGlobe Road Extraction dataset as the training dataset, the experimental results show that the best mloU score of the test set is 0.7099, which can improve the results by 0.1919 and 0.1596 in this paper compared with U-Net and D-LinkNet classical networks.

1. Introduction

Road extraction from satellite images is a necessary step for many applications, such as vehicle navigation urban planning intelligent transportation, image registration, GIS updates, land use detection, etc. Due to the complexity of the environment around roads in high-resolution images, a variety of focusing algorithms have emerged. Most of these methods can be divided into three categories: generation of road pixel-level markers [1,2], detection of road skeletons [3,4], and a combination of both [5,6].

Road area segmentation is the preliminary task of road extraction, which can be used as guidance of the topology delineation. In recent years, Deep learning methods have been increasingly applied to information extraction from high-resolution satellite images due to their good performance and generalization ability [7,8]. Since Mihi et al. [9] applied deep learning methods to road extraction, other deep learning models have been applied to road extraction research[10,11]. Zhong et al.[12] introduced the fully convolutional network into the segmentation of the road area and achieved dense end-to-end reasoning. But the simple linear interpolation upsampling of FCN made the model performance poor. Evolved from FCN, U-Net has a symmetric encoder-decoder structure, and the decoder uses parameter-learnable deconvolution, which makes the semantic segmentation more accurate. Zhang et al. [13] and Alexander et al. [14] proposed a semantic segmentation road extraction neural network, which uses unit construction with a similar architecture to U-Net, allowing the model to use fewer parameters but obtain better performance. Although the U-Net segmentation algorithm subsamples the road remote sensing image several times, making a large number of image features less, the network does not take into account the perceptual domain of the image and therefore has shortcomings for small target extraction.

To overcome the loss of spatial information and obtain more global context information, in the 2018 CVPR DeepGlobe Road Extraction Challenge, Zhou et al. [15] used dilated convolution to expand the receptive field and save a certain amount of spatial information while acquiring highly discriminative feature information, which improved the accuracy of the extraction and ultimately achieved excellent results in the competition. Although D-LinkNet uses Dilated convolution to control the receptive field by interval sampling, it has the same serious problem of subsampling as U-Net, and there is not much fusion of features to recover in the upsampling process, but simple feature splicing, which is not comprehensive for multi-scale feature fusion.

A more general and robust road segmentation method is still needed to improve the performance of road extraction from HR-RS images. In this paper, to reduce the loss of spatial information at multiple levels, strengthen the global context of FCNs and improve the accuracy of road extraction, we design a novel method based on Receptive Field and Improved Additive Margin Loss Function to extract roads from remote sensing imagery and achieve better performance. Firstly, simply cutting and calculating the energy function of the edge, then using a new initialization method of full connection layer weight in the backbone network of feature extraction layer is adopted.

2. Receptive Field

The receptive field is defined as the size of the area mapped on the original image by the pixel points on the feature maps output from each layer of CNN(Convolutional Neural Networks),as shown in figure 1, the pixel points in the higher layer feature maps are influenced by a certain region of the original image, this parameter is very critical for vision tasks, so when designing neural networks, the receptive field should be fully considered, which will achieve higher accuracy and robustness, generalization performance of the models.

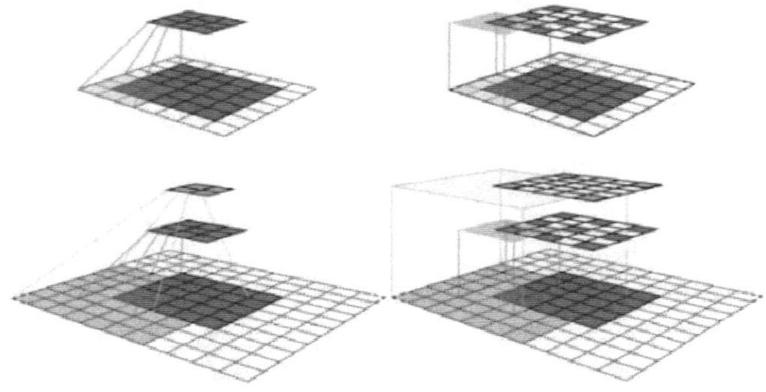

Figure 1. Schematic diagram of receptive field in DNN(Deep Neural Networks)

Calculation of the receptive fields in DNN is divided into two types, one is general convolution (Equation (1)) and the other is dilated convolution(Equation (2)):

$$F(i, j-1) = (F(i, j)-1) * \text{ stride } + \text{ kernelsize} \quad (1)$$

$$F(i, j-1) = (F(i, j)-1) * \text{ stride } + \text{ dilation } *(\text{ kernelsize } -1)+1 \quad (2)$$

where $F(i, j-1)$ represents the local receptive field of layer i to layer j, and the formula calculates the receptive field from top layer to bottom layer.

In summary, it is not difficult to understand that the problem of receptive field is the size of the region of the image observed when the field of view is fixed. For the image segmentation task, the larger size of the region of interest within the field of view, the more useful features are extracted, the better the performance of the deep network.

2.1. Cropping

In this paper, the size of the original image and semantic label mask is 512*512.Then the dataset is expanded by cropping, which also reduces the redundant interference pixels and expands the perceptual field, and a sliding window of 256*256 is used for extraction in cropping as shown in (3).

$$I_k = \sum_{i=1}^{n}\sum_{j=1}^{m} I_{(j-1)*256+1,\, j*256} \times H \tag{3}$$

Where I_k is the kth image after cropping, is a sliding window of size 256*256 and n, m is the quotient of the image length and width with 256.The before and after results of the cropped mask image are shown below:

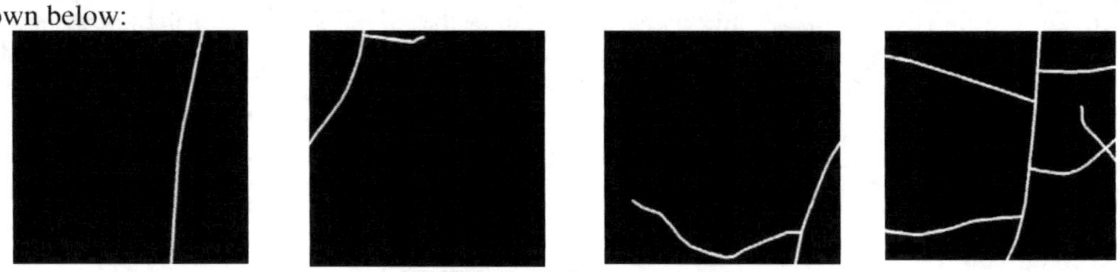

Figure 2. Before cropping

Figure 3. After cropping

After simple and efficient processing by cropping, although most of the images have expanded the perceptual field of the region of interest, but inevitably a small number of cropped images have fewer roads, which are not conducive to network training and therefore need to be eliminated, this paper proposes a method based on edge energy for fast and accurate elimination.

2.2. Image filtering based on edge energy

The amount of information in an image is reflected in the visual layer of the brain through edge information, and the edge of the binary mask image directly reflects the amount of information in the entire binary image. In this paper, the amount of information is reflected based on the edge energy of level set, so as to filter out images with less information. The level set algorithm uses implicit function to represent the curve, calculated by the magnitude of external energy, and approaches the edge of the road when the energy is maximum, because if the road is small in the image area and cannot approach the edge under the same parameters, then the energy is small.

The level set algorithm defines an external energy that shifts the zero level curve towards the object boundary. Let I be an image, and g is defined by:

$$g = \frac{1}{1 + |\nabla G_\sigma * \mathrm{I}|^2} \tag{4}$$

where G_σ is the Gaussian kernel of the standard operator σ. The function φ(x,y) is defined as the external energy term as follow:

$$\varepsilon_{g,\lambda,v}(\phi) = \lambda \mathrm{L}_g(\phi) + \mathrm{v}\mathrm{A}_g(\phi) \tag{5}$$

where λ >0 and v are constants, $\mathrm{L}_g(\phi)$ and $\mathrm{A}_g(\phi)$ are defined as follows.

$$\mathrm{L}_g(\phi) = \int_\Omega g\delta(\phi)\,|\nabla(\phi)|\,dxdy \tag{6}$$

and

$$\mathrm{A}_g(\phi)=\int_\Omega gH(-\phi)\mathrm{dxdy} \tag{7}$$

where δ is a single variable Dirac delta function and H is a Heaviside function. The energy term function is now defined as:

$$\varepsilon(\phi)=\mu\mathrm{P}(\phi)+\varepsilon_{g,\lambda,v}(\phi) \tag{8}$$

The external energy term $\varepsilon_{g,\lambda,v}$ is constantly moving towards the boundary, while the internal energy term $\mathrm{P}(\phi)$ is used to correct for the errors between the distance of symbols.

For further analysis of $\mathrm{L}_g(\phi)$, it is assumed that φ represents a differentiable parameterized curve c (p) , p $\in$ [0,1]. In (6), the energy function $\mathrm{L}_g(\phi)$ calculates the length of the metric function $\mathrm{ds}=\mathrm{g(C(p))}|\mathrm{C}'(\mathrm{p})|\mathrm{dp}$ formed by the zero level set curve φ. In (7), the energy function is used to accelerate the rate of expansion of the imaginary boundary of the entire curve. When the parameter of the g function is 1, then the energy function in equation (7) is shown as follows:

$$\Omega_\phi^-=\{(\mathrm{x,y})\,|\,\phi(\mathrm{x,y})<0\} \tag{9}$$

The energy function $\mathrm{A}_g(\phi)$ is the weighted Ω_ϕ^-. The coefficient v > 0 or v < 0 in A_g is determined by the phase position relationship between the initial contour and the region of interest. By means of the finite element variational method, the partial derivatives (first variational set) in the function e can be written as:

$$\frac{\partial\varepsilon}{\partial\phi}=\mu[\Delta\phi+\mathrm{div}(\frac{\nabla\phi}{|\nabla\phi|})]+\lambda\delta(\phi)\mathrm{div}(\mathrm{g}\frac{\nabla\phi}{|\nabla\phi|})+\mathrm{vg}\delta(\phi) \tag{10}$$

where Δ is Laplace. The Euler-lagrange equation $\frac{\partial\varepsilon}{\partial\phi}=0$,φ=0, and the maximum directional derivative of the minimization function g process is:

$$\frac{\partial\phi}{\partial t}=-\mu[\Delta\phi+\mathrm{div}(\frac{\nabla\phi}{|\nabla\phi|})]-\lambda\delta(\phi)\mathrm{div}(\mathrm{g}\frac{\nabla\phi}{|\nabla\phi|})-\mathrm{vg}\delta(\phi) \tag{11}$$

The second and third terms in the right-hand side of the (11) correspond to the energy gradient flows $\lambda\mathrm{L}_g(\phi)$ and $\mathrm{vA}_g(\phi)$, which cause the level set curve to move towards the boundary of the target region. The gradient flow of the internal energy function $\mu P(\phi)$ is as follows:

$$\Delta\phi-\mathrm{div}(\frac{\nabla\phi}{|\nabla\phi|})=div[(1-\mu\frac{1}{|\nabla\phi|})\nabla\phi] \tag{12}$$

where $1-\mu\frac{1}{|\nabla\phi|}$ is the diffusivity, that μ is the diffusivity gain coefficient controlling the overall diffusion rate. If $|\nabla\phi|>1$, then the diffusivity is positive, if $|\nabla\phi|<1$, the profile reverses the diffusion, thus increasing the influence of gradient flow. The iteration reaches the maximum value when the energy diffuses to the edge. If most of the value is 0, it cannot approach the edge to the maximum. The diffusion gain coefficient set in this paper is 0.05. In the process of controlling the entire gradient change, we need to find the maximum energy quickly, so we don't need to calculate all the pixels, because in a binary image, there are only two elements (0 and 1) in the binary image, so this paper uses small batch descent to optimise the original gradient calculation.

Let $h_\theta=\frac{\partial\phi}{\partial t}$ be the objective function to be solved. A small batch gradient descent doesn't put all the data in at once, but instead puts in a small batch of data to update the formula, which becomes:

for i=1, i+b, i+2b, i+3b,...:

$$\theta_j = \theta_j - \alpha \frac{1}{b} \sum_{k=i}^{i+b-1} (h_\theta(x^{(k)}) - y^{(k)}) x_j^{(k)} \tag{13}$$

where b is the number of samples in a small batch, and x is the pixel and y is the constant.

3. Initialization of Fully Connected Layers

The traditional parameter initialization methods include zero initialization and normal distribution initialization. This article refers to the idea of attention mechanism to modify the last fully connected layer parameters, and adds weights on the basis of normal distribution initialization, making important weights more important.

Initializing the weights to a small number is a general solution to break network symmetry. The idea is that the neurons are all random and unique at the beginning, so they compute different updates and integrate themselves into various parts of the entire network. An implementation of the weight matrix might look like W=0.01∗np.random.randn(D,H) ,where randn is sampled from a unit standard Gaussian distribution with mean zero. Through this formula, the weight vector of each neuron is initialized to a random vector sampled from a multidimensional Gaussian distribution, so the neurons point in random direction in the input space. Of course, decimals can also be randomly selected from a uniform distribution, but in practice it doesn't seem to have much effect on the final performance. Thus, this paper proposes a new initialization method Leaky-He:

$$\begin{aligned} & filterSize = [H, W] \\ & numChannels = K \\ & numIn = filterSize(1) * filterSize(2) * numChannels \\ & \mathrm{var}Weights = \frac{2}{((1 + scale^2) * numIn)} \\ & weights = rand(sz) * \sqrt{\mathrm{var}Weights} \end{aligned} \tag{14}$$

In (14), H, W is the length and width of the filter, K is the number of channels of the filter, scale is the scale factor, this paper is set to 0.01.

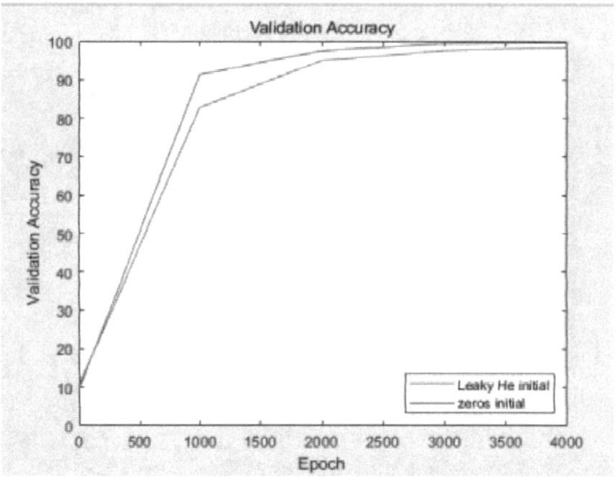

Figure 4. Verification Accuracy of Leaky-He

As shown in figure 4, when zero initialization and Leaky He initialization are applied to the same verification set, Leaky He initialization method obtains higher accuracy and converges slightly faster than zero initialization method.

4. Experiments

To verify the effectiveness of our approach, about 14000 cropped images of 256 * 256 * 3 size are trained to complete the road segmentation task. The comparison method is the classic network U-Net and the 2018 champion method D-LinkNet, and the quantitative comparison is mean IoU. The comparison result is displayed in table 1. Our method has higher averaged IoU compared with the U-Net and D-LinkNet, suggesting the effectiveness of our method in road extraction from HR-RS data.

Table 1. Quantitative comparison of different methods on the DeepGlobe testing dataset. Improved Deeplabv3+ network parameter model get higher score than the U-Net and D-LinkNet on the testing set.

Model	IoU on validation set
U-Net	0.5153
D-LinkNet	0.5788
Improved Deeplabv3+	0.7099

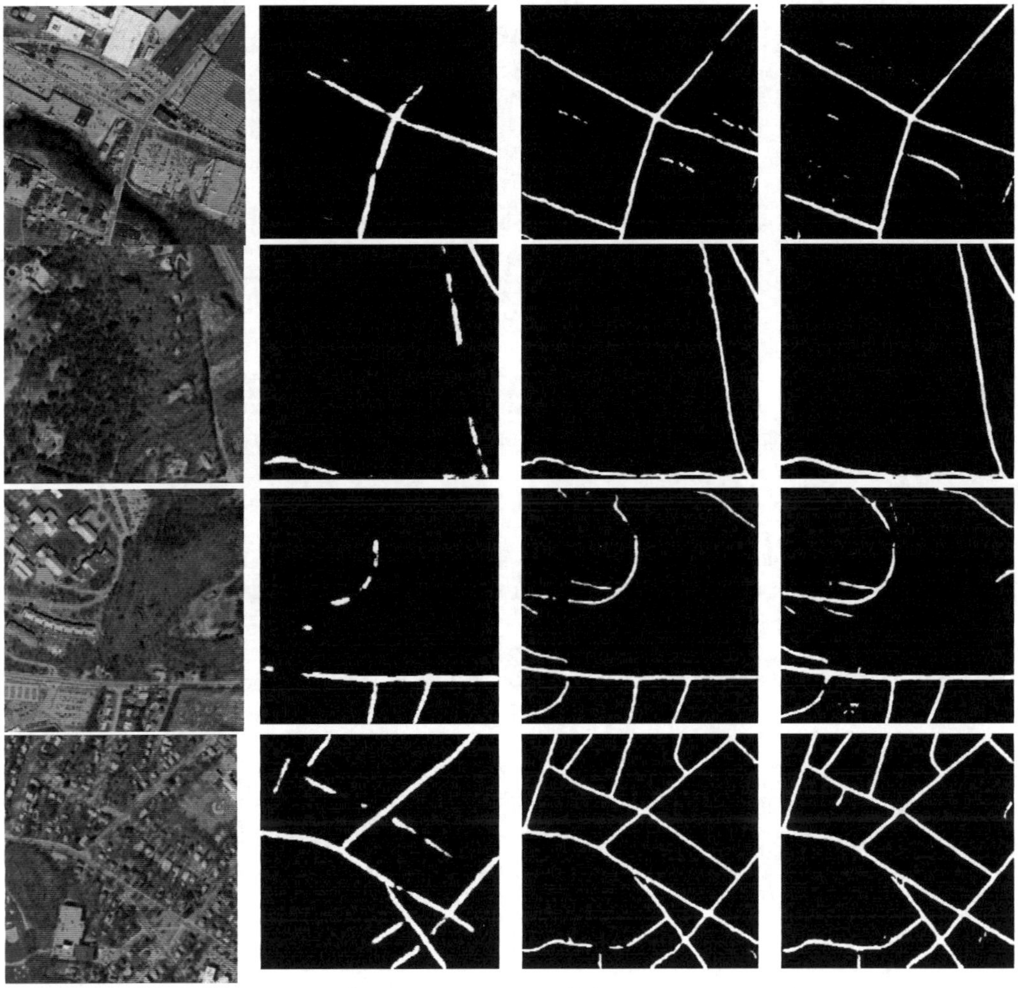

(a)Input (b) U-Net (c) D-LinkNet (d)Improved Deeplabv3+

Figure 5. Comparison of road extraction results on DeepGlobe road dataset: (a) Input Image. (b) Result from U-Net. (c) Result from D-LinkNet. (d) Result from improved Deeplabv3+. Improved Deeplabv3+ avoids weaknesses in U-Net and D-LinkNet, and makes better predictions.

5. Conclusion

In this paper, we propose a new method of road segmentation based on Receptive Field and Improved Deeplabv3+. We calculated important accuracy measurement IoU to evaluate the performance of the suggested technique in road extraction. The proposed method could achieve an mean IoU accuracy of 70.99% for DeepGlobe dataset confirmed that the model could obtain accurate road results and produce high-resolution segmentation map. Moreover, the proposed model is compared with other deep learning-based techniques, and the visual and quantitative outcomes prove the superiority of proposed method in road extraction from high-resolution remote sensing imagery.

6. References

[1] Zhang Z, Liu Q, Wang Y. Road extraction by deep residual u-net[J]. IEEE Geoscience and Remote Sensing Letters, 2018, 15(5): 749-753.
[2] Alshehhi R, Marpu P R. Hierarchical graph-based segmentation for extracting road networks from high-resolution satellite images[J]. ISPRS journal of photogrammetry and remote sensing, 2017, 126: 245-260.
[3] Liu B, Wu H, Wang Y, et al. Main road extraction from zy-3 grayscale imagery based on directional mathematical morphology and vgi prior knowledge in urban areas[J]. PloS one, 2015, 10(9): e0138071.
[4] Sujatha C, Selvathi D. Connected component-based technique for automatic extraction of road centerline in high resolution satellite images[J]. EURASIP Journal on Image and Video Processing, 2015, 2015(1): 1-16.
[5] Bastani F, He S, Abbar S, et al. Roadtracer: Automatic extraction of road networks from aerial images[C]//Proceedings of the IEEE Conference on Computer Vision and Pattern Recognition. 2018: 4720-4728.
[6] Máttyus G, Luo W, Urtasun R. Deeproadmapper: Extracting road topology from aerial images[C]//Proceedings of the IEEE International Conference on Computer Vision. 2017: 3438-3446.
[7] Zhong Z, Li J, Cui W, et al. Fully convolutional networks for building and road extraction: Preliminary results[C]//2016 IEEE International Geoscience and Remote Sensing Symposium (IGARSS). IEEE, 2016: 1591-1594.
[8] Xu Y, Xie Z, Feng Y, et al. Road extraction from high-resolution remote sensing imagery using deep learning[J]. Remote Sensing, 2018, 10(9): 1461.
[9] Computer Vision--ECCV 2010: 11th European Conference on Computer Vision, Heraklion, Crete, Greece, September 5-11, 2010, Proceedings, Part V[M]. Springer, 2010.
[10] Geng L, Sun J, Xiao Z, et al. Combining CNN and MRF for road detection[J]. Computers & Electrical Engineering, 2018, 70: 895-903.
[11] Wei Y, Wang Z, Xu M. Road structure refined CNN for road extraction in aerial image[J]. IEEE Geoscience and Remote Sensing Letters, 2017, 14(5): 709-713.
[12] Zhong Z, Li J, Cui W, et al. Fully convolutional networks for building and road extraction: Preliminary results[C]//2016 IEEE International Geoscience and Remote Sensing Symposium (IGARSS). IEEE, 2016: 1591-1594.
[13] Zhang Z, Liu Q, Wang Y. Road extraction by deep residual u-net[J]. IEEE Geoscience and Remote Sensing Letters, 2018, 15(5): 749-753.
[14] Buslaev A, Seferbekov S, Iglovikov V, et al. Fully convolutional network for automatic road extraction from satellite imagery[C]//Proceedings of the IEEE Conference on Computer Vision and Pattern Recognition Workshops. 2018: 207-210.
[15] Zhou L, Zhang C, Wu M. D-linknet: Linknet with pretrained encoder and dilated convolution for high resolution satellite imagery road extraction[C]//Proceedings of the IEEE Conference on Computer Vision and Pattern Recognition Workshops. 2018: 182-186.

Acknowledgments
The authors gratefully acknowledge the Foundation for Scientific Research Projects of Universities in Guangdong Province(No.2020KQNCX209,No.2021KTSCX287,No.2021KQNCX215,No.SZK2020Q01).

Journal of Physics: Conference Series **2320** (2022) 012022 doi:10.1088/1742-6596/2320/1/012022

One-Shot Face Recognition through a Region Inception ResNet with Modified Triplet Loss

Bo Sun, Ao Luo, Bojie Rong, Jun He* and Lejun Yu

School of Artificial intelligence, Engineering Research Center of Intelligent Technology and Educational Application, Ministry of Education, Beijing Normal University Beijing, China

Correspondence: hejun@bnu.edu.cn

Abstract. Face recognition has become a convenient method for identity verification, especially the one-shot task has much practical value. Some former works have achieved considerable results. However, they perform ill under the situation of COVID-19 with more and more people wearing a mask. This requires the extracted face features distinguishable and robust enough for classifying different masked people. It is difficult for the one-shot task with only one sample for training. To solve this problem, we designed a network called *Region Inception ResNet* with *Modified Triplet Loss*, which generates robust features. It keeps high accuracy even under masked condition. The networks are trained on the datasets of CASIA-Webface and Ms-Celeb-1M, tested on LFW (Labelled Faces in the Wild). Experiments in Section 6 show the effectiveness of our method.

1. Introduction

Face recognition plays an important role in many aspects as a verification method, such as mobile payment verification, employee punch-in and so on. Some works, [2], [5], [21-23], have contributed a lot in this area. In this paper, we focus on one-shot face recognition, whose testing face has only one photo for the network to record. In the training step, it is difficult to train networks by only these one-shot human categories. Normally, researchers train networks on normal categories which have multiple samples for every face; when testing, one-shot categories are mapped into a distinguishing feature space and classified by these trained networks. However, under the situation of COVID-19, people tend to verify their identities while wearing a mask, which hampers former methods to recognize different persons. Therefore, more distinguishable and robust face features are required.

In DeepID [2], Tang et al. trained 60 convolutional networks with a large face image dataset to extract robust features from 5 regions on each face image. [23] took a similar approach that splits face into 4*4 groups. However, the large volume restricted them for practical usage. On the other hand, many researchers focus on confining distances among different categories, such as triplet loss [5], RegularFace [22]. However, all of them are not robust enough to handle various one-shot categories, especially for faces wearing masks.

In this paper, we propose a **Region Inception ResNet** with **Modified Triplet Loss**, which extends the idea of multi-region and triplet loss, performing well on one-shot face recognition, especially on masked faces. It has 2 contributions:

- With the help of a **multi-region feature mapping** module, the Region Inception ResNet (RIR) splits extracted face feature maps into several parts, which improves robustness on masked faces while keeping the network lightweight.
- We introduce the **modified triplet loss** which consists of **modified triplet loss function and modified triplet mining strategies**, both of which fit well on the dataset consists of one-shot and normal samples (introduced in Section 3.2). For triplet loss function, different margins are applied on one-shot and normal categories; for triplet mining strategies, we generate more hard triplets containing one-shot samples in training phase.

The rest sections of this paper are organized as follows. Section 2 provides an overview of the related work on face recognition. The RIR network and modified triplet loss are demonstrated in Section 3. Section 4 illustrates experiments on RIR and modified triplet loss. Conclusion is presented in Section 5.

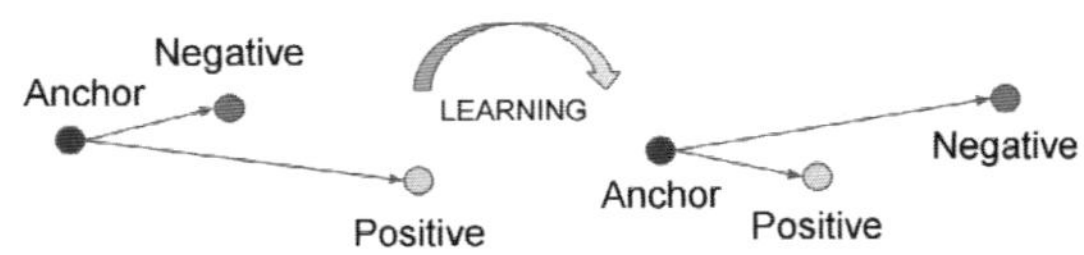

Figure 1. Training processes of triplet loss.

2. Related Works

In this section, we introduce some former works directly linked with our work. The Inception-ResNet-V1 [12] is taken as backbone of our network. Enlightened by DeepID [11], we introduce and improve the multi-region module in our network. Finally, triplet loss [13] is taken as our base loss function, which has considerable effect on face recognition.

2.1. Inception-ResNet-V1

We implement Inception-ResNet-V1 as our backbone network. It has an effective structure for extracting features, which is shown in Figure 1. The STEM in its first part extracts shallow features which are inputs of Inception modules. These Inception modules take advantage of ResNet, extracting robust features from shallow features.

However, the network is not robust enough for recognizing a person under multiple conditions (such as partial illumination or wearing a mask) in one-shot task.

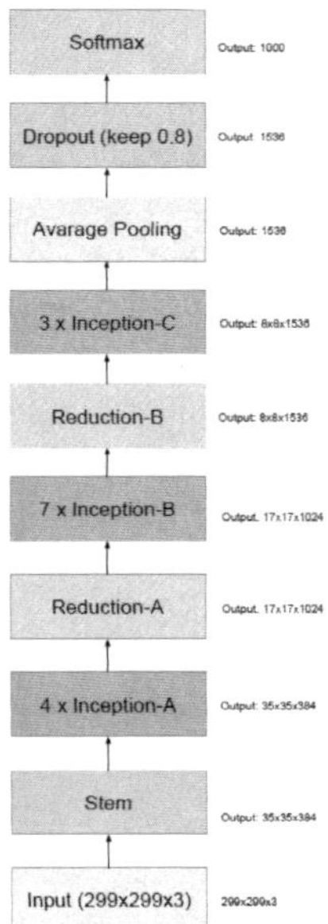

Figure 2. Inception-ResNet-V1.

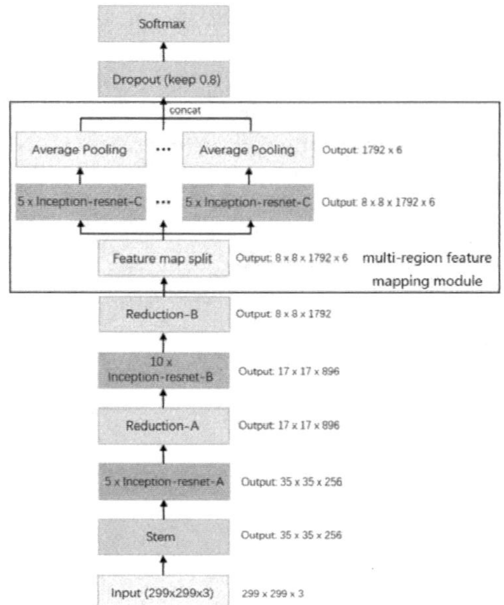

Figure 3. Region Inception ResNet(RIR).

2.2. DeepID

Sun Y [11] et al. proposed a feature extracting method, which detects five landmarks of a face (two eyes, nose tip, and two mouth corners). For high robustness, they had trained 60 CNN models, each of which outputs two 160-dimension vectors as features of those landmarks. It is a large-volume network with models trained separately which obtained the state-of-the-art result at that time.

Obviously, this method has a drawback that shallow features are extracted by homogeneous beginning layers in 60 models, which cause high computational redundancy, wasting time and storage space.

2.3. Triplet Loss

It is vital to extract distinguishable features in face recognition. Thus, Florian [13] et al. introduced Triplet Loss, which aims at a distinctive feature map with high Euclidean distances among different categories. It mainly consists of triplet loss function and triplet mining strategies.

2.3.1. Triplet loss function. As shown in Figure 2, a triplet consists of three samples, namely "Anchor", "Positive" and "Negative". "Anchor" is the baseline image, termed as x_i^a; "Positive" is the same-category sample of "Anchor", termed as x_i^p; "Negative" is a different-category sample against "Anchor", termed as x_i^n. Euclidean distance between "Anchor" and "Negative" is termed as d_n; Euclidean distance between "Anchor" and "Positive" is termed as d_p.

Triplet loss function aims to reduce d_p and increase $d_n$which are shown in equation (1), and the goal of triplet loss is shown in equation (2)

$$\begin{gathered} d_n = \|f(x_i^a) - f(x_i^n)\|_2^2 \\ d_p = \left\|f(x_i^a) - f(x_i^p)\right\|_2^2 \\ \forall \left(f(x_i^a), f(x_i^p), f(x_i^n)\right) \in T \end{gathered} \tag{1}$$

$$d_p + \alpha < d_n \tag{2}$$

$f(x)$ means mapping x into feature space. α is a margin that is enforced between d_p and d_n. T is the set of all possible triplets in the training set.

The **triplet loss function** is

$$loss = \sum_{i}^{N} [d_p - d_n + \alpha]_+ \tag{3}$$

The subscript "+" means to take larger one between result of the function and 0, N is number of triplets.
We can classify the triplets into three types:

- **Easy triplet**. Loss is 0.
- **Hard triplet**. d_n is shorter than d_p.
- **Semi-hard triplet**. d_n is larger than d_p, but shorter than $(d_p + \alpha)$.

When taking one-shot categories for training, data augment methods are applied on the single training sample so as to generate "Anchor" and "Positive", resulting in plenty of easy triplets.

2.3.2. Triplet mining strategies. If triplets are generated randomly, there would be plenty of easy triplets which have no benefit for training. Thus, triplet mining strategies have been applied to extract more semi-hard triplets and hard triplets. Those are discussed in Section 3.2.2.
Original triplet mining strategies perform well in normal face recognition. However, when it comes to one-shot training data, it faces a class-imbalance problem that the samples of normal categories are far more than those of one-shot ones. That is, one-shot categories have little impacts on calculated loss.

3. Region Inception ResNet with Modified Triplet Loss

3.1. Region Inception ResNet

We designed a **Region Inception ResNet** (RIR, shown in Figure 3) by introducing a **multi-region feature mapping** module (**MRFM**) to Inception-ResNet-V1 [12], which extracts well-performing features. The beginning part of RIR is similar with Inception-ResNet-V1, consisting of a Stem structure, 5 Inception-ResNet-As, a Reduction-A, 10 Inception-ResNet-Bs and a Reduction-B. Then we implement MRFM following Reduction-B. Finally, a dropout layer and a softmax layer are utilized to output classification results.

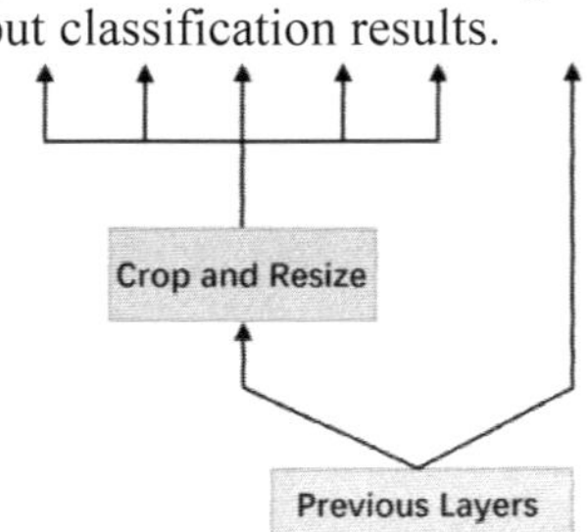

Figure 4. Feature map splitting operation.

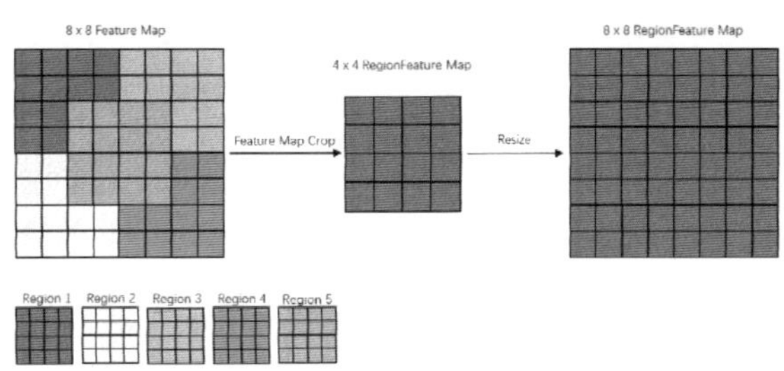

Figure 5. Patterns for splitting feature map.

The main difference between RIR and Inception-ResNet-V1 is the MRFM module, which plays an important role in improving robustness. At the beginning of MRFM, we generate 6 feature maps from Reduction-B module's output (8*8*1792), shown in Figure 4. To be concrete, this 8*8 input is cropped into 5 regions, namely left eye, right eye, left mouth corner, right mouth corner and nose tip, each of which is a 4*4 sub feature map. These 5 regions, along with the original output of Reduction-B module, are inputted into respective Inception-ResNet-C modules. To keep inputs of all Inception-ResNet-Cs same, we resize the 5 sub feature maps from 4*4 to 8*8, shown in Figure 5. After these 6 feature maps are processed by Inception-ResNet-C modules and an average pooling separately, we concatenate their outputs as a 6*1792 matrix and finally output to the next part. These 6 feature maps are independent with each other, so even the person is wearing a mask, the whole network can still keep robust.

In summary, RIR extracts feature maps with a single CNN, then splits these feature maps into multi-regions. In this way, it keeps small model size and obtains high robustness.

3.2. Modified Triplet Loss

To extract more robust and distinguishable features, we add some one-shot categories to training data. Though the original triplet loss performs well on normal datasets, it needs some modification to adapt to this mixed dataset. In Section 3.2.1, we illustrate the modified triplet loss function, which increases Euclidean distances among different categories for one-shot training data. On the other hand, we change triplet mining strategies in Section 3.2.2 to generate more triplets from one-shot categories.

3.2.1. Modified triplet loss function. The "Positive" samples of one-shot categories are generated by data augment. These "Positive" ones have low d_ps which result in many easy triplets and are useless for optimizing loss. To solve this problem, an additional margin is applied on triplets containing a one-shot-category "Anchor". The ***modified triplet loss function*** *is shown in equation (4).*

$$loss = \sum_{i}^{N_1}\left[d_{ip} - d_{in} + \alpha_1\right]_+ + \sum_{j}^{N_2}\left[d_{jp} - d_{jn} + \alpha_2\right]_+$$
$$where \begin{cases} d_{in} = \|f(x_i^a) - f(x_i^n)\|_2^2 \\ d_{ip} = \|f(x_i^a) - f(x_i^p)\|_2^2 \\ d_{jn} = \|f(x_j^a) - f(x_j^n)\|_2^2 \\ d_{jp} = \|f(x_j^a) - f(x_j^p)\|_2^2 \end{cases} \quad (4)$$

Where N_1, N_2 represent the cardinality of triplets with "Anchor" from normal categories or one-shot categories, α_1, α_2 are margins for normal or one-shot "Anchor" triplets respectively, $\alpha_1 \leq \alpha_2$. The experiments in Section 4 show that our modification is effective.

3.2.2. Modified Triplet Mining. There are two mining strategies.

- **Batch all**. It selects all hard triplets and semi-hard triplets in current mini-batch.
- **Batch hard**. For every sample in mini-batch, it finds a "Positive" with the max d_p and a "Negative" with the min d_n, to generate triplets. P triplets are generated when taking a mini-batch that contains P images.

We make some changes on mining strategies to raise the ratio of triplets with one-shot "Anchor" samples. There are four steps in our mining strategy.

- Take B samples from the training dataset to form a mini-batch, $B = P_1 \times K_1 + P_2 \times K_2$. P_1 and P_2 represent numbers of normal categories and one-shot categories, K_1 and K_2 are image numbers of each category.
- Take Every image in the mini-batch as an "Anchor", pick the positive sample makes d_p maximum, and the negative sample makes d_n minimum to form a triplet. We generate B triplets in this step.
- For $P_1 \times K_1$ normal-category images, pick a one-shot-category image who makes p_n minimum. Therefore, we generate $P_1 \times K_1$ more triplets.
- Take these $B + P_1 \times K_1$ triplets into equation (4) to calculate loss, and optimize it.

4. Experiments

4.1. Datasets

We take **CASIA-Webface** [14] and **Ms-Celeb-1M** [19] datasets as our training datasets, and test all the models on **LFW** (Labeled Faces in the Wild) [20].

CASIA-Webface dataset has 494,414 images of 10,575 persons, shown in Figure 6 (a). Ms-Celeb-1M dataset is a dataset for a competition [18], and contains 21000 images of famous persons, as shown in

Figure 6 (b). LFW dataset is used specifically for unconstrained face recognition, which has more than 13000 images of 5749 persons.

(a) CASIA-Webface dataset.

(b) Ms-Celeb-1M dataset.

Figure 6. Instances of two datasets.

4.2. Validating Standard

For Region Inception ResNet, we take **Accuracy** as the validating standard, in which N is number of all testing images, C is the count of images classified correctly, shown in equation (5).

$$\mathrm{Acc} = \frac{\mathrm{C}}{\mathrm{N}} \tag{5}$$

For experiments on modified triplet loss, we use RIR as the base network, take **Coverage** of human face under specific **Precision** as the validating criterion.

$$\mathrm{Precision} = \frac{\mathrm{C}}{\mathrm{M}} \tag{6}$$

$$\mathrm{Coverage} = \frac{\mathrm{M}}{\mathrm{N}} \tag{7}$$

Where M is the number of samples recognized as a human face by the network, C is the number of correctly classified samples and N is the total number of testing images. In face recognition, a relatively high precision is required for practical use, so we keep precision as a prerequisite and take coverage as the standard to valuate networks. Under the same precision, higher coverage means better performance. Here we take coverages under precisions of 99% and 99.9% in experiments.

4.3. Experiments on Region Inception ResNet

4.3.1. Performance of the whole network. At first, we compare accuracies of RIR and recent networks on LFW, shown in Table 1.

Table 1. Results trained on LFW (* means trained on non-public datasets).

Network	Model Num	Training Data	LFW Accuracy (%)	LFW Validation (%) @FAR=1e-3
DeepFace [7]	3	4M*	97.35	N/A
Deep FR [15]	1	2.6M	98.95	N/A
DeepID2+ [11]	25	300K*	98.70	N/A
FaceNet [5]	1	200M*	99.65	N/A
Baidu [16]	1	1.3M*	99.13	N/A
Inception-ResNet-V1	1	Webface	98.85	96.700±0.0185
Inception-ResNet-V1	1	Msceleb	98.93	96.633±0.0158
RIR	1	Webface	99.05	96.900±0.0154

Table 1 shows the effectiveness of our method. It shows that RIR achieves a high accuracy on LFW at 99.20%, and is the best method except FaceNet, which is trained on a non-public dataset of 200 million face images. The proposed network is trained on public dataset with less data, it is less time-consuming and more robust. The comparisons of different dataset volume is shown in 4.3.2.

4.3.2. Ablations on datasets of different scale. We train two RIR models on datasets with different size of 500K and 3.2M images. Table 2 shows that the whole network tends to achieve higher accuracy when trained on larger datasets.

Table 2. Influence of different dataset scale.

Network	Training Data	Data Size	LFW Accuracy (%)	LFW Validation (%) @FAR=1e-3
RIR	Webface	500K	99.05	96.900+-0.0154
RIR	MsCeleb	3.2M	**99.20**	**97.066+-0.0164**

4.3.3. Ablations on masked samples. We validate the robustness of our network with two shielding methods from [17]. As shown in Figure 7, shielding method 1 covers a sample from the bottom up, which simulates the masked faces; shielding method 2 uses squares of different sizes to shield an image randomly.

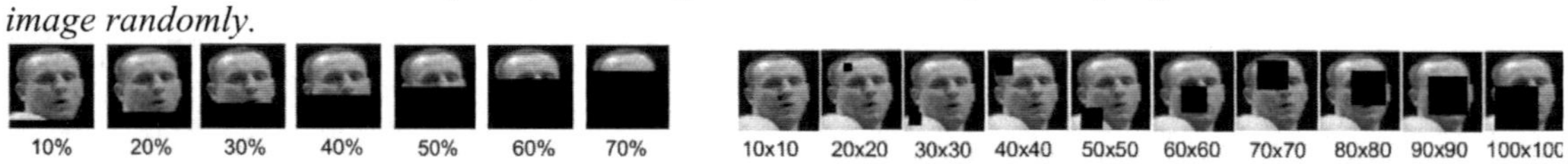

(a) Shielding with different portions. **(b)** Shielding with different block sizes.

Figure 7. Two kinds of shielding methods.

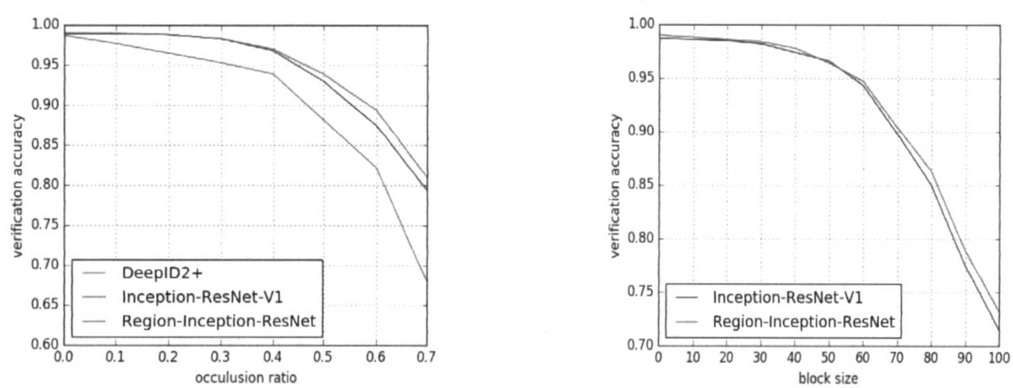

(a) Shielding method 1. **(b)** Shielding methods 2.

Figure 8. Experiment results of shielding methods.

Table 3, 4 and Figure 8 show that our method is effective under both shielding methods.

In the experiments of shielding method 1 (shown in Table 3), RIR achieves the highest accuracy. In Figure 8 (a), we notice that when shielded part reaches 40%, accuracy of DeepID2+ starts dropping quickly, and when higher than 50%, RIR performs better than Incetption-ResNet-V1 and DeepID2+. Even the human face only remains the forehead, RIR still has an accuracy of 89.4%.

In experiments of shielding method 2 (shown in Table 4), we compare Inception-ResNet-V1 and RIR. The result shows that under most shielding conditions, RIR performs better than Inception-ResNet-V1.

Table 3. Accuracies with shielding method 1.

Network	Shielding Ratio						
	10%	20%	30%	40%	50%	60%	70%
DeepID2+	97.7%	96.5%	95.3%	93.9%	88.2%	82.2%	68.0%
Inception-ResNet-V1	98.9%	98.8%	98.3%	96.8%	93.0%	87.5%	79.4%
RIR	**99.0%**	**98.8%**	**98.3%**	**97.0%**	**93.9%**	**89.4%**	**81.0%**

Table 4. Accuracies with shielding method 2.

Network	Shielding Block Size									
	10x10	20x20	30x30	40x40	50x50	60x60	70x70	80x80	90x90	100x100
Inception-ResNet-V1	98.6%	98.5%	98.2%	97.4%	96.6%	94.3%	89.8%	85.0%	77.5%	71.5%
RIR	**98.8%**	**98.6%**	**98.4%**	**97.8%**	**96.4%**	**94.7%**	**90.4%**	**86.3%**	**78.9%**	**73.2%**

4.4. Experiments on Modified Triplet Loss

4.4.1. Ablations on different margin settings. Different margins are applied for normal categories (α_1) and one-shot categories (α_2) to generate less easy triplets. The result is shown in Table 5.

Table 5. Ablations on different margins.

Method	C@99%	C@99.9%
triplet loss ($\alpha_1 = 0.2, \alpha_2 = 0.2$)	74.28%	18.13%
triplet loss ($\alpha_1 = 0.2, \alpha_2 = 0.3$)	74.96%	20.38%
triplet loss ($\alpha_1 = 0.2, \alpha_2 = 0.4$)	**76.53%**	**22.24%**
triplet loss ($\alpha_1 = 0.2, \alpha_2 = 0.5$)	76.32%	21.60%

Table 6. Ablations on mining strategies.

Method	C@99%	C@99.9%
Hard triplet mining only	76.53%	22.24%
Hard + one-shot triplet mining	**79.34%**	**26.53%**

When α_2 increases from 0.2 to 0.4, the coverage is improved; when α_2 is larger than 0.4, the coverage goes down, so we set α_2 to 0.4. This experiment proves effect of modified triplet loss function.

4.4.2. Ablations on mining strategies. We test on different mining strategies separately, shown in Table 6. By adding triplets with one-shot categories as negative samples, the coverage increases by 2.81% under precision of 99% and 4.29% under precision of 99.9%. It shows that the modified mining strategy has a better performance.

5. Conclusion

In this paper, we proposed a **Region Inception ResNet with Modified Triplet Loss**, which got 99.20% accuracy on LFW dataset. The robustness of RIR is shown in shielding experiments, under most of the shielding conditions, our method performed better than former networks. On the other hand, with the help of one-shot categories, the modified triplet loss function keeps Euclidean distances among different categories highly distinguishable. The modified mining strategies increases portion of one-shot categories, contributing to improvement of accuracy. In conclusion, with less training time and storage, our work is easier to apply one-shot face recognition on servers and classify masked faces.

6. References

[1] Krizhevsky A, Sutskever I, Hinton G E. Imagenet classification with deep convolutional neural networks[C]//Advances in neural information processing systems. 2012: 1097-1105.

[2] Sun Y, Wang X, Tang X. Deep learning face representation from predicting 10,000 classes[C]//Proceedings of the IEEE Conference on Computer Vision and Pattern Recognition. 2014: 1891-1898.

[3] Yang J, Zhang D, Frangi A F, et al. Two-dimensional PCA: a new approach to appearance-based face representation and recognition[J]. IEEE transactions on pattern analysis and machine intelligence, 2004, 26(1): 131-137.

[4] Szegedy C, Liu W, Jia Y, et al. Going deeper with convolutions[C]//Proceedings of the IEEE conference on computer vision and pattern recognition. 2015: 1-9.

[5] Schroff F, Kalenichenko D, Philbin J. Facenet: A unified embedding for face recognition and clustering[C]//Proceedings of the IEEE conference on computer vision and pattern recognition. 2015: 815-823.
[6] Simonyan K, Zisserman A. Very deep convolutional networks for large-scale image recognition[J]. arXiv preprint arXiv:1409.1556, 2014.
[7] Taigman Y, Yang M, Ranzato M A, et al. Deepface: Closing the gap to human-level performance in face verification[C]//Proceedings of the IEEE conference on computer vision and pattern recognition. 2014: 1701-1708.
[8] Sun Y, Chen Y, Wang X, et al. Deep learning face representation by joint identification-verification[C]//Advances in neural information processing systems. 2014: 1988-1996.
[9] Hoffer E, Ailon N. Deep metric learning using triplet network[C]//International Workshop on Similarity-Based Pattern Recognition. Springer, Cham, 2015: 84-92.
[10] Wang J, Song Y, Leung T, et al. Learning fine-grained image similarity with deep ranking[C]//Proceedings of the IEEE Conference on Computer Vision and Pattern Recognition. 2014: 1386-1393.
[11] Sun Y, Wang X, Tang X. Deeply learned face representations are sparse, selective, and robust[C]//Proceedings of the IEEE conference on computer vision and pattern recognition. 2015: 2892-2900.
[12] zegedy C, Ioffe S, Vanhoucke V, et al. Inception-v4, inception-resnet and the impact of residual connections on learning[C]//Thirty-First AAAI Conference on Artificial Intelligence. 2017: 4278-4284.
[13] Schroff F, Kalenichenko D, Philbin J. Facenet: A unified embedding for face recognition and clustering[C]//Proceedings of the IEEE conference on computer vision and pattern recognition. 2015: 815-823.
[14] Yi D, Lei Z, Liao S, et al. Learning face representation from scratch[J]. arXiv preprint arXiv:1411.7923, 2014.
[15] Parkhi O M, Vedaldi A, Zisserman A. Deep Face Recognition[C]//BMVC. 2015, 1(3): 6-18.
[16] Liu J, Deng Y, Bai T, et al. Targeting ultimate accuracy: Face recognition via deep embedding[J]. arXiv preprint arXiv:1506.07310, 2015.
[17] Zoph B, Vasudevan V, Shlens J, et al. Learning transferable architectures for scalable image recognition[C]//Proceedings of the IEEE conference on computer vision and pattern recognition. 2018: 8697-8710.
[18] Kemelmacher-Shlizerman I, Seitz S M, Miller D, et al. The megaface benchmark: 1 million faces for recognition at scale[C]//Proceedings of the IEEE Conference on Computer Vision and Pattern Recognition. 2016: 4873-4882.
[19] Guo Y, Zhang L, Hu Y, et al. Ms-celeb-1m: A dataset and benchmark for large-scale face recognition[C]//European Conference on Computer Vision. Springer, Cham, 2016: 87-102.
[20] Huang G B, Mattar M, Berg T, et al. Labeled faces in the wild: A database forstudying face recognition in unconstrained environments[C]. 2008.
[21] Zhong Y, Deng W, Wang M, et al. Unequal-training for deep face recognition with long-tailed noisy data[C]//Proceedings of the IEEE Conference on Computer Vision and Pattern Recognition. 2019: 7812-7821.
[22] Zhao K, Xu J, Cheng M M. Regularface: Deep face recognition via exclusive regularization[C]//Proceedings of the IEEE Conference on Computer Vision and Pattern Recognition. 2019: 1136-1144.
[23] Shi Y, Yu X, Sohn K, et al. Towards Universal Representation Learning for Deep Face Recognition[J]. arXiv preprint arXiv:2002.11841, 2020.

Acknowledgments

This work is supported by the National Natural Science Foundation of China (Grant No. 62077009, 62177006), and partial supported by the Guangdong Provincial Natural Science Foundation (Grant No.

2214050002868) and Zhuhai Science and Technology Planning Project (Grant No. ZH22036201210161PWC).

Deep Supervised Cross-modal Hashing for Ship Image Retrieval

Jiaen Guo, Haibin Wang, Bo Dan, and Yu Lu

Naval Aviation University, Yantai Shandong, 264001, China

guojiaen@163.com; hesonwhb@163.com; lovelyn19841204@163.com; 17664113162@163.com

Abstract. The retrieval of multimodal ship images obtained by remote sensing satellites is an important content of remote sensing data analysis, which is of great significance to improve the ability of marine monitoring. In this paper, We propose a novel cross-modal ship image retrieval method, called Deep Supervised Cross-modal Hashing(DSCMH). It consists of a feature learning part and a hash learning part used for feature extraction and hash code generation separately, both two parts have modality-invariant constraints to keep the cross-modal invariability, and the label information is also brought to supervise the above process. Furthermore, we design a class attention module based on the cross-modal class center to strengthen class discrimination. The experiment results show that the proposed method can effectively improve the cross-modal retrieval accuracy of ship images and is better than several state-of-the-art methods.

1. Introduction

With the repaid development of remote sensing technology, remote sensing images are getting massive, multi-source, and high-resolution. The abundant observation information needs powerful data processing method to obtain its full utilization[1]. As an important branch, remote sensing image retrieval is being widely studied[2,3]. Fast and accurate retrieval of remote sensing images can help researchers locate relevant information in massive data and provide important support for target recognition. The current research for remote sensing image retrieval are mainly focusing on remote sensing scenes[4,5], and most of them are single modality.

It's worth noting that there are few pieces of research on cross-modal ship remote sensing image retrieval. However, in a multi-sensor networking system, the cross-sensor correlation of ship targets can effectively improve the ability of target tracking and recognition. Thus we mainly focus on the cross-modal retrieval of ship images to make up for the blank of the research and provide support for ship target recognition.

In order to achieve fast and accurate retrieval of multimodal ship images, in this paper, we proposed a novel cross-modal retrieval method DSCMH, which is a supervised deep neural network built in an end-to-end manner. As is shown in Fig.1, the feature learning part and hash learning part are used to learn discriminate features and generate high-quality hash codes, and the labels are used as supervisory information deeply participating in the training of DSCMH. Specifically, we design a cross-modal class center to supervise the feature learning process, it can also integrate the class information into the learned deep convolutional features in an attention way, which effectively enhances the association between feature learning and class information keeping the intra-discrimination simultaneously. The hash learning part directly learns hash codes of the features learned from the feature learning part to ensure semantic invariance and intra-discrimination. To further eliminate the modality difference, we add modality-invariant constraints to both two parts.

The major contributions of this work can be summarised as follows:

- A deep supervised cross-modal hash method is proposed for ship image retrieval, the learning of hash codes and modality features are integrated into the same end-to-end learning framework, which can simultaneously learn discriminate features and compact hash codes in a supervised way;
- Based on the labels of the samples, we proposed a cross-modal class center guided learning mechanism to ensure the semantic invariability and structure preservation between the feature learning part and hash learning part, the cross-modal class center can be also used to integrate the label information into the learned features;
- The proposed method is evaluated on two public bimodal ship image datasets. The experiment results show that DSCMH has more advantages in retrieval accuracy than several existing state-of-the-art methods, which indicates the great advantages of our proposed method on cross-modal ship image retrieval tasks.

2. The Proposed Method

In this section, we will introduce the proposed DSCMH. Firstly, we give the formulation of the cross-modal retrieval problem, then, we will introduce the framework of DSCMH in detail, after that, the feature learning part and the hash learning part will be introduced separately.

2.1. Problem Formulation

Assume we have *m* modalities of ship images, and there are *k* groups of matched images, so we have *k*m* images in total. Without losing generality, we focus on the bimodal ship image retrieval tasks. Taking panchromatic and multispectral bimodal images as examples, the paired bimodal ship images can be donated as $\Phi = \left\{\left(x_i^\alpha, x_i^\beta\right)\right\}_{i=1}^k$, where x_i^α is the panchromatic modality, x_i^β is the multispectral modality. Besides, the image pairs have a label matrix $L = \left\{L_i\right\}_{i=1}^k$, and we assign each pair of images a class label vector L_i, donated as $L_i = \left\{l_{i1}, l_{i2}, \cdots, l_{ic}\right\}$,where *c* is the amount of ship image classes, L_i is defined in the form of one-hot coding, if the sample *i* belongs to the *j* class, $l_{ij} = 1$, otherwise 0.

Given all the above training data Φ and L, the goal of the feature learning part is learning a feature transform function $f_i^\alpha = F_\alpha\left(x_i^\alpha, \theta_{F_\alpha}\right)$ and $f_i^\beta = F_\beta\left(x_i^\beta, \theta_{F_\beta}\right)$, which are two deep neural networks for feature extraction, where $f^\alpha \in \mathbb{R}^d$ and $f^\beta \in \mathbb{R}^d$ are the extracted features from different modalities, θ_{F_α} and θ_{F_β} are the parameters of the networks. The networks transform the original images into high-level convolutional features so that samples from different modalities can be measured for similarities in a common metric space. The conventional method of similarity measurement is calculating the distance between different modality feature vectors directly, and the distance is used to retrieve similar images.

To meet the needs of large-scale image fast retrieval, we binary encode the feature vector to lower the storage and processing cost. The hash learning part uses hash layers $h_i^\alpha = H_\alpha\left(f_i^\alpha, \theta_{F_\alpha}\right)$ and $h_i^\beta = H_\beta\left(f_i^\beta, \theta_{F_\beta}\right)$ to generate the binary codes, where θ_{F_α} and θ_{F_β} are the parameters of the hash layer, h_i^α and h_i^β the generated hash codes. The challenge of the hash learning part is trying to keep the similarities from the feature vectors and reduce the quantitative loss as much as possible.

2.2. Framework of DSCMH

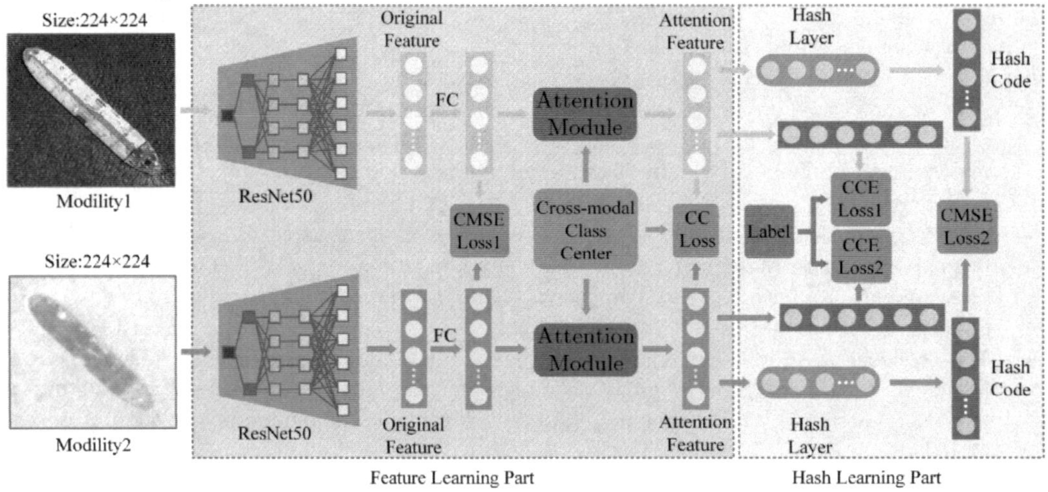

Figure 1. Framework of DSCMH

The framework of the DSCMH is shown in Fig.1. As mentioned above, DSCMH is divided into two parts: feature learning and hash learning, feature learning part consist of two subnetworks used to extract deep discriminant features from bimodal ship images, while the hash learning part generates hash codes through the hash layer. As is shown in Fig.1, the paired bimodal ship images are simultaneously fed into the Resnet50 network pre-trained on ImageNet, after the linear layer, the original images are processed into feature vectors with a dimension of 1024.

The cross-modal class center is the common center of all modalities which is randomly initialized and updated with the training process. The original feature is weighted by the cross-modal class center in an attention way to obtain an enhanced feature representation. Then the attention feature is fed into the hash learning part. The hash layer reduces the attention feature to the setting dimension and then transforms it into binary codes. Another important role of the cross-modal class center is expanding the distance of different classes in the form of cross-modal center loss. The class attention module and cross-modal center loss will be introduced in 2.3.

Modality-invariability constraint is introduced in both parts which can eliminate the gap between different modalities and realize the convergence in the common metric space. In addition, hashing learning part uses the label as supervision to ensure class discrimination.

2.3. Cross-modal Class Center Guided Feature Learning

The feature learning part is divided into two subnetworks, and every subnetwork is used to learn the corresponding modility's high-level convolutional features. To obtain more discriminative information, we use the Resnet50 network pre-trained on ImageNet as the backbone, and take the output of the last layer as the final feature vector. To make better use of the supervisory information, we design a cross-modal class center $C=\left\{C_j\right\}_{j=1}^{c}, C\in\mathbb{R}^{c\times d}$ for feature generation, where $C_j\in\mathbb{R}^{d}$ is *j*th class center. The cross-modal class center can fuse the class information into the generated feature vectors through an attention mechanism, so the feature is more distinguishable. What's more, it is the common center of all modalities, so the modal consistency is further improved after weighting. The cross-modal class center is set according to the class of the images *c* and the dimension of output feature vectors *d*. The common class center is learned iteratively, so that different modal features can approach the center of each class in the common space with training, and intra-class compactness is stronger.

Fig.2 shows the framework of the class attention module, we can see that the attention module generates a new attention feature by fusing the original feature and the weighted cross-modal center in an attention way.

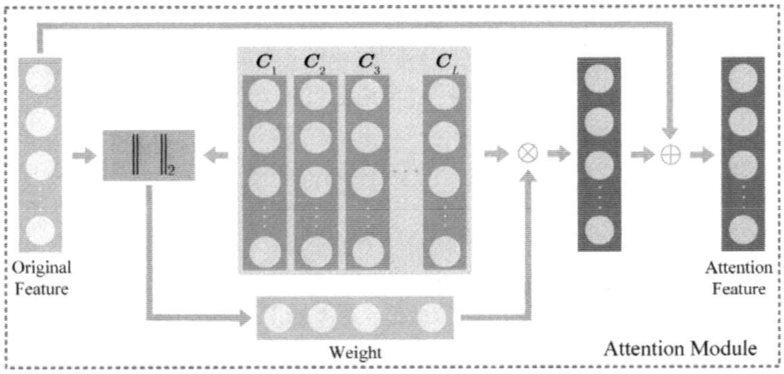

Figure 2. The framework of the attention module

As is shown in Fig. 2, the cross-modal center C is weighted by a couple of weights before fusing, which is obtained by calculating the distance between the original feature and the cross-modal class center, here we use the Frobenius norm to measure their similarities. Then cross-modal class center and weight are both used to generate the weighted feature. Take the original feature f^{α} as an example, the process is demonstrated as follows:

$$\hat{f}^{\alpha} = \frac{\sum_{j=1}^{c}\left(\left\|f^{\alpha} - \boldsymbol{C}_j\right\|_2^2 \cdot \boldsymbol{C}_j\right)}{\sum_{j=1}^{c}\left\|f^{\alpha} - \boldsymbol{C}_j\right\|_2^2} \tag{1}$$

The weighted feature $\hat{f}^{\alpha} \in \mathbb{R}^d$ obtained by weighting the cross-modal class center is then used to generate the attention feature together with the original feature, which is demonstrated as follows:

$$\tilde{f}^{\alpha} = \lambda \cdot f^{\alpha} + (1-\lambda) \cdot \hat{f}^{\alpha} \tag{2}$$

Where $\tilde{f}^{\alpha} \in \mathbb{R}^d$ donates the attention feature, λ is used to control the attention to the cross-modal class center.

As a result, the attention feature both has the original feature and the corresponding class information, which strongly keeps the semantic relevance preservation between feature representation and its label.

To narrow the modality gap and make the features from different modalities closer in common metric space, we need a discrimination loss in training to optimize the DSCMH. Here we use the mean square error loss to measure the distance between features. Mean square error loss is the most commonly used regression loss function, which is defined as the sum of squares of the distance between the predicted value and the target. To enhance cross-modal feature similarity and reduce cross-modal loss of intra-class samples, Cross-modal Mean Square Error (CMSE) loss is designed for network optimization based on mean square error loss, which is defined as follows:

$$L_{CMSE1} = \frac{1}{k}\sum_{i=1}^{k}\left\|f_i^{\alpha} - f_i^{\beta}\right\|_2^2 \tag{3}$$

The cross-modal mean square error loss focuses on the feature distance between cross-modal paired samples, which aims to reduce the distance between different modal features corresponding to the same target so that it can effectively reduce the modal difference between paired samples.

In addition to the attention weighting for feature vectors, another function of the cross-modal class center is to be used in the cross-modal center loss. Center loss is used to reduce the intra-class difference of samples belonging to the same class. By pre-setting the center position of each class and optimizing it in the training process, different sample features belonging to the same class can approach their cross-modal class center continuously. Similarly, to meet the needs of cross-modal retrieval, Cross-modal Center (CC) loss is introduced to improve the original center loss function in multi-modal network optimization, which is defined as follows:

$$L_{CC} = \frac{1}{2}\sum_{i=1}^{k}\left\| f_i^{\alpha} - \boldsymbol{C}^i \right\|_2^2 + \frac{1}{2}\sum_{i=1}^{k}\left\| f_i^{\beta} - \boldsymbol{C}^i \right\|_2^2 \tag{4}$$

$\boldsymbol{C}^i$ is the corresponding class center of sample i. Different modal samples belonging to the same class have the same class center in the common feature space, therefore, the cross-modal center loss function can help the network learn more modal invariant features and increase the compactness within the class so that it has stronger adaptability to different modalities. Different from the original cross-modal class center updating method, we use the gradient descent method to update it:

$$\boldsymbol{C} \leftarrow \boldsymbol{C} - \eta \frac{\partial L_{CC}(\boldsymbol{C})}{\partial \boldsymbol{C}} \tag{5}$$

2.4. Deep Hash Learning

To maintain the similarity of the original feature vectors in the process of hash code generation, the hash learning part needs to restrict the generation of hash code. The output feature vector generated by the feature learning part is fed into the hash learning part and divided into two branches. One branch reduces the dimension of the attention feature vector and outputs the prediction feature vector $\hat{f}_i^{\alpha}, \hat{f}_i^{\beta} \in \mathbb{R}^c$. The Cross-modal Cross-Entropy(CCE) loss is designed to measure the classification ability of DSCMH, which is defined as follows:

$$L_{CCE} = -\frac{1}{N}\left(\sum_{i=1}^{2} L_i \cdot \log\left(\hat{f}_i^{\alpha}\right)\right) - \frac{1}{N}\left(\sum_{i=1}^{N} L_i \cdot \log\left(\hat{f}_i^{\beta}\right)\right) \tag{6}$$

The other branch feeds the attention feature $\hat{f}_i^{\alpha}, \hat{f}_i^{\beta}$ into the hash layer to generate hash code $h_i^{\alpha}, h_i^{\beta}$. The hash layer first reduces the dimension of the attention feature according to the required length and then generates hash codes through the sign function. The sign function is defined as follows:

$$sign(x) = \begin{cases} +1, x > 0 \\ -1, x \le 0 \end{cases} \tag{7}$$

The generation of hash code will lose some information and bring quantization loss, to reduce the difference between hash codes of different modalities, modal invariance constraints are imposed on the generated hash codes, which is also measured by the cross-modal mean square error loss:

$$L_{CMSE2} = \frac{1}{k}\sum_{i=1}^{k}\left\| h_i^{\alpha} - h_i^{\beta} \right\|_2^2 \tag{8}$$

The joint loss function is a combination of L_{CC}, L_{CMSE} and L_{CCE}, which is defined as follows:

$$L_{joint} = \alpha \cdot \left(L_{CCE1} + L_{CCE2}\right) + \beta \cdot L_{CC} + \gamma \cdot \left(L_{CMSE1} + L_{CMSE2}\right) \tag{9}$$

3. Experiments

3.1. Datasets and Baselines

3.1.1. Datasets

The experiment selects the bimodal remote sensing ship image dataset MPSC [32] to verify the retrieval effect of the proposed algorithm and uses the shore-based bimodal ship image dataset VAIS [33] to verify the DSCMH's ability for cross-modal retrieval. MPSC is a panchromatic (PAN) and multispectral (MS) bimodal remote sensing ship image dataset, including 6 types of ships and 2632 paired images; VAIS is an infrared spectroscopy (IR) and visible light (VIS) optical bimodal ship image dataset, including 6 types of ships and 1088 paired images. Both two datasets have been divided into training dataset and test dataset in advance. The class information of the two datasets is given in Table 1.

3.1.2. Baselines

To verify the effectiveness of DSCMH, we compare our DSCMH with six state-of-the-art cross-modal retrieval methods, including AGAH[6], DADH[7], DCMH[8], DSCMR[9], MAN[10], and SDML[11]. The first three methods are hashing methods, and the others are conventional methods. It should be noted that these methods are not applied to ship image retrieval initially, we get the source codes from the authors and apply them to cross-modal ship image retrieval. For a fair comparison, all methods use Resnet50 pre-trained on ImageNet as feature extractor and train with maximal epoch 300.

Table 1. Bimodal ship dataset information

MPSC	Class	Destroyers	Frigates	Combat boats	Bulk carriers	Container ships	Oil tankers
	Amount	99	179	261	919	387	787
VAIS	Class	Merchant	Medium-other	Medium-Passenger	Sailing	Small	Tug
	Amount	146	138	117	284	353	50

3.2. Implementation Details and Evaluation Metric

3.2.1. Implementation Details

we build the feature learning part with two pre-trained networks, then we remove the last 2048×1000 layer and use the penultimate layer to output the 2048-dimensional feature vector, following this, the linear layer outputs a 512-dimensional feature vector. In the hash learning part, sigmoid activation is used to output the predicted label, and tanh is used to transform the attention feature into hash codes. We deploy DSCMH on an NVIDIA GTX 3090 GPU via Pytorch and use Adam as the optimizer with a learning rate of 10^{-3}. The $\alpha, \beta, \gamma, \lambda$ is set to 1, 0.0001, 0.01 and 0.8.

3.2.2. Evaluation Metric

As for the result of the experiment, we use mean Average Precision(mAP) as evaluation metric, which is a commonly used evaluation criterion on cross-modal retrieval. The mAP metric jointly considers the ranking information and precision, and we put it on four different cross-modal ship image retrieval tasks:
1) retrieving PAN samples using MS queries(MS2PAN)
2) retrieving MS samples using PAN queries(PAN2MS)
3) retrieving IR samples using VIS queries(VIS2IR)
4) retrieving VIS samples using IR queries(IR2VIS)

3.3. Performance Comparisons

Firstly, we compare DSCMH with all state-of-the-art methods and give the mAP results in Table 2.

Table 2. mAP comparison results(all methods)

Method		AGAH[6]	DADH[7]	DCMH[8]	DSCMR[9]	MAN[10]	SDML[11]	DSCMH
MPSC	MS2PAN	0.444	0.458	0.400	0.627	0.262	0.269	**0.637**
	PAN2MS	0.457	0.461	0.370	0.608	0.287	0.271	**0.602**
VAIS	VIS2IR	0.401	0.398	0.404	0.433	0.255	0.273	**0.447**
	IR2VIS	0.390	0.392	0.368	0.420	0.263	0.274	**0.436**

From Table 2. we can see that DSCMH outperforms all the other six state-of-the-art methods on the two benchmark datasets. The main reason is that DSCMH brings supervisory information to the retrieval task, both the feature learning part and hash learning part use labels to supervise the generation of feature and hash codes. Besides, the cross-modal class center narrows the modality gap, and the modal

invariance constraint further removes the difference among modalities, so that the same class samples from different modalities can get closer in the matrix space, which leads to a higher mAP.
To give a better comparison between DSCMH and other hash methods, we conduct experiments on different hash code lengths and give the mAP results in Table 3.

Table 3. mAP comparison results(hash methods)

Method	**MPSC**						**VAIS**					
	MS2PAN			PAN2MS			VIS2IR			IR2VIS		
	16 bits	32 bits	64 bits	16 bits	32 bits	64 bits	16 bits	32 bits	64 bits	16 bits	32 bits	64 bits
AGAH[6]	0.426	0.437	0.444	0.452	0.446	0.457	0.374	0.390	0.401	0.345	0.369	0.390
DADH[7]	0.444	0.455	0.458	0.449	0.453	0.461	0.391	0.389	0.398	0.366	0.386	0.392
DCMH[8]	0.380	0.378	0.400	0.355	0.346	0.370	0.390	0.401	0.404	0.379	0.384	0.368
DSCMH	**0.624**	**0.634**	**0.637**	**0.587**	**0.608**	**0.602**	**0.433**	**0.431**	**0.447**	**0.409**	**0.433**	**0.436**

Table 3. shows the retrieval results of all the hash methods. We can see that longer hash codes bring higher retrieval accuracy, and our DSCMH achieves the highest mAP result among all the hash methods. This is because we also apply modal invariance constraint to the hash learning part, which makes the hash codes we used to retrieve more similar.

4. Conclusion

Aiming at the cross-modal ship image retrieval problem, we proposed a deep supervised hashing method to achieve an accurate and fast retrieval. Design of cross-modal class center and application of hash coding makes the cross-modal ship image retrieval easier. The experiment results show that our proposed method can achieve higher mean average precision on two widely-used ship image datasets than most of the cross-modal retrieval methods, and performs better than several hash methods, which proves the effectiveness of DSCMH on cross-modal ship image retrieval tasks.

5. Reference

[1] K. Walter, M. J. Gibson and A. Sowmya, "Self-Supervised Remote Sensing Image Retrieval," IGARSS 2020 - 2020 IEEE International Geoscience and Remote Sensing Symposium, pp. 1683-1686, 2020.
[2] Y. Li, Y. Zhang, X. Huang and J. Ma, "Learning Source-Invariant Deep Hashing Convolutional Neural Networks for Cross-Source Remote Sensing Image Retrieval,"IEEE Transactions on Geoscience and Remote Sensing, " vol. 56, no. 11, pp. 6521-6536, 2018.
[3] Y. Li, Y. Zhang, X. Huang, H. Zhu and J. Ma, "Large-Scale Remote Sensing Image Retrieval by Deep Hashing Neural Networks," IEEE Transactions on Geoscience and Remote Sensing, vol. 56, no. 2, pp. 950-965, 2018.
[4] J. Ma, D. Shi, X. Tang, X. Zhang, X. Han and L. Jiao, "Cross-Source Image Retrieval Based on Ensemble Learning and Knowledge Distillation for Remote Sensing Images," 2021 IEEE International Geoscience and Remote Sensing Symposium IGARSS, pp. 2803-2806, 2021.
[5] B. Chaudhuri, B. Demir, L. Bruzzone and S. Chaudhuri, "Region-Based Retrieval of Remote Sensing Images Using an Unsupervised Graph-Theoretic Approach," IEEE Geoscience and Remote Sensing Letters, vol. 13, no. 7, pp. 987-991,2016.
[6] W. Gu, X. Gu, J. Gu, B. Li, Z. Xiong, and W. Wang. "Adversary Guided Asymmetric Hashing for Cross-Modal Retrieval". Proceedings of the 2019 on International Conference on Multimedia Retrieval, 159-167, 2019.
[7] C. Bai, C. Zeng, Q. Ma, J. Zhang and S. Chen. "Deep Adversarial Discrete Hashing for Cross-Modal Retrieval, " Proceedings of the 2020 International Conference on Multimedia Retrieval, 525–531, 2020.

[8] Q. Jiang and W. Li, "Deep Cross-Modal Hashing," 2017 IEEE Conference on Computer Vision and Pattern Recognition (CVPR), pp. 3270-3278, 2017.
[9] L. Zhen, P. Hu, X. Wang and D. Peng, "Deep Supervised Cross-Modal Retrieval," 2019 IEEE/CVF Conference on Computer Vision and Pattern Recognition (CVPR), pp. 10386-10395, 2019.
[10] P. Hu, D. Peng, X. Wang and Y. Xiang. "Multimodal adversarial network for cross-modal retrieval." Knowledge-Based Systems 180.SEP.15:38-50, 2019.
[11] P. Hu, L. Zhen, D. Peng, and P. Liu. "Scalable Deep Multimodal Learning for Cross-Modal Retrieval". Proceedings of the 42nd International ACM SIGIR Conference on Research and Development in Information Retrieval, 635–644, 2019.

Acknowledgements
This work was supported by The National Defense Science and Technology Excellence Youth Talent Fund (2017-JCJQ-ZQ-003) and the Taishan Scholar Engineering Special Fund (ts 201712072)

An Image Encryption Scheme based on Adaptive Bit-Plane Division

Jun Gu, Feng Xu*, Tao Zeng and Xin Lyu

College of Computer and Information, Hohai University, Nanjing 211100, China.

*Correspondence: xufeng@hhu.edu.cn

Abstract. With the application and popularization of digital images, the security of images has been paid more and more attention. The bit-level image encryption method has become one of the commonly used methods. It has the characteristic of changing the pixel value while disturbing the position of the pixel point, but at the same time it is accompanied by a higher computational complexity. In order to improve the efficiency of the encryption algorithm, according to the characteristic that the higher 4 of the 8 bit-planes of the grayscale image contain 94.125% of the information of the entire image, the encryption structure of higher 4 bit-planes and lower 4 bit-planes is widely used. But for different images, this characteristic is not accurately applicable. In response to this problem, we design a method of adaptive bit-plane division, which uses mutual information and Jensen-Shannon divergence to measure the amount of information of bit-planes and then divides bit-planes into two parts. The results of bit-plane division are different for different images. Based on this, we propose an image encryption scheme based on adaptive bit-plane division. Compared with the traditional higher 4-bits and lower 4-bits encryption method, this scheme reduces the encryption cost and improves the encryption efficiency. Experimental results show that the proposed scheme can effectively resist the chosen plain-image attacks and has sufficient security.

1. Introduction

In recent years, digital image encryption algorithms mainly focus on pixel-level image encryption algorithms and bit-level image encryption algorithms. Pixel-level encryption algorithms usually perform scrambling-diffusion operations and multiple iterations on the pixels of images[1][2]. The bit-level encryption algorithms are to perform binary decomposition of the image pixel value to obtain the bit-planes of the image and then encrypt the bit-planes, so that the pixel value can be changed while changing the pixel position, which can improve the encryption efficiency of the algorithm[3][4][5].

Zhu and Zhang[6] first proposed Bit-level Permutation (BLP) in 2011. Pixel values of a grayscale image from 0 to 255 were decomposed into 8 bits in binary and the bits of the same level of the pixels were arranged to change the pixel value while disturbing the pixel position. The scrambling position of the bit was restricted to the same bit-plane which cannot change the number of 0 and 1 bits in the bit-plane. Zhang[7] proposed a scrambling method between different bit-planes based on an "expansion-contraction" mechanism. Decomposing a grayscale image of N×N size into 8 bit-planes and merging two adjacent bit-planes to obtain four groups, the four groups of image matrices were combined into a 2N×2N matrix for scrambling. Then 8 new bit-planes were split by the scrambling result according to the inverse process of the combination and were merged to complete the encryption process. Although this algorithm significantly reduced the high correlation of the bit-planes, it required an additional 4

 2320 (2022) 012024 doi:10.1088/1742-6596/2320/1/012024

times the storage space. Just for encrypting images of size N×N, this strategy cannot be used for M×N images. In order to solve these two problems, Wang[8] proposed a new bit-level scrambling strategy, which divided the M×N grayscale image into N columns, and performed binary conversion on each pixel in the N column. Then they used Arnold transform to map 8×N bits to new positions and performed cyclic shift. At present, in addition to various methods of bit-level scrambling, the characteristic that the higher 4 bit-planes of the 8 bit-planes of grayscale images contain 94.125% of the information of the entire image had attracted the attention of researchers. Liu[9] scrambled the entire image and then decomposed the 8 bit-planes into higher 4-bits and lower 4-bits, then scrambled and diffused the higher 4-bits, and finally merged the bit-planes to obtain the cipher image. Lin[10] merged the lower 4-bits of the bit-planes into an image matrix and juxtaposed with the higher 4-bits so that five image matrices were encrypted, which not only reduced the calculation and time cost, but also ensured the security of the cipher image. For some special scenarios, such as teleconferences, surveillance cameras and other real-time applications and computing environments with limited resources, it was usually directly used to encrypt the higher 4-bits or even the higher 2-bits, while the rest parts were not encrypted to achieve real-time and reduce resource utilization by sacrificing certain security[11][12][13].

But for different images, the conclusion that the higher 4-bits in the image bit-planes account for 94.125% of the entire image information is not always applicable. In response to this problem, we propose a method of adaptive bit-planes division by measuring the similarity of bit-planes based on mutual information and Jensen-Shannon divergence. In addition, we propose an image encryption scheme based on the adaptive division method. The division method is used to divide the image into two parts, high-significant bit-planes and low-significant bit-planes, and then different encryption methods are used to encrypt the two parts. Experimental results show that this scheme can significantly improve efficiency while ensuring safety.

The remainder of this paper is organized as follows. Section 2 introduces some preliminary materials including bit-plane decomposition and similarity measures. Section 3 presents the proposed bit-plane adaptive division in detail. Section 4 gives an efficient image encryption scheme based on bit-planes adaptive division and two existing encryption algorithms. Experiment results and security analyses are stated in Section 5. Finally, section 6 concludes this paper.

2. Related works

2.1. Bit-plane decomposition

In grayscale images, the range of pixel values is [0,255], so each pixel can be represented by an 8-bit binary sequence. Therefore, the grayscale image can be divided into 8 bit-planes through binary bit-plane decomposition. There are two commonly used coding rules for bit-plane decomposition[14]:

2.1.1. Binary bit-plane decomposition. The grayscale value of each pixel p(i,j) of the standard test grayscale image X of size M×N ranges from [0,255], where i,j represent the position of the pixel value in the image, and it can be represented by 8-bit binary. If $b_{i,j,0}$, $b_{i,j,1}$,……, $b_{i,j,7}$ mean the binary representation of each pixel, there are:

$$b_{i,j,k} = \left\lfloor \frac{p_{i,j}}{2^k} \right\rfloor \bmod 2, k = 0,1,2,\text{L}\ 7 \tag{1}$$

$$p_{i,j} = \sum_{k=0}^{7} b_{i,j,k}\, \text{g}2^k \tag{2}$$

2.1.2. Gray Code bit-plane decomposition. If any two connected numbers only have one binary number different in the coding of a group of numbers, this kind of coding is called Gray Coding. The method of converting the corresponding n-bit binary code to n-bit Gray code is as follows:

$$\begin{cases} G(i) = B(i), & i = n-1 \\ G(i) = B(i) \oplus B(i+1), & 0 \le i < n-1 \end{cases} \tag{3}$$

where G(i) represents Gray code, and B(i) represents binary.

The original image P is decomposed into 8 bit-planes, which are P_0, P_1, P_2, P_3, P_4, P_5, P_6, and P_7. The weight of the information carried by each bit-plane is different[15]. The calculation method of the percentage of pixel information in each bit-plane is as follows, and the results are listed in Table 1:

$$p(i) = \frac{2^i}{\sum_{i=0}^{7} 2^i} \tag{4}$$

Table 1. Percentage of pixel information in each bit plane.

i th	Percent *p(i)* (%)	*i* th	Percent *p*(i) (%)
1	0.3922	5	6.275
2	0.7843	6	12.55
3	1.5686	7	25.10
4	3.1373	8	50.20

However, for different grayscale images, the information distribution of the 8 bit-planes is different and the image encryption method that always selects the higher 4-bits for protection is not flexible enough, so that when facing a massive image data encryption scene, the algorithm does not have sufficient encryption efficiency.

2.2. Similarity measures

2.2.1. Mutual Information. Mutual information is a measure of the correlation between two random variables[16]. From the perspective of probability, the mutual information of two discrete random variables X and Y are respectively described in equation (5):

$$\begin{cases} I(X;Y) = \sum_{y \in Y} \sum_{x \in X} p(x,y) \log\left(\frac{p(x,y)}{p(x)p(y)}\right) \\ p(x,y) = p\{X \le x, Y \le y\} \\ p(x) = p\{X \le x, Y \le \infty\} = \sum_{y} p(x,y) \end{cases} \tag{5}$$

Where p(x,y) represents the joint probability distribution of X and Y, and p(x)、p(y) represent the marginal probability distribution. The greater the mutual information value, the greater the similarity between X and Y.

For 8 bit-planes of grayscale images, the minimum value of mutual information between every two adjacent bit-planes is always concentrated in the lowest two bit-planes (P_0 and P_1), as shown in Figure 1(a) and it is obviously showed that the result of this division is not accurate.

2.2.2. Jensen-Shannon divergence. Kullback-Leibler divergence is also known as relative entropy[16]. For the same random variable x, there are two separate probability distributions P(x) and Q(x). KL divergence can be used to measure the difference between these two distributions, the calculation method is respectively described in equation (6).

$$D_{KL}(p \| q) = \sum_{i=1}^{n} p(x_i) \log(p(x_i) / q(x_i)) \tag{6}$$

Jensen-Shannon divergence, a variant based on KL divergence, measures the similarity of two probability distributions[17], and solves the problem of KL divergence asymmetry. Generally, the JS divergence is symmetric, and its value is between 0 and 1. The smaller the JS divergence, the stronger the fitting degree of P_1 and P_2, and the greater the similarity. It is defined as follows:

$$JS(P_1 \| P_2) = \frac{1}{2} KL(P_1 \| \frac{P_1 + P_2}{2}) + \frac{1}{2} KL(P_2 \| \frac{P_1 + P_2}{2}) \quad (7)$$

For the 8 bit-planes of grayscale images, the maximum value of JS divergence between every two adjacent bit-planes is always concentrated in the two highest bit-planes (P_6 and P_7), as shown in Figure 1(b), and it is obviously showed that the result of division is also not accurate.

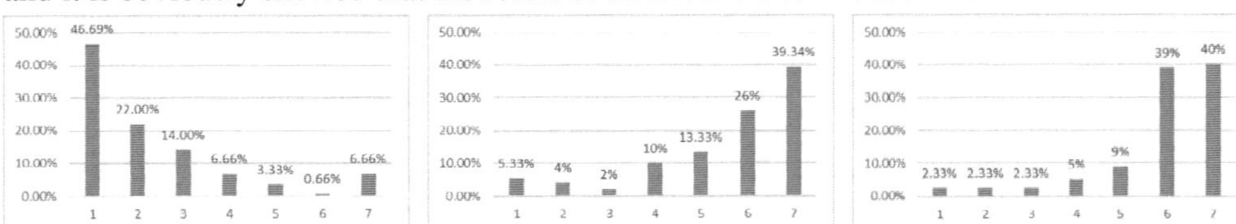

(a) (b) (c)

Figure 1. (a) The results of dividing the bit-planes using mutual information; **(b)**The results of dividing the bit-planes using Jensen-Shannon divergence; **(c)**The results of dividing the bit-planes using Jensen-Shannon divergence and mutual information.

3. Bit-plane adaptive division

Aiming at the shortcomings of the above-mentioned similarity metrics and the inflexibility of the traditional image encryption methods based on the higher 4 bit-planes and lower 4 bit-planes, we propose a new similarity metric to divide the bit-planes.

3.1. Calculation of the correlation of bit-planes

The grayscale image P is decomposed into 8 bit-planes using Gray coding, and the mutual information $Score_{MI}$ and JS divergence $Score_{JS}$ of every two adjacent bit-planes are calculated with the length of seven.

Take the reciprocal of $Score_{MI}$, and perform logsumexp operation between $Score_{JS}$ and it to get the comprehensive division index $Score_{MI\_JS}$, with the length of seven.

$$Score_{MI\_JS} = LogSumExp(1 / Score_{MI}, Score_{JS}) = \log(e^{1/Score_{MI}} + e^{Score_{JS}}) \quad (8)$$

3.2. Bit-planes division

According to the characteristic that the higher bit-planes contains more information than the lower bit-planes and the similarities between every two adjacent bit-planes, the two bit-planes with the lowest similarity can be found to divide all bit-planes into two parts: high-significant bit-planes and low-significant bit-planes. The details are as follows:

According to the two bit-planes P_i, P_{i+1} which have the maximum value of $Score_{MI\_JS}$, the 8 bit-planes are divided into two parts: high-significant bit planes P_{i+1},...,P_6,P_7, and low-significant bit planes P_0, P_1,..., P_i.

Figure 1(c) shows the division results of various images. It can be seen that the 8 bit-planes are divided into higher one bit-plane and lower seven bit-planes for about 40%, and divided into higher 2 bit-planes and lower 6 bit-planes for about 39%. The division into higher 3 bit-planes and lower 5 bit-planes accounts for about 9%, and 7% of the cases are divided at the lower 3 bit-planes.

4. The proposed encryption scheme

Aiming at the shortcomings of the above-mentioned similarity metrics and the inflexibility of the traditional image encryption methods based on the higher 4-bits and lower 4-bits, this section proposes a new similarity metric to divide the bit-planes. We propose an efficient image encryption scheme based on bit-planes adaptive division. This scheme performs different encryption algorithms for the two divided parts by bit-plane adaptive division. The flowchart of the encryption scheme is shown in Figure 2. This scheme effectively saves time and cost while ensuring the security of the encryption result.

4.1. High-significant bit-planes encryption algorithm

Figure 3 shows the framework of the algorithm using 2D-LSCM chaotic mapping, row-column synchronization scrambling and diffusion operations to encrypt high-significant bit-planes[18].

Assume that the high-significant bit-planes are P_6 and P_7 respectively, and the size of each bit-plane is M×N. Detailed steps are as follows:

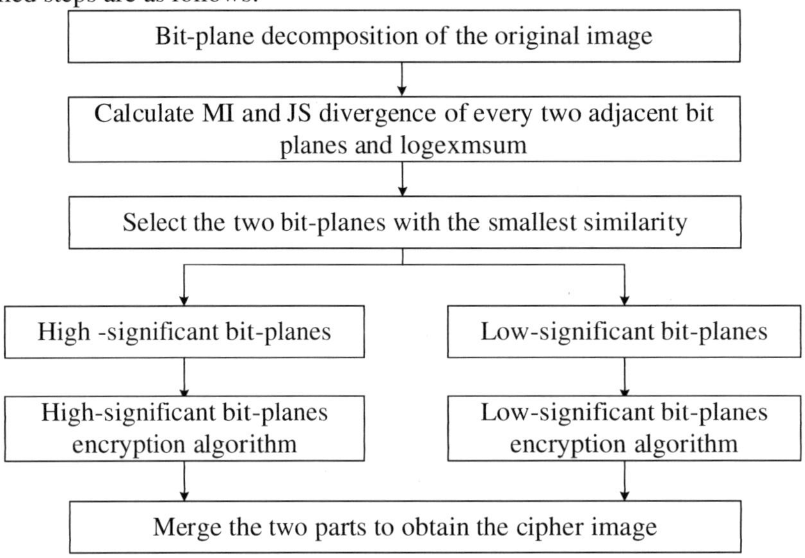

Figure 2. Flow of encryption algorithm.

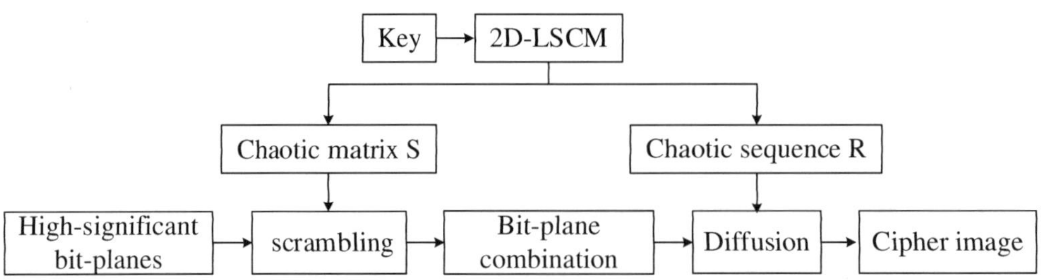

Figure 3. Encryption algorithm of high-significant bit-planes.

4.1.1. Chaotic matrices and sequence generation. This section uses the two-dimensional chaotic system Logistic-Sine-coupling-map (2D-LSCM), which can be defined as follows:

$$\begin{cases} x_{i+1} = \sin(\pi(4rx_i(1-x_i) + (1-r)\sin(\pi y_i))) \\ y_{i+1} = \sin(\pi(4ry_i(1-y_i) + (1-r)\sin(\pi x_{i+1}))) \end{cases} \tag{9}$$

Where r is the control parameter, and $r \in (0,1)$.

Use x_0, y_0, r_1 as the initial value to iterate the chaotic system 2D-LSCM for m+2MN times, discard the former m values to avoid harmful effects. The chaotic system generates two chaotic matrices of size M×N, namely S_6, S_7. Do the same operation on x_0, y_0, r_2 to get the chaotic sequence R of size M×N.

4.1.2. Bit scrambling. The row and column positions of the bit-planes (P_6, P_7) are simultaneously scrambled and detailed steps are as follows:

- *Step 1*: Sort each column of matrix S_6 to obtain index matrix O;
- *Step 2*: Set the row index m=1;
- *Step 3*: Sort the values of the pixels at the position $\{(O_{m,1},1),(O_{m,2},2),\ldots,(O_{m,N},N)\}$ in S_6 to obtain the index vector v;
- *Step 4*: Select the pixels at position $\{(O_{m,1},1),(O_{m,2},2),\ldots,(O_{m,N},N)\}$ in P_6 and confuse them by v;
- *Step 5*: Iterate *Step 3* to *Step 4* for m=2~M to get the scrambled matrix T_6;
- *Step 6*: Repeat *Step 1* to *Step 5* to obtain the scrambled matrix T_7 of P_7.

4.1.3. Diffusion. Perform a diffusion operation on the matrixes T_6 and T_7 and detailed steps are as follows:

- *Step 1*: Combine T_6 and T_7 to obtain T, and the combination formula is as shown in equation (2);
- *Step 2*: Convert T into a sequence TD with length of M×N, R is a chaotic sequence with length of M×N generated by 2D-LSCM with an initial value of x_0, y_0, r_2;
- *Step 3*: Calculate the sequence TC according to equation (10) for R and TD;

$$TC=\begin{cases}((TD_1+TD_{M\times N}+TD_{M\times N-1})+\lfloor R_i\times 2^{32}\rfloor)\bmod F,\ if\ i=1\\((TD_2+TD_1+TD_{M\times N})+\lfloor R_i\times 2^{32}\rfloor)\bmod F,\ if\ i=2\\((TD_i+TD_{i-1}+TD_{i-2})+\lfloor R_i\times 2^{32}\rfloor)\bmod F,\ if\ i\in[3,M\times N]\end{cases}\tag{10}$$

Where F is the number of pixel values allowed in the plaintext image. For example, when TD is a 2-bits image, then F=4, and the operation ⌊X⌋ is the largest integer less than or equal to X.

- *Step 4*: The sequence TC is transformed into the M×N matrix HC, which is the encrypted image of the high-significant bit-planes.

4.2. Low-significant bit-planes encryption algorithm

Perform XOR operation on the pixel value according to the parity of the pixel value to encrypt the low-significant bit-planes. Finally, the two parts of the encryption results are combined to obtain the complete cipher image. The framework of the algorithm is shown in Figure 4.

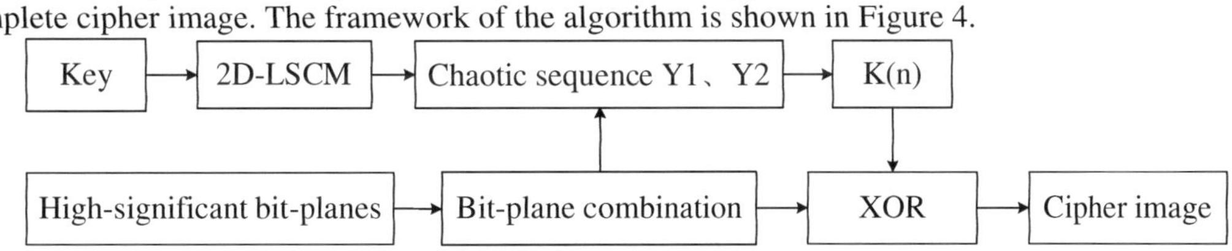

Figure 4. Encryption algorithm of low-significant bit-planes.

Assume that the low-significant bit-planes are $P_1,P_2,\cdots,P_5$ respectively, and the size of each bit-plane is M×N. Detailed steps are as follows:

4.2.1. Chaotic matrices and sequence generation. Use x_0, y_0, r_3 and x_0, y_0, r_4 as the initial value to respectively iterate the chaotic system 2D-LSCM for m+MN times, discard the former m values to avoid harmful effects. The chaotic system generates two chaotic sequences of size M×N, namely Y_1, Y_2.

4.2.2. XOR operation. Detailed steps are as follows:

- *Step 1*: Combine P_0, P_1, P_2, P_3, P_4, and P_5 according to equation (2) to obtain an image LP of size M×N; then convert LP into a sequence L with length of 1×MN;
- *Step 2*: Take the n-th pixel A(x,y) in the image and if the sequence number of the point n is odd, the encryption key k(n) is constructed from the real chaotic sequence $Y_1(n)$ according to equation (11); if else, $Y_2(n)$ is used to construct k(n).

$$k(n)=\begin{cases}\mathrm{mod}(\lfloor Y_1(n)\times 10^{15}\rfloor,256), if\ \mathrm{mod}(n,2)=1\\ \mathrm{mod}(\lfloor Y_2(n)\times 10^{15}\rfloor,256), if\ \mathrm{mod}(n,2)=0\end{cases} \tag{11}$$

- *Step 3*: Use A(x, y) and the key value k(n) generated in *Step 2* to perform bit-XOR operations to obtain the encrypted pixel value A'(x, y).
- *Step 4*: Repeat *Step 2* to *Step 3* until all pixels are encrypted to obtain the encrypted image LC.
- *Step 5*: Combine the encrypted image HC and the encrypted image LC according to equation (2) to obtain the cipher image C of the original image P.

5. Experimental results and performance analysis

This section provides the simulation results and performance analysis of the encryption scheme. In addition, we study two similar encryption algorithms[9][15] and three bit-level encryption algorithms[19][20][21] proposed in recent years, and compare the performance of these several image encryption schemes. The experimental data come from the USC-SIPI dataset[1], LLVIP dataset[2], DIV2K dataset[3] and DOTA dataset[4]. Specifically, we select 39 images of the USC-SIPI dataset, including 256×256, 512×512, and 1024×1024 images; the LLVIP dataset is a collection of infrared images(IR), and 30 images are selected for experiment; the DIV2K dataset includes high-definition images in 2K size, and we select 30 images for experiment; 10 images in 4K size are selected for experiment from the DOTA dataset which is a collection of remote sensing images. Experimental environment: Intel(R) Core(TM) i5-8250U CPU@1.8GHz, Windows10. The compilation platform is MATLAB R2018b.

5.1. Encryption results

Figure 5 shows the encryption results of images of different sizes and the corresponding pixel distribution histograms. The second column and the fourth column are the histogram of the plain image and the histogram of the cipher image. We can observe that the pixel distribution of the cipher image tends to be uniform, and the plain image cannot be identified after encryption, which means the algorithm can effectively resist statistical attacks. Our encryption algorithm can totally reconstruct the plain-images by using the correct key.

5.2. Security analysis

Image security is the most important indicator of image encryption algorithm. This section analyzes the security of the proposed encryption scheme from the following aspects.

5.2.1. Key space and sensitivity analysis. The key space is the collection of all legal keys in the entire encryption and decryption process. Alvarez[22] researched and found that when the key space is greater than 2^{100}, the algorithm can effectively resist brute force attacks. The key of the algorithm in this paper is $Key=\{x_0,y_0,r_1,r_2,r_3,r_4\}$, where x_0,y_0 are the initial values of the 2D-LSCM chaotic map, and r_1,r_2,r_3,r_4 are the control parameters of the 2D-LSCM chaotic map. Set them to 15 decimal numbers,

[1] https://sipi.usc.edu/database/database.php

[2] https://bupt-ai-cz.github.io/LLVIP/

[3] https://data.vision.ee.ethz.ch/cvl/DIV2K/

[4] https://captain-whu.github.io/DOTA/dataset.html

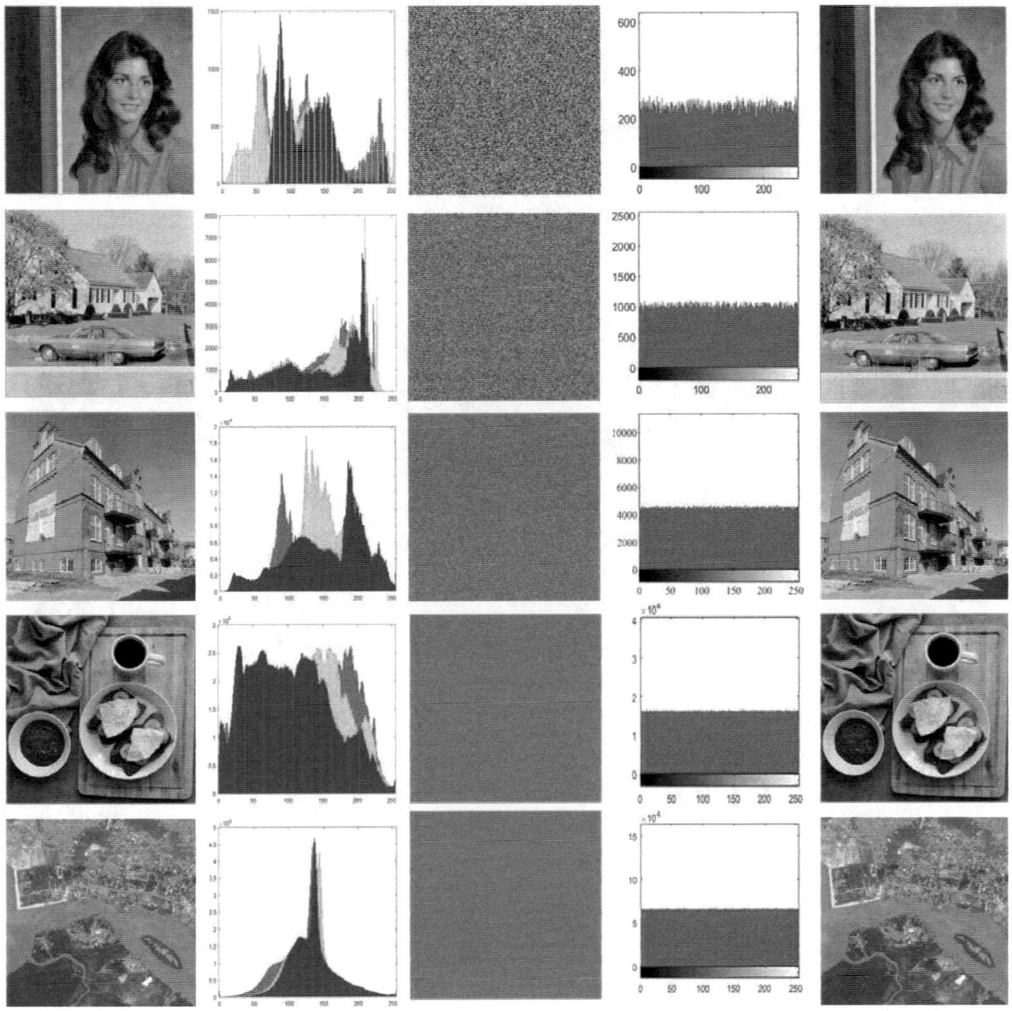

(a) (b) (c) (d) (e)

Figure 5. Simulation results of different images. **(a)**Plain-images; **(b)**histograms of the plain-images; **(c)**cipher-images; **(d)**histograms of the cipher-images; **(e)**decrypted images.

and the key space is 10^{90}. 2^{100} is smaller than 10^{90} which means that our encryption algorithm is sufficient to resist current brute force attacks.

Key sensitivity refers to the degree of influence of the correct key on the correctness of the encryption and decryption results, that is, when other conditions remain unchanged, a small change to the key will cause a huge difference between the result and the correct result. The bit values of different bits of the correct key K1 are respectively changed to obtain keys K2 and K3. Figure 6 shows the key sensitivity of the image named Lena during encryption and decryption. It can be seen that the encryption results of different keys K1 and K2 are very different; only the correct key can restore the original image, and when two keys K2 and K3 with small differences are used to decrypt the cipher image, the decryption results also vary greatly.

(a)Plain image P | (b)Cipher image $C_1 = Enc(P, K_1)$ | (c) Cipher image $C_2 = Enc(P, K_2)$ | (d) Differences between (b) and (c)

(e) Decrypted image $D_1 = Dec(C_1, K_1)$ | (f) Decrypted image $D_2 = Dec(C_1, K_2)$ | (g) Decrypted image $D_3 = Dec(C_1, K_3)$ | (h) Differences between (f) and (g)

Figure 6. Key sensitivity analysis.

5.2.2. Information entropy analysis. Information entropy is an index to measure the uncertainty of information[23]. The greater the information entropy of the cipher image, the greater the randomness of the image, and the higher the security of the encryption result. For a 256-level grayscale image, the ideal value of information entropy is 8. The calculation formula of information entropy is as follows:

$$H(m) = -\sum_{i=0}^{N-1} P(m_i) \log P(m_i) \tag{12}$$

Table 2 shows the information entropy of the cipher image obtained by the three algorithms under the adaptive division and higher 4-bits and lower 4-bits division. It can be seen that the results of the cipher image information entropy of the two divisions are not much different, and are approximately 8. It shows that the self-adaptively divided cipher image has sufficient security.

Table 2. Shannon entropy test of cipher images of different sizes.

size	Proposed -ad	Proposed -4+4	Liu's[9]-ad	Liu's[9]-4+4	Teng's[15]-ad	Teng's[15]-4+4
256	7.997149±0.0003	7.997127±0.0002	7.998704±4E-01	7.998293±0.7073	7.997015±4E-01	7.997333±0.7073
512	7.999264±5E-05	7.999288±6E-05	7.997409±3E-01	7.996728±0.1513	7.997614±3E-01	7.998800±0.1513
1024	7.999816±5E-06	7.999837±5E-06	7.998410±6E-01	7.999320±0.0062	7.997410±5E-01	7.997631±0.0063
IR	7.999859±1E-05	7.999856±1E-05	7.996174±4E-01	7.998231±0.1509	7.997618±6E-01	7.997852±0.1509
2K	7.999934±8E-06	7.999325±7E-06	7.997650±2E-01	7.998640±0.0261	7.997651±3E-01	7.997641±0.0261
4K	7.999989±6E-07	7.999989±6E-07	7.998331±0.2570	7.998565±0.0713	7.997333±0.2569	7.997566±0.0713

5.2.3. Correlation analysis of adjacent pixels. The correlation of adjacent pixels reflects the degree of correlation between the grayscale values of pixels in adjacent positions in the image[24]. The pixels of digital images have a strong correlation in the three directions of horizontal, vertical, and diagonal[25]. One of the functions of the encryption algorithm is to reduce the correlation between adjacent pixels, as far as possible to achieve ideal state which is 0. The calculation formula for the correlation coefficient of N pairs of adjacent pixels in the image is as follows:

$$\begin{cases} E(u)=\dfrac{1}{N}\sum_{i=1}^{N}u_i \\ D(u)=\dfrac{1}{N}\sum_{i=1}^{N}(u_i-E(u))^2 \\ Cov(u,v)=\dfrac{1}{N}\sum_{i=1}^{N}(u_i-E(u))(v_i-E(u)) \\ r_{u,v}=\dfrac{Cov(u,v)}{\sqrt{D(u)}\times\sqrt{D(v)}} \end{cases} \tag{13}$$

Among them, u and v are the grayscale values of adjacent pixels in the image, $E(u)$ is the mean value of u, $D(u)$ is the variance of u, $Cov(u,v)$ is the covariance of u and v, and $r_{u,v}$ is the correlation coefficients of u and v.

We randomly select 4000 pairs of adjacent pixels in three directions from plain images and cipher images by the proposed scheme and then plot their gray levels as x-values and y-values in a 2D plane, as shown in Figure 7.

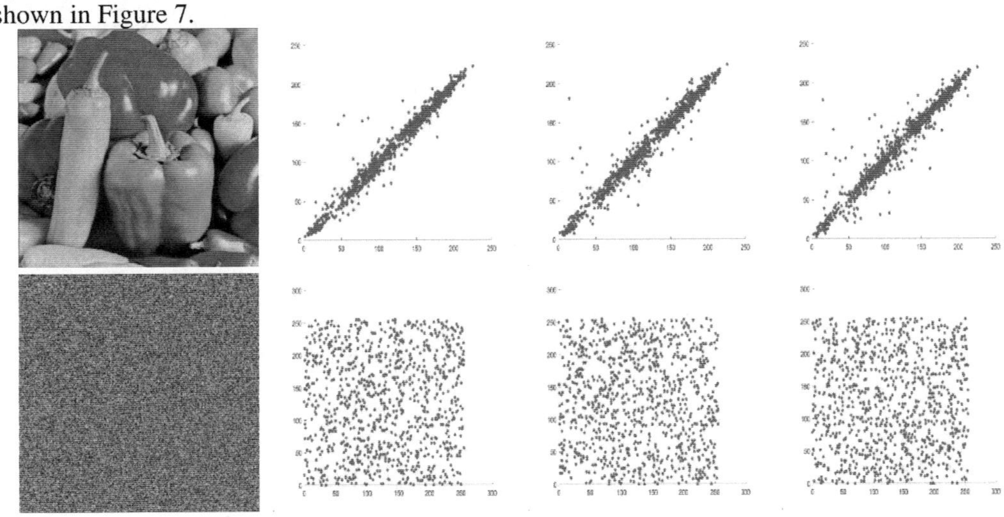

Figure 7. Correlations of plain images and corresponding cipher images in three directions.

Table 3. Pixel correlations of cipher images of different sizes in horizontal direction.

size	Proposed -ad	Proposed -4+4	Liu's[9]-ad	Liu's[9]-4+4	Teng's[15]-ad	Teng's[15]-4+4
256	0.003308±0.0322	0.002077±0.0291	-0.01345±0.0351	-0.00367±0.0234	0.006745±0.0247	-0.00677±0.0339
512	0.003345±0.0194	-0.00035±0.0227	-0.00470±0.0297	-0.00494±0.0272	-0.00305±0.0289	-0.00493±0.0225
1024	-0.00319±0.0027	-0.01102±0.0140	0.011845±0.0046	-0.01061±0.0121	0.001018±0.0030	0.009163±0.0204
IR	0.000920±0.0206	-0.00166±0.0228	0.006385±0.0144	-0.00529±0.0155	0.002189±0.0155	0.010146±0.0132
2K	0.001748±0.0174	-0.00172±0.0149	-0.00021±0.0105	0.002014±0.0087	0.000728±0.0155	-0.00481±0.0127
4K	-0.00082±0.0083	0.001848±0.0086	-0.00385±0.0097	0.004910±0.0091	0.004806±0.0098	-0.00134±0.0082

Table 4. Pixel correlations of cipher images of different sizes in vertical direction.

size	Proposed -ad	Proposed -4+4	Liu's[9]-ad	Liu's[9]-4+4	Teng's[15]-ad	Teng's[15]-4+4
256	-0.01036±0.0336	-0.00236±0.0298	0.000884±0.0239	0.003261±0.0272	0.004371±0.0431	0.017132±0.0363
512	-0.00660±0.0236	-0.00716±0.0206	-0.00821±0.0203	0.004188±0.0327	-0.00473±0.0233	0.008786±0.0288
1024	-0.01133±0.0126	-0.00788±0.0256	0.006888±0.0209	-0.00781±0.0102	-0.00805±0.0132	0.001798±0.0127
IR	0.004842±0.0190	0.000124±0.0198	0.006295±0.0313	0.000204±0.0114	0.002154±0.0194	-0.00064±0.0194
2K	0.007458±0.0140	-0.00081±0.0144	0.007987±0.0312	0.004943±0.0131	0.001903±0.0157	-0.00258±0.0130
4K	0.004415±0.0095	0.004429±0.0087	-0.00274±0.0103	0.001072±0.0105	0.001042±0.0093	-0.00503±0.0105

Table 5. Pixel correlations of cipher images of different sizes in diagonal direction.

size	Proposed -ad	Proposed -4+4	Liu's[9]-ad	Liu's[9]-4+4	Teng's[15]-ad	Teng's[15]-4+4
256	0.001019±0.0028	0.000794±0.0019	0.000675±3E-03	0.000493±0.0026	0.000122±2E-03	-0.00097±0.0035
512	-0.00025±0.0017	-0.0003±0.0016	-0.00015±2E-03	0.000505±0.0013	-0.00011±1E-03	8.87E-05±0.0011
1024	-0.00026±0.0004	-0.00013±0.0002	-0.00337±3E-03	0.000220±0.0002	0.000129±5E-04	6.05E-05±0.0004
IR	-3.2E-05±0.0006	-0.00013±0.0005	-0.00729±5E-05	0.000132±0.0005	0.000222±6E-04	0.000347±0.0005
2K	8.10E-05±0.0004	-1.3E-05±0.0004	0.002000±2E-05	0.000164±0.0009	-0.00019±3E-04	-0.00014±0.0004
4K	-7.4E-05±0.0002	-2.4E-05±0.0002	-0.00189±0.0031	2.16E-05±0.0001	3.98E-05±0.0001	4.07E-05±0.0001

Tables 3, 4, and 5 respectively show the correlation coefficients of adjacent pixels in the horizontal, vertical, and diagonal directions of the cipher image obtained by the three algorithms under adaptive division and higher 4-bits and lower 4-bits division. It can be seen that the pixel correlation results of two divisions of the cipher image are not much different, and are close to 0 indicating that the adaptive division of the cipher image has sufficient security.

5.2.4. Differential attack analysis. Differential attack analysis, also known as plaintext sensitivity analysis, usually uses two indicators, Number of Pixel Change Rate(NPCR) and Unified Average Changing Intensity(UACI), to evaluate the anti-differential attack capability of image encryption algorithms[26]. NPCR is the ratio of the number of pixels which changes in the cipher image to the total number of pixels in the image, and UACI is the average intensity of the grayscale value difference between two cipher images. The calculation method is defined as follows:

$$D(i,j)=\begin{cases}0 & C_1(i,j)=C_2(i,j)\\ 1 & C_2(i,j)\neq C_1(i,j)\end{cases} \tag{14}$$

$$NPCR=\frac{\sum_{i,j}^{M,N}D(i,j)}{M\times N}\times 100\% \tag{15}$$

$$UACI=\frac{\sum_{i,j}^{M,N}\left|C_1(i,j)-C_2(i,j)\right|}{M\times N\times 255}\times 100\% \tag{16}$$

Among them, C_1、C_2 are the pixel matrices of the two cipher images and D is a two-dimensional matrix. D and C_1、C_2 have the same size. $C_1(i,j)$、$C_2(i,j)$ represent the pixel value of the position (i,j) in the two cipher images. M and N represent the length and width of the image respectively.

Table 6. NPCR results of diffident image encryption algorithms using different divisions.

size	Proposed -ad	Proposed -4+4	Liu's[9]-ad	Liu's[9]-4+4	Teng's[15]-ad	Teng's[15]-4+4
256	99.61802±0.0186	99.61792±0.0216	99.44244±0.2909	99.31559±1.0058	99.51945±0.2075	99.30959±1.0597
512	99.60765±0.0158	99.60580±0.0154	99.58870±0.1284	99.56258±0.0822	99.56830±0.0736	99.56494±0.0766
1024	99.60556±0.0072	99.61141±0.0065	99.57790±5E-01	99.60515±0.0084	99.60575±1E-03	99.60559±0.0038
IR	99.60912±0.0051	99.60954±0.0045	99.62906±9E-01	99.55357±0.1221	99.55297±1E-01	99.55285±0.1225
2K	99.60914±0.0029	99.60989±0.0032	99.47155±1E-01	99.6018±0.01516	99.60271±1E-02	99.60109±0.0173
4K	99.60979±0.0013	99.60974±0.0013	99.50619±0.3211	99.58006±0.0509	99.57966±0.0509	99.57950±0.0512

This paper calculates the NPCR and UACI of the cipher image obtained by the three algorithms under the adaptive division and higher 4-bits and lower 4-bits division, as shown in Table 6 and Table 7. We can observe from Table 6 that the NPCR of the encryption results of the three algorithms under adaptive division is closer to the ideal value of 99.6094% than the traditional higher 4-bits and lower 4-bits division. Table 7 also shows the UACI results of the encryption results of two divisions are not much different, and are all close to the ideal value of 33.4635%. However, Liu's[9] only scrambles the lower bit-planes so that the encryption result is too dependent on the randomness of the lower bit-planes and lower bit-planes obtained by adaptive division is generally more than higher bit-planes. In that case, the ideal value of UACI is lower than 33.4635%, so the UACI result of the encryption result under adaptive division is relatively small. The results show that the encryption algorithm proposed in this paper under adaptive division can effectively resist differential attacks.

Table 7. UACI results of diffident image encryption algorithms using different divisions.

size	Proposed -ad	Proposed -4+4	Liu's[9]-ad	Liu's[9]-4+4	Teng's[15]-ad	Teng's[15]-4+4
256	33.46621±0.1086	33.45184±0.0964	33.09462±1.0221	33.48655±0.0789	33.49528±0.2676	33.42294±0.0872
512	33.44568±0.0461	33.44838±0.0282	32.39709±0.5131	33.46085±0.0460	33.48166±0.2883	33.46481±0.0351
1024	33.44262±0.0094	33.45811±0.0174	32.62205±2.5193	33.44267±0.0182	33.50560±0.2303	33.46023±0.0108
IR	33.46141±0.0203	33.46201±0.0190	33.22703±1.1929	33.47400±0.0174	33.52549±0.5035	33.46416±0.0240
2K	33.46850±0.0129	33.46590±0.0139	32.80728±1.1314	33.46568±0.0145	33.48563±0.4162	33.45814±0.0143
4K	33.44139±0.1604	33.46427±0.0065	33.27393±0.5864	33.46429±0.0509	33.43651±0.1622	33.46132±0.0102

5.3. Efficiency analysis

The encryption scheme proposed in this article has a faster encryption speed because it has the following characteristics: 1) Compared with the traditional higher 4-bits and lower 4-bits encryption structure, the encryption scheme proposed in this article distinguishes the high significant bit-planes and the low significant bit-planes according to the actual situation of different images. The high significant bit-planes adopt a relatively high-complexity encryption method to ensure security, and the low significant bit-planes sacrifice a certain amount of safety and adopt a faster encryption algorithm. Compared with the traditional methods of using the same algorithm for all bit-planes, the scheme proposed in this paper saves certain computing resources in the part of low significant information and has higher efficiency; In selective encryption methods, the part that is not encrypted greatly increases the possibility of plaintext information leakage. Our scheme encrypts all plaintext information, which has higher security.

Table 8. Execution time of different image encryption algorithms using two different divisions. '-ad' represents that the encryption algorithm uses an adaptive division scheme, and '-4+4' represents that the encryption algorithm uses higher 4-bits and lower 4-bits division scheme. The time unit is seconds.

Algorithm	256×256	512×512	1024×1024	IR	2K	4K
Liu's[9]	0.508333±0.024	2.033854±0.062	8.229167±0.073	10.17012±0.359	22.10938±3.070	141.6761±2.170
Liu's[9]-ad	0.300000±0.116	0.756230±0.294	4.570313±0.290	5.774023±0.597	12.26445±2.242	83.84730±21.89
Teng's[15]	0.666994±0.397	2.297915±0.059	9.504384±0.094	12.15185±0.511	26.34693±3.692	104.45881±5.114
Teng's[15]-ad	0.486523±0.278	1.663111±0.255	6.990453± 0.607	9.089827±0.765	19.68961±3.121	81.56392±14.01
Proposed-4+4	0.841146±0.051	3.979539±0.147	20.34375±0.111	27.34355±2.551	78.79063±19.16	326.3594±3.050
Proposed-ad	0.494792±0.205	2.180804±0.691	8.473958±1.842	20.74941±15.80	36.30273±17.29	199.3551±31.81

This section selects two encryption algorithms based on higher 4-bits and lower 4-bits in Liu's[9] and Teng's[15] to conduct comparative experiments to verify the encryption efficiency of our encryption scheme as listed in Table 8. It can be seen that compared with the traditional encryption algorithms based on higher 4-bits and lower 4-bits, the encryption algorithm under the adaptive division scheme has higher efficiency. At the same time, Table 9 compares the encryption time of several image encryption algorithms[19][20][21] proposed in recent years using images of different sizes. We can observe that compared with other bit-level algorithms, our encryption scheme has higher efficiency.

Table 9. Execution time of several advanced image encryption algorithms using two. The time unit is seconds.

Algorithm	256×256	512×512	1024×1024
Chai's[19]	2.3545	10.8232	38.9186
Wang's[20]	2.0423	6.1855	-
Wang's[21]	0.6689	1.0671	11.5876
Proposed-ad	0.494792±0.205	2.180804±0.691	8.473958±1.842

6. Conclusion

According to the characteristic of higher bit-planes with more information and lower bit-planes with less information, we propose a method of adaptive bit-plane division in which the mutual information and JS divergence of adjacent bit-planes are calculated to measure the similarity, and then the higher bit-planes and lower bit-planes are divided. Based on this and two existing encryption algorithms, we also propose an image encryption scheme. The encryption scheme first effectively divides the image bit-planes into two parts adaptively, then selects a more secure encryption method for the higher bit-planes and a more efficient encryption method for the rest. Finally, the two parts of the encryption result are merged into a plane to obtain the final cipher image. After experimental simulations and comparative analysis of encryption efficiency, key space, information entropy, adjacent pixel correlation, NPCR and UACI, the scheme has higher efficiency and sufficient security.

7. References

[1] Wang X, Chen S, Zhang Y. A chaotic image encryption algorithm based on random dynamic mixing[J]. Optics and Lasers in Engineering,2011,181(6):1171-1186.
[2] Mansouri A, Wang X. A novel one-dimensional chaotic map generator and its application in a new index representation-based image encryption scheme[J]. Information Sciences, 2021,563:91-110.
[3] Zhou W, Wang X, Wang M, Li A. A new combination chaotic system and its application in a new Bit-level image encryption scheme[J]. Optics and Lasers in Engineering,2021,149.
[4] Teng L, Wang X, Yang F. Color image encryption based on cross 2D hyperchaotic map using combined cycle shift scrambling and selecting diffusion[J]. Nonliner Dynamics, 2021,105:1859-1876.
[5] Raza S, Stapute V. A novel bit permutation-based image encryption algorithm[J]. Nonliner Dynamics,2019,95(2):859-873.
[6] Zhu Z, Zhang W, Wong KW, Yu H. A chaos-based symmetric image encryption scheme using a bit-level permutation[J]. Information Sciences,2011,181(6):1171-1186.
[7] Zhang W, Wong KW, Yu H, Zhu Z. A symmetric color image encryption algorithm using the intrinsic features of bit distributions[J]. Communications in Nonlinear Science and Numerical Simulation,2013,18(3):584-600.
[8] Zhang Y, Wang X. A symmetric image encryption algorithm based on mixed linear–nonlinear coupled map lattice[J]. Information Sciences,2014,273:329-351.
[9] Liu J, Yang D. A digital image encryption algorithm based on bit-planes and an improved logistic map[J]. Multimedia Tools and Applications,2018,77(8):10217-10233.
[10] Lin Z, Ni X, Chen Z. Image encryption algorithm of adaptive parameter based on bitplane[J]. Computer Engineering and Design,2016, 37(03):597-601+642.
[11] Wang X, Luan D. A novel image encryption algorithm using chaos and reversible cellular automata[J]. Communications in Nonlinear Science and Numerical Simulation, 2013,18(11): 3075–3085.
[12] Li L, Yao Y, Chang X. Plaintext-dependent selective image encryption scheme based on chaotic maps and DNA coding[C]//International Conference on Dependable Systems and Their Applications(DSA), 2017,pp.57-65.
[13] Ren H, Dai L, Zhang J. A digital image encryption algorithm based on bit-plane transformation[J]. Computer Engineering,2013,39(06):185-189.
[14] Zhou Y, Cao W. Image encryption using binary bitplane[J].Signal Processing, 2014,100:197-207.
[15] Teng L, Wang X. A bit-level image encryption algorithm based on spatiotemporal chaotic system and self-adaptive[J].Optics Commu-nications,2012,285(20):4048-4054.
[16] Cover T, Thomas J. Elements of Information Theory, Second Edition[M]. John wiley &Sons,2006:13-55.
[17] Majtey A, Lamberti P. Jenson-Shannon divergence as a measure of distinguishability between mixed quantum states[J]. Physical Review A,2005,72(5):762-776.

[18] Hua Z, Jin F, Xu B. Logistic-Sine-coupling map for image encryption[J]. Signal Processing,2018,149:148-161.
[19] Chai X, Gan Z, Yuan K, Chen Y.A novel image encryption scheme based on DNA sequence operations and chaotic systems[J]. Neural Computing&Application,2019,31: 219-237.
[20] Wang X, Chen S, Zhang Y. A chaotic image encryption algorithm based on random dynamic mixing[J].Optics and Laser Techn-ology,2021,138.
[21] Wang X, Guan N, Zhao H. A new image encryption scheme based on coupling map lattices with mixed multi-chaos[J]. Scientific Reports,2020,10(1).
[22] Alvarez G, Li S. Some basic crypto-graphic requirements for chaos-based cryptosystems[J]. International Journal of Bifurcation and Chaos, 2006,16(8): 2129-2151
[23] Wu Y, Zhou Y. Local Shannon entropy measure with statistical tests for image randomness[J]. International Journal of Bifurcation and Chaos, 2013,222:323-342.
[24] Sha K, Wang Z. Research On Image Encryption Algorithm based on Chaos[J].Audio Engineering, 2019,43(01):64-67.
[25] Wu Y, Zhou Y. Design of image cipher using latin squares[J]. Information Sciences.2014.264:317-339.
[26] Wu Y, Noonan J. NPCR and UACI Randomness Tests for Image Encryption[J].Cyber Journals: Multidiplinary Journals in ence and Technology, Journal of Selected Areas in Telecommunications, 2011.

Acknowledgments
This research was supported in part by the National Key Research and Development Program of China (Grant No. 2018YFC0407105), the Water Conservancy Science and Technology Project of Jiangsu Province of China (Grant No. 2021080), Project of Water Science & Technology of Jiangsu Province (Grant No. 2021080).

Detection of Self-explosive Insulators in Aerial Images Based on Improved YOLO v4

Li Yi[1], Li Jinguo[1], Zhai Yongjie[2] and Meng Ping[1]

[1] College of Computer Science and Technology, Shanghai University of Electric Power, Shanghai, 200000, China
[2] School of Control and Computer Engineering, North China Electric Power University, Hebei, Baoding, 071000, China

YukoAmamiya@mail.shiep.edu.cn

Abstract. In view of the low detection efficiency of traditional aerial image-based self-explosive insulators and the need to manually extract the features of self-explosive insulators, a self-explosive insulator detection algorithm based on the improved YOLO v4 is proposed: We use the hybrid data augmentation method to increase the number of defective samples of self-explosive insulators and increase the diversity of self-explosive insulators to improve the detection accuracy of the algorithm. In addition, the channel attention mechanism is combined with the YOLO v4 algorithm to improve the feature extraction capability of algorithms to self-explosion insulators. The proposed algorithm is verified using the expanded self-explosive insulator dataset and the traffic signs dataset TT100K. The experimental results show that the proposed algorithm has a mAP of 92.0 % on the self-explosive insulator dataset and has an F1-score of 78.70% on TT100K dataset, which has positive significance for the research on self-explosive insulator detection.

1. Introduction

As an important component of transmission lines, insulators play the role of electrical insulation and mechanical fixation. The integrity of the insulators structure has a significant impact on the safety of power grid operation [1]. Since insulators work in outdoor environments, the failures such as breakage and bursting of insulator strings are unavoidable [2]. Therefore, it is essential to accurately detect the self-explosion area in insulator strings to ensure the safety of transmission lines.

The existing traffic sign detection algorithms are divided into two classes:

(1) The first class of insulator detection algorithms is based on traditional machine learning methods. The methods used to extract insulator features include threshold segmentation according to the color feature of the insulator and edge extraction according to the contour feature of the insulator [3]-[5]. However, traditional methods are mainly used to identify insulators in special scenes, so their applicability is not satisfactory.

(2) The second class of insulator detection algorithms is based on a convolutional neural network, and the two most representative insulators detection algorithms are one-stage insulator detection algorithm and two-stage insulators detection algorithm [6]-[7].

To achieve real-time and high-accurate detection of self-explosive insulators, we propose a one-stage self-explosive insulator detection algorithm, based on the one-stage target detection algorithm

YOLO V4 [8] and combined with channel attention mechanism (Squeeze-and-Excitation Networks) [9]. By enhancing the feature of self-explosive insulators and suppressing irrelevant background information, the detection speed and accuracy of self-explosive insulators are optimized.

The second part introduces the related research on insulators detection. The third part and the fourth part introduces this experiments in detail, and the fifth part analyses the experimental results of the proposed algorithms.

2. Related works

In this section, we will summarize and analysis prior works concerning general object detection and insulators detection.

2.1. General object detection

In 2012, with the proposed convolutional neural network algorithm, computer vision tasks entered the era of deep learning. Girshick R et al. proposed convolutional neural networks (CNN), which extract target features to complete the object detection task and successfully applied the deep learning method to the target detection task for the first time. In 2015, Girshick R et al. proposed a Fast R-CNN algorithm [10] as the R-CNN algorithm is time-consuming in the region proposal stage and used the ROI pooling layer to speed up feature extraction. In the same year, He Kaiming et al. improved the Fast R-CNN algorithm and proposed a Faster R-CNN algorithm [11] where RPN was used to improve the detection speed and accuracy. Fast R- CNN and Faster R-CNN are classic two-stage target detection algorithms with excellent, but the detection speed is slow, and the balance between detection speed and accuracy cannot be achieved.

The one-stage object detection algorithm can balance detection speed and accuracy, and the representative ones are the YOLO (You Only Look Once) [12]-[14] series algorithm proposed by Joseph Redmon et al. and SSD (Single Shot MultiBox Detector) [15] algorithm proposed by Liu W et al. Because of the smaller size and less information of small objects, and the network structure limitation of YOLO algorithm, YOLO and YOLO 9000 algorithms cannot be applied to small object detection tasks. In 2018, Joseph Redmon proposed the YOLO V3 algorithm by reconstructing the structure of the YOLO algorithm. It successfully solved the above problem of the YOLO series algorithm and improved the detection accuracy. In 2020, Alexey Bochkovskiy et al. proposed the YOLO V4 algorithm, the latest algorithm of the YOLO series algorithm, and its detection accuracy is greatly improved, and the detection speed is the same with YOLO V3, and the mAP of YOLO V4 in the coco dataset is 43.5%.

2.2. Insulator detection

In 2019, Cheng et al. [16] used the Faster R-CNN algorithm to detect insulators in aerial images after parameter adjustments and optimizations. The detection accuracy of the Faster R-CNN algorithm for insulators in aerial images reached 90.5 %. In 2019, Zhao et al. [17] added an attention module to the Faster R-CNN algorithm to improve its insulator feature extraction ability, and the detection accuracy of insulators is improved to 94.3% by fusing the proposed frames based on the attention mechanism in the network. In 2020, Yang et al. [18] added the Inception-Resnet module to the YOLO v3 algorithm to increase its ability to extract insulator features while replacing the loss function of YOLO v3. The detection accuracy of insulators is 71.93%. In 2021, Gao et al. [19] used the YOLOv4 algorithm to improve the detection accuracy of burst insulators by re-clustering the prior frame, with the detection accuracy for burst insulators being 91.8%. So, we decided to use YOLO V4 algorithm as the basic algorithm of our method and improved the performance of YOLO V4 in our task.

3. Hybrid data augmentation methods

The self-explosive insulator detection algorithm based on deep learning requires a large number of self-explosive insulator samples for training[20]-[21]. However, the proportion of positive and negative samples is seriously imbalanced in the aerial image dataset, which makes it difficult for the algorithm to accurately extract the feature of the self-explosive insulators. Thus we proposed a hybrid data augmentation method, which effectively expands defective samples through supervised and unsupervised data augmentation methods to balance the defective sample dataset.

3.1. Supervised data augmentation methods

In the data augmentation methods, supervised data augmentation methods are divided into two classes: geometric transformation data augmentation methods and color transformation data augmentation methods. In this paper, the rotation, image brightness changing and image fogging methods in the supervised data augmentation method are selected to simulate the insulator images taken by drones at different angles, different times of the day and different weather conditions. This move is to increase the diversity of self-explosive insulator data in the dataset. Figure 1 shows samples of self-explosive insulators obtained using the supervised data augmentation methods in this paper.

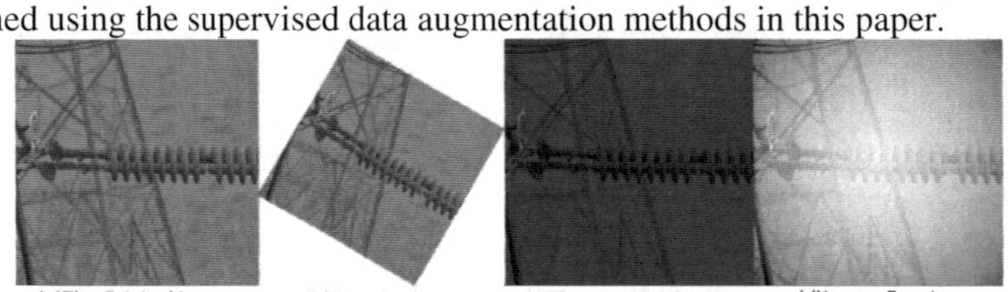

(a)The Original Image (b)Rotating Image (c)Changing Brightness (d)Image Fogging

Figure 1. Different types of supervised data augmentation methods.

3.2. Unsupervised data augmentation methods

The unsupervised data augmentation method uses a game between the generator and the discriminator in the unsupervised data augmentation algorithm to make the algorithm learn the target distribution in the training set. Then a false image consistent with the distribution of the real data is obtained by feeding noise into the generator. Finally, the discriminator will judge the authenticity of the image generated by the generator, the unsupervised data augmentation algorithms including GAN (Generative Adversarial Network), DCGAN (Deep Convolution Generative Adversarial Networks) and CycleGAN [22]-[24].

GAN and DCGAN require a large amount of data and high computational resources for more complex image generation tasks. However, CycleGAN does not require a large number of one-to-one sample training and realizes the transfer of image style by inputting the generated original and target images into the algorithm to learn the data distribution. Therefore, the CycleGAN algorithm was used to generate simulated self-explosive insulator images to increase the diversity of self-explosive insulator data in the dataset.

Step 1: In this paper, we selected some self-explosive insulator images of the dataset as the data to be input to the CycleGAN algorithm for training. The selected images are then blurred in the region where the self-explosive insulators are located. Finally, the blurred images are used as test A and the original images are used as test B. Then test A and test B are input into the CycleGAN algorithm to learn the distribution of the self-explosive insulators.

Step 2: After the CycleGAN algorithm was used to learn the distribution of self-explosive insulators in the images, another part of the images is selected from the dataset and the self-explosive insulator areas in this part of the image data are also blurred and then input into the CycleGAN algorithm as Test A to obtain a simulated self-explosive insulator image. Figure 2 shows the process of using the CycleGAN algorithm to generate simulate self-explosive insulator images.

In Figure 2, G represents the generator in the Cycle GAN algorithm; F represents the discriminator in the Cycle GAN algorithm; D_X is used to determine whether $F(Y)$ is the real data and D_Y is used to determine whether $G(X)$ is the real data. The loss function used by CycleGAN in learning the distribution of targets is shown in Equation 1:

$$L(G,F,D_X,D_Y) = L_{GAN}(G,D_Y,X,Y) + L_{GAN}(F,D_X,Y,X) + \lambda L_{cyc}(G,F) \tag{1}$$

where $L_{GAN}(G,D_Y,X,Y)$ and $L_{GAN}(F,D_X,Y,X)$ represent the adversarial loss function in the CycleGAN algorithm and $L_{cyc}(G,F)$ represents the cyclic consistency loss function in the CycleGAN algorithm.

Finally, we selected some simulated images generated in Step 2 and added them to the dataset to increase the diversity of the self-explosive insulators.

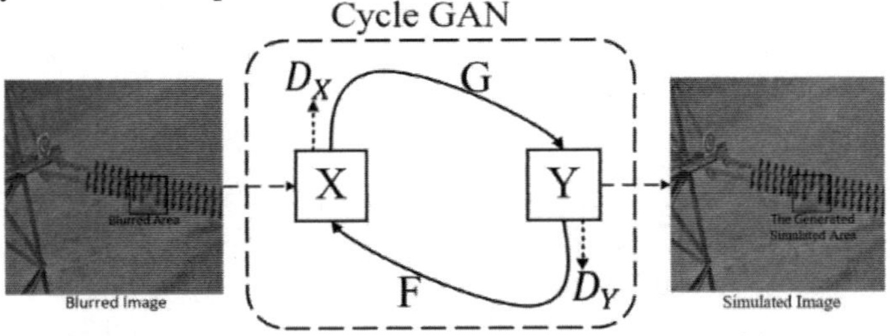

Figure 2. The generation process of the simulated image.

4. The improved YOLO V4

4.1. Preliminaries of YOLO V4

The YOLO v4 algorithm proposed by Alexey Bochkovskiy is an improved target detection algorithm developed from YOLO v3. The YOLO v4 algorithm divides the images which are input into the network by using several grids and then uses candidate boxes of different sizes on each grid to predict the class of objects in the grid. Its detection speed remains essentially the same as that of the YOLO v3 algorithm, but the detection accuracy is improved compared to the YOLO v3 algorithm. Therefore, we decided to use the YOLO v4 algorithm to detect self-explosive insulators.

The YOLO v4 algorithm uses the CSPDarknet53 network as the feature extraction network of the algorithm. CSPDarknet53 network has a residual structure consists of 1×1 and 3×3 sized convolutional layers, and the residual structure used in the CSPDarknet 53 network is shown in Figure 3.

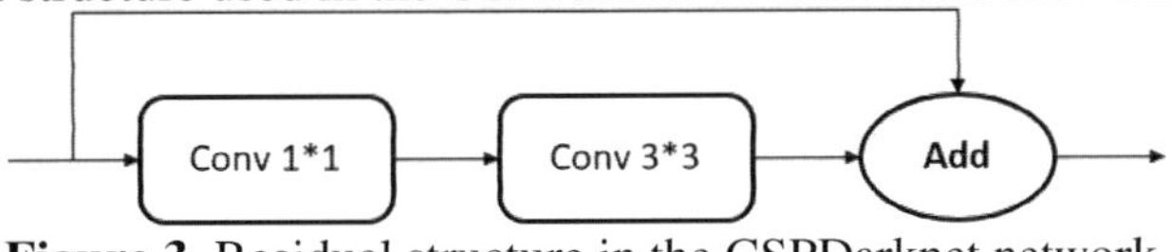

Figure 3. Residual structure in the CSPDarknet network.

Compared with the Darknet53 network, the CSPDarknet53 network adds a Cross Stage Partial structure to each convolutional block to reduce the computational volume, improves the feature extraction capability of the network, and replaces the Leaky_ReLU activation function with the Mish activation function. The CSPDarknet53 network is easier to train than the Darknet53 network and can better obtain small target features.

Compared with the YOLO v3 algorithm that uses FPN as the neck of the algorithm, the YOLO v4 algorithm uses SPP [25] and FPN+PAN as the neck of the algorithm. The combination of FPN+PAN can enhance the feature extraction network capability to extract features, while SPP can improve the target locating ability of the feature extraction network. The neck structure in the YOLO v4 algorithm is shown in Figure 4.

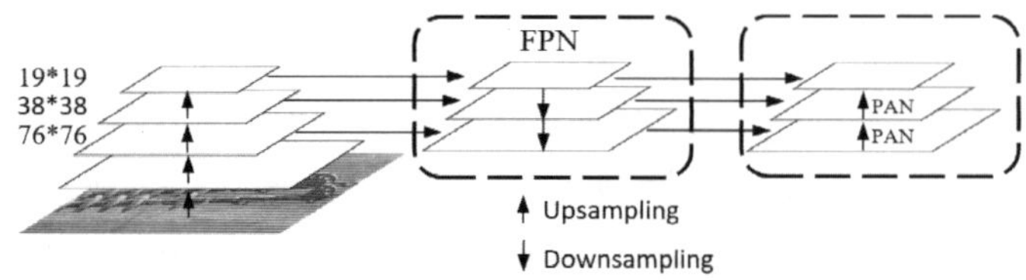

Figure 4. Neck structure of the YOLO v4 algorithm.

The YOLO v4 algorithm improves the detection capability compared to the YOLO v3 algorithm, but there is a large amount of residual structure in the YOLO v4 algorithm. The use of jump connections in the residual structure can increase network depth and make the network easier to be optimized [26]. Therefore, we decided to improve the YOLO v4 algorithm to enhance the detection of self-explosive insulators in the aerial images.

4.2. The improved YOLO V4

SENet is a simple but effective channel attention mechanism module. By embedding SENet into the feature extraction network of YOLO v4, the receptive field of the feature extraction network can be expanded, and its ability to extract self-explosive insulator features can be improved. SENet itself does not bring a lot of computation to the network, causing too much negative impact on the network, so we decided to add SENet to the YOLO v4 algorithm to improve the detection accuracy of self-explosive insulators in aerial images.

The core operational steps of SENet are the squeeze and the excitation operations. The squeeze operation is performed by compressing the input feature map U_k into a real number Y_k by using global average pooling. Y_k contains all the information of the feature map U_k and Y_k is calculated as shown in Equation (2):

$$Y_K = \frac{1}{W \times H} \sum_{i=1}^{W} \sum_{j=1}^{H} U_k(i,j) \qquad k = 1,2,...,C \tag{2}$$

where U_k represents the feature map obtained after convolution, C represents the number of channels of U , $W \times H$ and represents the spatial dimension of U .

Then we used the excitation operation to capture the correlation between the channels in the feature map U_k . By feeding the global feature Y_k into the excitation operation, the weight of each channel of the feature map U_k can be obtained. Then the weights of the feature map U_k are multiplied with the values of each channel of the feature map Y_k to enhance the self-explosive insulator information and suppress irrelevant background information in the feature map U_k . The structure of the SENet is shown in Figure 5, where Y_k is the feature map obtained by augmenting the self-explosive insulator information in U_k .

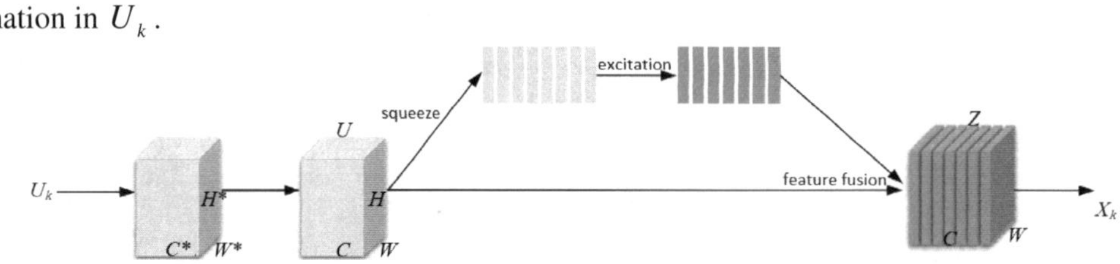

Figure 5. Structure of SENet.

Since the YOLO v4 algorithm network can be divided into two parts: the feature extraction network and the detection network, SENet is added to the feature extraction and the detection networks of the YOLO v4 algorithm network to improve the detection capability of the algorithm for self-explosive insulators.

Step 1: The feature extraction network of the YOLO v4 algorithm uses the CSPDarknet 53 network as the backbone network of the algorithm. Because there are a large number of residual structures in the backbone network of the YOLO v4 algorithm, the SENet is combined with the residual structures in the network to form a new attentional residual structure by adding SENet: since the residual structure of the backbone network consists of two convolutional and residual layers with a size of 1×1 and 3×3, we add the SENet to the last convolutional and residual layers of the residual structure to form a new attentional residual structure. Compared with the original residual structure, the improved one enhances the feature extraction ability of the backbone network to self-explosive insulators without much increase in computation through adding of the channel attention mechanism.

Step 2: The detection network of the YOLO v4 algorithm uses feature fusion before YOLO Head to obtain a deep layer feature map with strong semantic information and a shallow layer feature map with target location information. Thus, we can get a feature map with strong semantic information and target location information. After the fusion of the feature maps, the algorithm adds some convolutional layers after the feature fusion layers to achieve a better fusion of the feature maps. Therefore, SENet is added between the penultimate convolutional layer and the penultimate third convolutional layer before the YOLO Head, which can enhance the feature information and location information of the self-explosive insulators in the feature map after feature fusion and improve the detection capability of the YOLO head detection module for self-explosive insulators.

By using SENet to improve the YOLO v4 algorithm, we enhanced its detection capability for self-explosive insulators. Figure 6 shows the structure of the improved YOLO v4 algorithm.

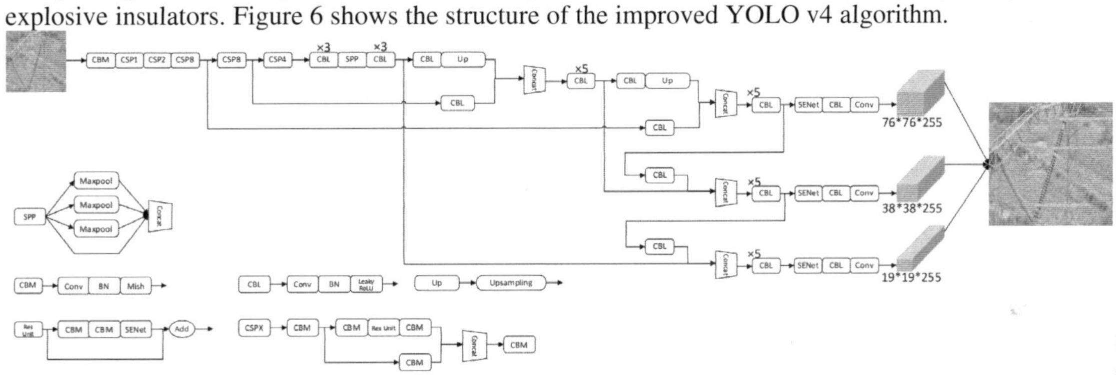

Figure 6. Structure of improved YOLO v4.

5. Experimental results and analysis

5.1. Datasets and evaluation metrics

The self-explosive insulator images used in this experiment were taken by drones during the power inspection in a city power supply bureau, with a resolution of 2144×1424 and JPG image format. The insulator dataset used contains different types of self- explosive insulators such as glass and ceramic. The resolution of self-explosive insulators images was adjusted to 1024×1024 to better train the algorithm, and the dataset was expanded using supervised and unsupervised data augmentation methods to increase the diversity of the self-detonation insulator dataset. A total of 1616 images of self-explosive insulators were obtained, of which 1152 images were randomly selected as the training set and 464 images were used as the test set. To better validate the effectiveness of our method, the TT100K dataset produced by Wei Zhu et al. [27] and published by Tsinghua University with Tencent, was chosen. The images in TT100K datasets have complex background information, and traffic signs account for a small proportion of the images, which makes it difficult for the algorithm to detect them in TT100K datasets. Therefore, we consider that experiments on TT100K dataset can better validate our method. There are a total of 9176 images with a resolution of 2048×2048 in the TT100K dataset, of which 6105 images are used for training images and the other 3071 images for testing images.

The common evaluation metrics for algorithms include precision, recall and F1-score. Precision is a measure of the proportion of samples judged to be positive by the algorithm out of the total number of positive samples. Recall is a measure of the proportion of positive samples determined by the algorithm to the total number of samples, and F1-score is calculated by combining precision and recall. Because this paper aims to detect self-explosive insulators in insulator strings, and mAP is an average of the multiple accuracies obtained from the calculations, we considered that the use of mAP on the self-explosive insulator dataset can help to measure the performance of our proposed algorithm better. For most researchers using the TT100K dataset, the F1-score is used as the evaluation metric [28]-[35]. Thus, we also adopted it in the same way when validating the proposed algorithm on the TT100K dataset.

5.2. Experimental environment and parameter settings

We use the Pytorch deep learning framework to realize the improved YOLO v4 algorithm proposed in this paper, and the experimental environment is shown in Table 1. In training the improved YOLO v4 algorithm using the self-explosive insulator dataset and the TT100K dataset, the image sizes that are input to algorithm were set to 1024×1024 and 2048×2048 respectively. The values of the hyperparameters set in this algorithm are shown in Table 2. Figure 7 shows the results obtained by setting different Initial_learning_rate when training the improved YOLO v4 algorithm using the self-explosive insulator dataset. As can be seen in the left panel of Figure 7, when setting the initial learning rate of the algorithm to 0.01 and the results will be unstable in the early stages of training. As can be seen in the right panel of Figure 7, when the initial learning rate is set to 0.001, the algorithm does not show unstable training results, and the final training result is better than that of the initial learning rate of 0.01.

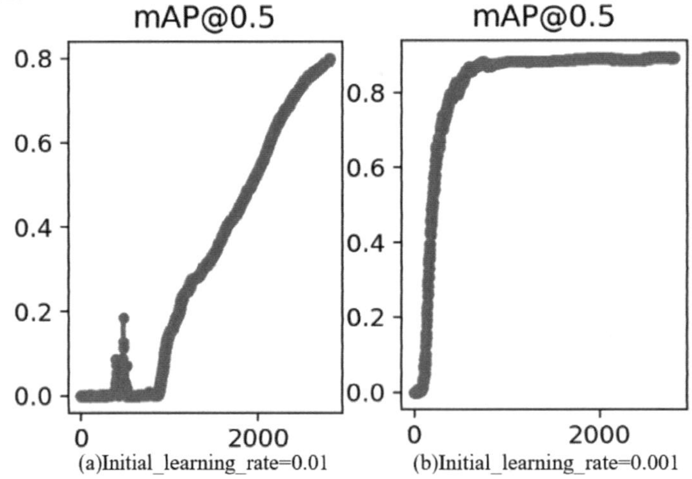

Figure 7. The training effects of different initial_learning_rates in the self-exploding insulator datasets

Table 1. System environment

System environment	Hardware environment
Ubuntu Linux	CPU: Intel i9-9900K RAM:32GB GPU: NVIDIA RTX 2080 8G

Table 2. Hyper parameter setting

Parameters setting	Parameters values
Batch_size	2

Initial_learning_rate	0.001
Final_learning_rate	0.0005

5.3. Experimental results and analysis

To verify the effectiveness of our method, the YOLO v4 algorithm was used on the self-explosive insulator dataset produced in this paper and the TT100K dataset, respectively to compare the performance of our method. Figure 8 shows the detection results of our method on the self-explosive insulator dataset, and Figure 9 shows those on the TT100K dataset.

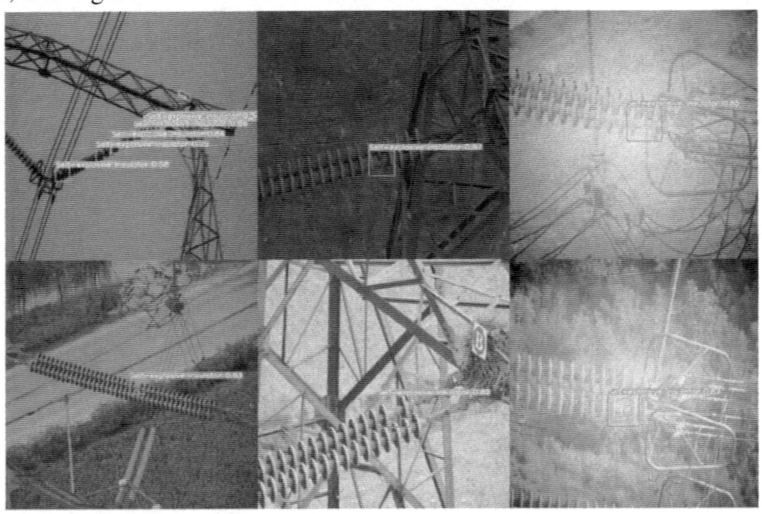

Figure 8. The detection results of our algorithm in the self-explosive insulators datasets.

As the result shows that the mAP of the YOLO v4 algorithm in the self-explosive insulator dataset is 86.80%, while that of our method is 92.00%. Thus, compared to the YOLO v4 algorithm our method has a 5.20% improvement in mAP. The F1-score of the YOLO v4 algorithm in the TT100K dataset is 76.20%, while that of our method is 78.70%, with a 2.50% improvement in the F1-score compared to the YOLO v4 algorithm, this result is due to the addition of SENet to the YOLO v4 algorithm's backbone network and YOLO Head, which improves the algorithm's ability to extract features for self-explosive insulators and traffic signs. At the same time, we used the methods of Cheng et al. [16] and Liu et al. [28] in the self-explosive insulator dataset and TT100K dataset, respectively, to compare with our method and the mAP obtained from our method and different method tested in the self-explosive insulator dataset are shown in Table 3, and the F1-score for each category tested in the TT100K dataset are shown in Table 4.

Figure 9. The detection results of our algorithm in the TT100K datasets.

Table 3. Detection results of different methods in the self-explosive insulator datasets

Method	mAP/%
Faster R-CNN [16]	78.84%
SSD	88.50%
YOLO v4	86.80%
Ours	92.00%

Table 4. Detection results of different methods in the TT100K datasets

Method(F1/%)	all	i2	i4	i5	il100	il60	il80	io	ip
FR [28]	62	53	57	60	55	71	72	66	51
SSD [28]	59	48	56	57	51	70	68	67	50
YOLO V4	76	74	88	90	90	87	78	74	85
Ours	79	81	90	91	87	89	83	76	86

Method(F1/%)	p10	p11	p12	p19	p23	p26	p27	p3	p5
FR [28]	54	49	58	67	71	63	82	52	71
SSD [28]	56	48	54	61	65	61	77	55	77
YOLO V4	75	84	61	74	88	83	96	75	88
Ours	80	87	69	74	90	86	93	74	87

Method(F1/%)	p6	pg	ph4	ph4.5	ph5	pl100	pl120	pl20	pl30
FR [28]	66	84	64	69	50	78	73	49	53
SSD [28]	66	76	60	69	50	77	69	50	77
YOLO V4	57	90	55	77	62	91	87	66	80

Ours	64	93	66	82	92	94	94	74	82

Method(F1/%)	pl40	pl5	pl50	pl60	pl70	pl80	pm20	pm30	pm55
FR [28]	61	61	50	60	67	61	63	65	69
SSD [28]	69	57	54	57	61	58	63	63	66
YOLO V4	78	78	74	78	74	79	64	66	70
Ours	80	83	79	81	76	83	63	67	64

Method(F1/%)	pn	pne	po	pr40	w13	w32	w55	w57	w59
FR [28]	63	65	45	85	47	58	51	60	53
SSD [28]	56	64	46	79	51	55	57	60	58
YOLO V4	90	86	70	68	67	68	70	83	79
Ours	90	89	71	82	55	82	70	80	44

Method(F1/%)	wo
FR [28]	45
SSD [28]	40
YOLO V4	48
Ours	44

Another evaluation metric used to compare our method with other different methods is the detection speed used in the self-explosive insulator dataset and the TT100K dataset, which is shown in Table 5 and that in the TT100K dataset is shown in Table 6. Table 5 and Table 6 show that the detection speed of our method is slower than that of the YOLO v4 algorithm since our method is improved by introducing multiple SENet to the network to improve the detection accuracy by enhancing the target features. Due to the addition of SENet, our method requires a lot of squeezing and excitation operations compared to the YOLO v4 algorithm, which increases the amount of computation to a certain extent and the detection speed is decrease.

Table 5. Detection speed of different methods in the self-explosive insulator datasets

Method	Speed/ms
Faster R-CNN [16]	142.1ms
SSD	12.0ms
YOLO v4	27.6ms
Ours	29.1ms

Table 6. Detection speed of different methods in the TT100K datasets

Method	Speed/%
FR [28]	154.3ms
SSD [28]	13.6ms
YOLO v4	28.2ms
Ours	29.1ms

It can be seen from Figure 8 that our method can detect the self-explosive insulators in the image when there are complex backgrounds and multiple self-explosive insulators in the image. Figure 9 shows that our method has a better detection ability of traffic signs in complex backgrounds. This paper also conducts experimental tests on adding channel attention mechanism SENet to different structures of the YOLO v4 algorithm. The result shows that the effect of adding SENet to the YOLO v4 algorithm's backbone network and YOLO Head proposed in this paper is better than that of SENet alone added to the backbone network and YOLO Head. Table 7 shows the results obtained in this paper by adding SENet to different positions in the YOLO v4 algorithm. As can be seen from Table 7, when using the Residual Block in the SENet backbone network and combining it with the YOLO Head will improve the ability of the backbone network to extract self-explosive insulators, which in turn will improve the ability of the backbone network to extract self-explosive insulator features, and when using the SENet in combination with the YOLO Head will improve the accuracy of the algorithm's detection module to detect and locate self-explosive insulators. The addition of SENet to both the algorithm's backbone network and YOLO Head improves the ability of the YOLO v4 algorithm to detect and locate self-explosive insulators in aerial images. We use the YOLO v4 algorithm and the improved YOLO v4 algorithm to make statistics on the detection results obtained using the self-explosive insulators dataset produced in this paper. The statistical results in Table 8 show that our method detected more self-explosive insulators than the YOLO v4 algorithm. As can be seen from Table 8, our method is able to detect more self-explosive insulators in self-explosive test set compared to the original YOLO v4 algorithm, thus our method is better able to achieve the detection of self-exploding insulators in aerial images. Therefore, the improved YOLO v4 algorithm proposed in this paper is better than the YOLO v4 algorithm in general.

Table 7. Detection results of different addition methods of SENet in the self-explosive insulator datasets

Method	mAP/%
YOLO v4	86.80%
YOLO v4(Backbone)	87.20%
YOLO v4(Head)	88.40%
Ours	92.00%

Table 8. Statistical results

Method	Defects	Positive	Negative
YOLO v4	575	463	112
Ours	575	476	99

6. Conclusion

In this paper, an improved YOLO v4 self-explosive insulator detection algorithm is proposed and we produced a self-explosive insulators dataset based on insulator images taken by drones during power inspection in a city power supply bureau. The experiments are conducted on the dataset using the improved YOLO v4 algorithm. The results show that our method has a mAP of 92.0% and a detection speed of 29.1ms, which are better than YOLO v4, SSD and Faster R-CNN algorithm in terms of accuracy, and can detect self-explosive insulators in aerial images with complex backgrounds and multiple self-explosive insulators. However, there is still some room for improvement in the detection speed of our method.

7. Reference

[1] Zhang Qian, Wang Jianping, Li Weitao. Insulator State Detection of Convolutional Neural Networks Based on Feedback Mechanism[J]. Transactions of China Electrotechnical Society, 2019,34(16): 3311-3321 (in Chinese).

[2] Lv Zhining. A survey of common faults analysis and detection methods for transmission line[J]. Automation & Instrumentation, 2020(01):161-164+168(in Chinese).
[3] Lin Jucai, Han Jun, Chen Fangming, Xu Xiong, Wang Yaping. Defects Detection of Glass Insulator Based on Color Image[J]. Power System Technology, 2011,35(01):127-133(in Chinese).
[4] Zhao Zhenbing, Jin Sixin, Liu Yachun. Aerial insulator image edge extraction method based on NSCT[J]. Chinese Journal of Scientific Instrument, 2012,33(09):2045-2052(in Chinese).
[5] Yan Kai, Xiong Ying. Study on the method of aerial insulator image detection and recognition[J]. Guizhou Electric Power Technology, 2015,18(10):51-53(in Chinese).
[6] Yi Jiyu, Chen Cifa, Gong guoqiang. Aerial Insulator Detection of Transmission Line Based on Improved Faster RCNN[J]. Computer Engineering,2021,47(06):292-298+304.DOI:10.19678/j.issn.1000-3428.0059872(in Chinese).
[7] He Wenqi, Li Jian, Zhao Wenhao. Insulator Anomaly Detection Based on an Improved Faster RCNN[J]. ZheJiang Electric Power, 2021,40(08):40-46.DOI:10.19585/j.zjdl.202108006(in Chinese).
[8] Alexey B, Chien-Yao Wang, Hong-Yuan, et al. YOLOv4: Optimal Speed and Accuracy of Object Detection[DB/OL]. (2020-4-24)[2020-09-21].
[9] Jie H, Li S, Gang S, et al. Squeeze-and-Excitation Net-works[J]. IEEE Transactions on Pattern Analysis and Machine Intelligence, 2017, PP(99).
[10] GIRSHICK R. Fast R - CNN [C]// Proceedings of Interna-tional Conference on Computer Vision. Santiago: IEEE, 2015: 1440-1448959.
[11] S. Ren, K. He, R. Girshick and J. Sun, "Faster R-CNN: Towards Real-Time Object Detection with Region Pro-posal Networks," in IEEE Transactions on Pattern Analysis and Machine Intelligence, vol. 39, no. 6, pp. 1137-1149, 1 June 2017, doi: 10.1109/TPAMI.2016.2577031.
[12] J. Redmon, S. Divvala, R. Girshick and A. Farhadi, "You Only Look Once: Unified, Real-Time Object Detection," 2016 IEEE Conference on Computer Vision and Pattern Recognition (CVPR), 2016, pp. 779-788, doi: 10.1109/CVPR.2016.91.
[13] J. Redmon and A. Farhadi, "YOLO9000: Better, Faster, Stronger," 2017 IEEE Conference on Computer Vision and Pattern Recognition (CVPR), 2017, pp. 6517-6525, doi: 10.1109/CVPR.2017.690. Conf. Circuits and Systems Theory, New York, 1994, pp. 8–16.
[14] Redmon J, Farhadi A. YOLOv3: An Incremental Im-provement[J]. arXiv E-prints, 2018.
[15] Wei Liu, Dragomir Anguelov, Dumitru Erhan, Christian Szegedy, Scott Reed, Cheng Yang Fu, and Alexander C. 554 Berg, "Ssd: Single shot multibox detector, " pp. 21-37, 2016.
[16] Cheng Haiyan, Zhai Yongjie, Chen Rui. Faster R-CNN based recognition of insulators in aerial images[J]. Modern Electronic Technology, 2019,42(02):98-102(in Chinese).
[17] Zhao Wenqing, Chen Xingfu, Zhao Zhenbing, Zhai Yongjie. Insulator recognition based on attention mechanism and Faster RCNN[J]. CAAI Transactions on Intelligent Systems, 2020,15(01):92-98(in Chinese).
[18] Yang Yanfei, Cao Yang. Detection of Glass Insulator in Images Taken by Drones Based on Improved YOLOv3[J/OL]. Computer Engineering and Applications:1-11[2021-05-28] (in Chinese).
[19] Gao Jianchen, Zhang Jiahong, Li Yingna, Li Chuan. Research on insulator burst fault identification based on YOLOv4[J/OL]. Laser & Optoelectronics Progress:1-13[2021-05-28] (in Chinese).
[20] Edisanter Lo, "Target Detection Algorithms in Hyperspectral Imaging Based on Discriminant Analysis," Journal of Image and Graphics, Vol. 7, No. 4, pp. 140-144, December 2019. doi: 10.18178/joig.7.4.140-144
[21] Teow Wee Teo and Mohd Zaid Abdullah, "Solar Cell Micro-Crack Detection Using Localised Texture Analysis," Journal of Image and Graphics, Vol. 6, No. 1, pp. 54-58, June 2018. doi:10.18178/joig.6.1.54-58

[22] Goodfellow, I., Pouget-Abadie, J., Mirza, M., Xu, B., Warde-Farley, D., Ozair,S., Courville, A., Bengio, Y.: Generative adversarial nets. In: Advances in neural in-formation processing systems. pp. 2672–2680 (2014).

[23] A. Radford, L. Metz, and S. Chintala, "Unsupervised representation learning with deep convolutional generative adversarial networks," 4th Int. Conf. Learn. Represent. ICLR 2016 - Conf. Track Proc., pp. 1– 16,2016.

[24] J.-Y. Zhu, T. Park, P. Isola, and A. A. Efros, ''Unpaired image-to-image translation using cycle-consistent adversarial networks,'' in Proc. IEEE Int. Conf. Comput. Vis., Oct. 2017, pp. 2223–2232.

[25] K. He, X. Zhang, S. Ren and J. Sun, "Spatial Pyramid Pooling in Deep Convolutional Networks for Visual Recognition," in IEEE Transactions on Pattern Analysis and Machine Intelligence, vol. 37, no. 9, pp. 1904-1916, 1 Sept. 2015, doi: 10.1109/TPAMI.2015.2389824.

[26] K. He, X. Zhang, S. Ren and J. Sun, "Deep Residual Learning for Image Recognition," 2016 IEEE Conference on Computer Vision and Pattern Recognition (CVPR), 2016, pp. 770-778, doi: 10.1109/CVPR.2016.90.

[27] Z. Zhu, D. Liang, S. Zhang, X. Huang, B. Li, and S. Hu, ''Traffic-sign detection and classification in the wild,'' in Proc. IEEE Conf. Comput. VisPattern Recognit. (CVPR), Jun. 2016, pp. 2110–2118.

[28] L. Wei, C. Xu, S. Li and X. Tu, "Traffic Sign Detection and Recognition Using Novel Center-Point Estimation and Local Features," in IEEE Access, vol. 8, pp. 83611-83621, 2020, doi: 10.1109/ACCESS.2020.2991195.

[29] Q. Tang, G. Cao and K. -H. Jo, "Integrated Feature Pyramid Network With Feature Aggregation for Traffic Sign Detection," in IEEE Access, vol. 9, pp. 117784-117794, 2021, doi: 10.1109/ACCESS.2021.3106350.

[30] H. Zhang et al., "Real-Time Detection Method for Small Traffic Signs Based on Yolov3," in IEEE Access, vol. 8, pp. 64145-64156, 2020, doi: 10.1109/ACCESS.2020.2984554.

[31] Florian Spiess, Lucas Reinhart, Norbert Strobel, Dennis Kaiser, Samuel Kounev, and Tobias Kaupp, "People Detection with Depth Silhouettes and Convolutional Neural Networks on a Mobile Robot," Journal of Image and Graphics, Vol. 9, No. 4, pp. 135-139, December 2021. doi: 10.18178/joig.9.4.135-139.

[32] Fitri Utaminingrum, Renaldi P. Prasetya, and Rizdania, "Combining Multiple Feature for Robust Traffic Sign Detection," Journal of Image and Graphics, Vol. 8, No. 2, pp. 53-58, June 2020. doi:10.18178/joig.8.2.53-58

[33] Rafflesia Khan, Tarannum Fariha Raisa, and Rameswar Debnath, "An Efficient Contour Based Fine-Grained Algorithm for Multi Category Object Detection," Journal of Image and Graphics, Vol. 6, No. 2, pp. 127-136, December 2018. doi: 10.18178/joig.6.2.127-136

[34] Ryo Hasegawa, Yutaro Iwamoto, and Yen-Wei Chen, "Robust Japanese Road Sign Detection and Recognition in Complex Scenes Using Convolutional Neural Networks," Journal of Image and Graphics, Vol. 8, No. 3, pp. 59-66, September 2020. doi: 10.18178/joig.8.3.59-66

[35] Teerapat Chaloeivoot and Suebskul Phiphobmongkol, "Building Detection from Terrestrial Images," Journal of Image and Graphics, Vol. 4, No. 1, pp. 46-50, June 2016. doi:10.18178/joig.4.1.46-50.

Acknowledgments
This work was supported by grants from the National Natural Science Foundation of China(61702321, U1936213), the National Natural Science Foundation of China(U21A20486).

Lossless Compression and Optimization Method of Data Flow Efficiency of Infrared Image Depth Learning Model of Substation Equipment

Guangwei Liu[1,2], Zhen Hao[1,2], Zheng Niu[1], Ka Mu[1,2], Jixian Ma[1,2] and Wenzhe Zhang[3]

[1]State Grid Jibei Electric Power Co., Ltd, Beijing, China
[2]North China Electric Power Research Institute Co., Ltd, Beijing, China
[3]School of Electrical Engineering, Xi'an JiaoTong University, Xi'an, China

gwliu2016@qq.com

Abstract. Infrared detection of substation equipment has become one of the important means for live detection of power grid equipment. The insulation fault and equipment defect of electrical equipment in operation can be found by using infrared thermal imaging technology. Infrared image deep learning model usually needs a very large computational cost and storage space, which greatly limits the application of deep learning model in embedded terminal. In order to apply the deep learning model well in embedded devices with limited resources, this paper proposes a lossless compression method of infrared image deep learning model of substation equipment, which can reduce the size of the deep learning network model by 35 to 49 times without loss of recognition accuracy. A hybrid sparse matrix storage format based on recursion and a cache blocking method based on multi-core/many-core processors are proposed to optimize the data flow efficiency between processors.

1. Introduction

The operation quality of the substation will directly affect the operation safety and stability of the whole power grid. Therefore, regular inspection of the substation system is an effective guarantee to ensure the normal operation of the whole substation.[1-2]. During the patrol inspection of the substation system, the patrol inspectors are required to have good professional quality and work ability. In the process of routine inspection, the three forms of visual inspection, ear hearing and hand touch are mainly used. However, this inspection method is often easy to ignore some hidden dangers in the process of equipment operation. For example, if the equipment has a heating problem, it is impossible to directly observe it only with the naked eye of the human body.

Through the infrared temperature measurement technology, it can effectively monitor the operation status of various equipment, and it can also implement more scientific and reasonable equipment inspection work, so as to improve the operation stability of the whole substation system. For example, when the substation equipment has abnormal temperature, it cannot be observed directly with eyes. However, if the fault is not handled in time, the maintenance of the whole substation equipment may be delayed. In this case, with the help of infrared temperature measurement technology, this situation can be well avoided to ensure the operation stability of the substation equipment. In recent years, scholars at home and abroad have conducted in-depth research on front-end intelligent technology based on edge computing and neural network acceleration technology.

Data compression is a technology that uses as few numbers as possible to represent the original signal. In the field of infrared detection of substation equipment, the owner of infrared image data has a rich amount of information. If the original image is used for in-depth learning, the image processing efficiency is low, the time-consuming is long, and there are high requirements for the storage space of the front-end device. Therefore, while preserving the useful information of the image, the smaller the storage space of the image as far as possible, so that the processing efficiency can be improved. Generally, when selecting image compression coding algorithm, the principle is to use the algorithm with large compression ratio, good reconstructed image quality and fast operation speed. Lossless image compression coding means that the reconstructed image is exactly the same as the original image, and it removes redundant information. Lossless compression coding can be generally divided into three aspects: (1) predictive coding; (2) Dictionary based coding method; (3) Method based on probability statistics. The nonlinear predictor in reference [3] used BP neural network. This predictor can obtain low-resolution wavelet coefficients and predict wavelet coefficients with consistent direction and high resolution. Reference [4] adopted a fully connected recurrent neural network nonlinear predictor, which is also combined with the speech coding algorithm. This method can obtain a very low transmission coding speed and improve the coding effect compared with the previous ordinary predictor. Reference [5] further compressed the model weights after pruning and quantization by using Huffman coding, and achieved a storage saving of 20%.

The research group has carried out relevant research on intelligent detection technology of infrared image of substation equipment based on deep learning algorithm. On the basis of the above research, this paper starts with the edge technology and front-end intelligent technology of infrared video intelligent diagnosis device, further explores the technology of deep learning in real-time recognition and full precision maintenance of embedded equipment, the optimization method of data flow efficiency between multi-core and multi-core.

2. Research on intelligent detection technology of infrared image of substation equipment based on deep learning algorithm

The research group constructed an accurate identification model of infrared images of substation equipment based on the improved yolov3 algorithm, realized automatic real-time positioning of infrared image target images, and proposed a complete set of fault diagnosis solutions for current and voltage overheating defects [6-7]. The algorithm mainly includes the following parts.

- Infrared Image Preprocessing Technology.
- Construction of infrared image database of typical substation equipment. The typical equipment infrared image is shown in Figure 1.
- Research on infrared image enhancement algorithm based on visible and infrared image dual view fusion technology.
- Infrared image recognition technology of substation equipment based on convolutional neural network. The schematic diagram of improved YOLOv3 target detection algorithm is shown in Figure 2.
- Construction of infrared image defect diagnosis model of substation equipment.

Figure 1. Typical equipment infrared image.

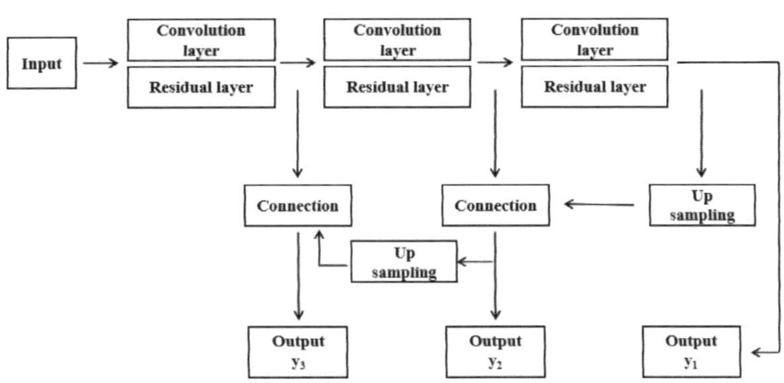

Figure 2. Schematic diagram of improved YOLOv3 target detection algorithm.

3. Research on real-time recognition and full accuracy maintenance technology of deep learning in embedded devices

Deep learning models usually contain millions or even tens of millions of parameters and more than ten or even dozens of layers of networks, which usually requires a very large computational cost and storage space. This makes the application of deep learning model in embedded terminal greatly limited. In order to apply the deep learning model well in embedded devices with limited resources, it is necessary to compress and accelerate the deep learning model while maintaining the recognition accuracy of deep learning. The compression and acceleration flow chart of deep learning network model is shown in Figure 3.

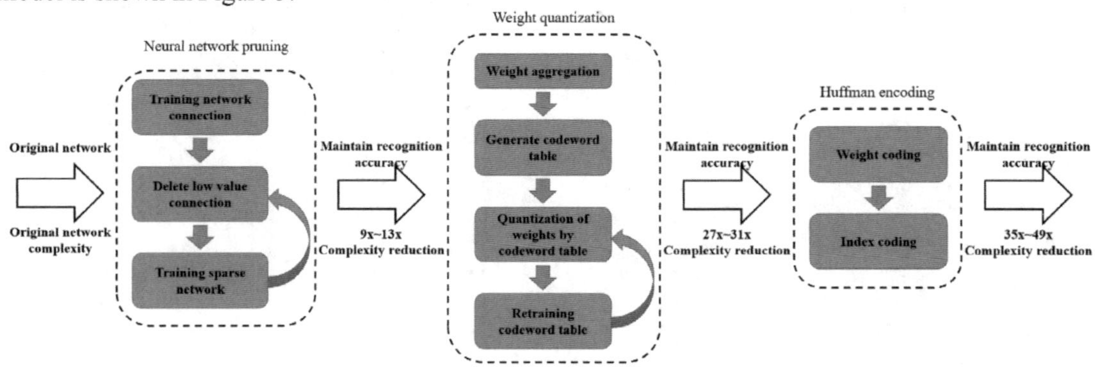

Figure 3. Compression and acceleration flow chart of deep learning network model.

As shown in the figure above, the proposed compression method of deep learning model with no loss of recognition accuracy is mainly divided into three steps: neural network pruning, weight quantization and Huffman coding, which can compress the size of deep learning network model 35 to 49 times with no loss of recognition accuracy.

3.1. Neural network pruning.

In large-scale deep learning networks, the activation of most neurons tends to zero, and these neurons with activation of 0 are redundant. Removing them can greatly reduce the size and computation of the model without affecting the performance of the model. Neural network pruning is mainly divided into three steps:

- Conducting normal network training.
- Deleting all connections whose weight is less than certain constraints.
- Retraining the sparse connection network obtained above.

The entropy value of the *j*-th characteristic diagram is calculated as follows:

$$H_j =-\sum_{i=1}^{m} p_i \log p_i$$

3.2. Weight quantification

There are two main research directions in the quantification of deep learning network model, one is weight simplification, and the other is weight sharing. This paper integrates the two methods, using weight reduction based on integer arithmetic reasoning and weight sharing based on clustering.

3.2.1 Weight reduction. In order to reduce the storage space of the model and accelerate the compression of the model, the weight reduction uses fewer bit bits to represent the actual floating-point weight of the model. In this paper, the weight reduction method based on integer arithmetic reasoning is used to quantify the weight and input value of the deep learning network model from float32 to uint8, the offset value from float32 to unit32, and the accumulator of unit32 to unit8. The weight reduction of deep learning model is shown in Figure 4.

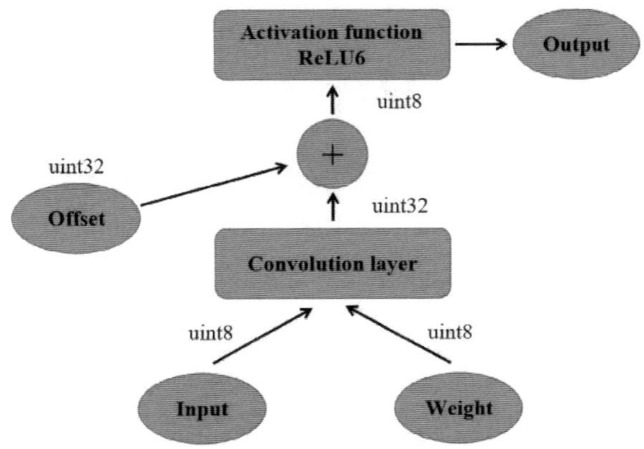

Figure 4. Weight reduction of deep learning model.

3.2.2. Weight sharing. Weight sharing is an application of information compression technology in neural network. The purpose is to use smaller space to store the original neural network. The weight matrix of each layer is clustered into several clusters by K-means clustering algorithm, and the cluster center value of each cluster is used to represent the weight of the cluster. Since the weight of the same cluster shares a weight size, only the index of the cluster storing the weight can be obtained by looking up the table.

When updating the weight, all gradient information is grouped according to the previous group of the weight matrix. Assuming that the centroid of group j is updated for the i-th time, the update process is as follows:

$$\text{centroids}_j^{i+1} = \text{centroids}_j^{i} - \text{lr} \times \left(\sum\nolimits_{k \in j} gradient_k\right)$$

3.2.3. Huffman code. Huffman coding is a common lossless coding technology. It carries out variable length coding according to the probability of symbol occurrence, and can further compress the deep learning network by using the biased distribution of effective weight.

4. Optimization method of deep learning data flow efficiency between multi-core/many-core processors

When the deep learning algorithm is running, a large number of sparse matrix storage and data cache blocking operations are involved between multi-core/many-core processors. This paper studies the hybrid sparse matrix storage format(HYB-R, recursive hybrid) based on recursion and the cache blocking method based on multi-core/many-core processors to optimize the data flow efficiency between multi-core/many-core processors.

4.1. Recursive hybrid storage format

The algorithm based on the traditional HYB format has the following two problems:

- For some sparse matrices, a large number of non-zero elements are divided into COO schemes.
- The matrix partition method used in HYB format is an empirical partition, which is not necessarily the best partition method.

Aiming at the problems of HYB format, this paper studies a new sparse matrix storage format, recursive hybrid sparse matrix storage format. The goal of HYB-R format is to store as many non-zero elements as possible in ELL format, so as to make full use of the advantages of ELL format in multi-core/many-core processors.

The following figure describes the process of creating HYB-R format. The matrix is first divided into two parts, ELL and COO. The COO part is then recursively divided into the ELL part and the COO part until there are few elements in the COO part. The HYB-R format is shown in Figure 5.

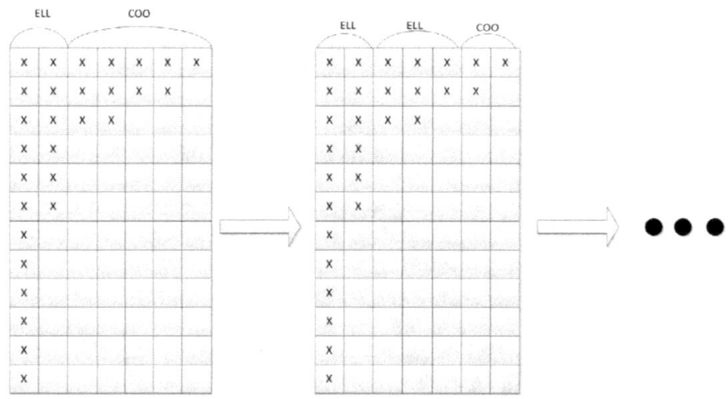

Figure 5. HYB-R format.

Firstly, the matrix and vector stored in HYB-R format are transferred from the host memory to the global memory, and then the kernel is started one by one. After the calculation, the target vector Y is transferred from the global memory to the host memory. During execution, the number of threads started by the ELL kernel gradually decreases as the number of remaining matrix rows becomes smaller and smaller.

4.2. Cache blocking method based on multi-core/many-core processor

This paper studies the cache blocking method based on multi-core / many-core processor to optimize the sparse matrix vector multiplication, in which the sparse matrix is divided into many sub blocks, and each sub block is stored in CSR format. Through the cache blocking method, the corresponding part of the source vector x can be reused in the processor's on-chip cache, so the access time for vector x can be greatly reduced.

The following figure describes how to assign tasks to threads of multi-core / multi-core processors. On the multi-core / multi-core processor of Fermi architecture, threads in the same block share an L1 cache. Therefore, we allocate a sub block row to a block, and all threads in the block calculate the same number of rows. In order to ensure that the corresponding part of vector x is reused in the cache, each thread calculates the sub block one by one. The task allocation diagram of SpMV algorithm based on BCSR formais shown in Figure 6.

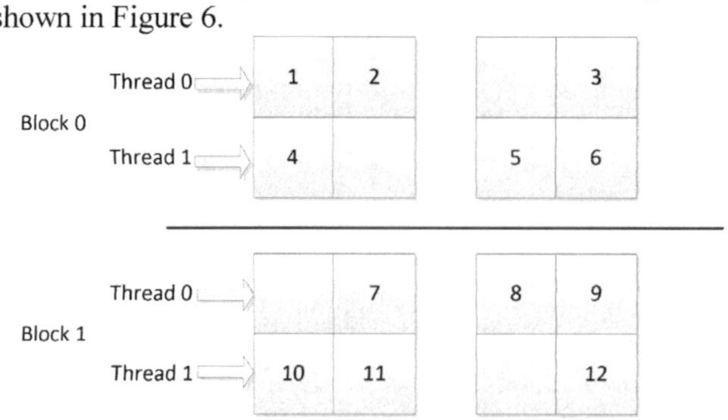

Figure 6. Task allocation diagram of SpMV algorithm based on BCSR format.

5. Conclusion

Infrared detection of substation equipment has become one of the important means for live detection of power grid equipment. The insulation fault and equipment defect of electrical equipment in operation can be found by using infrared thermal imaging technology. In recent years, intelligent detection has attracted more and more attention. In order to apply the deep learning model well in embedded devices with limited resources, this paper proposes a lossless compression method of infrared image deep learning model of substation equipment, and proposes a hybrid sparse matrix storage format based on recursion and a cache blocking method based on multi-core/many-core processors to optimize the data flow efficiency between processors. The research results of this paper can greatly improve the efficiency of deep learning network model without loss of recognition accuracy.

6. References

[1] M Lin, JG Yang, YY Jia et al., Research and Application on Detection Technology of SF6 Leak in Electrical Equipment [J], Jiangsu Electrical Engineering, 2014, 33(4): 27-33.

[2] M. Shawal Jadin, S. Kabir and S. Taib, Thermal Imaging for Qualitative-based Measurements of Thermal Anomalies in Electrical Components[C], 2011 Electronics Communications and Photonics Conference , 2011.

[3] Rosin P L. A simple method for detecting salient regions[J]. Pattern Recognition, 2009, 42(11): 2363-2371.

[4] Ma N and WEI G. A code-excited nonlinear predictive speech coding algorithm based on recurrent neural networks[J]. Journal of China Institute of Communications,2000， 21(10):31-37.

[5] Han S, Mao H and Dally W J. Deep compression: Compressing Deep Neural Networks with Pruning, Trained Quantization and Huffman coding[J]. Fiber, 2015.

[6] X. ZHANG, Z R LV, et al., Intelligent Detection Technology of Infrared Image of Substation Equipment Based on Deep Learning Algorithm[C], 2021 IEEE Sustainable Power and Energy Conference, 2021, pp. 3855-3860.

[7] Smirnov E A, Timoshenko D M and Andrianov S N. Comparison of Regularization Methods for ImageNet Classification with Deep Convolutional Neural Networks [J]. Aasri Procedia,2014, 6(1):89 - 94.

Image Point Set Registration: Phased Gaussian Mixture Model

Xin Hu* and Hao Qin

School of Electronics and Control Engineering, Changan University, Xian, China

*huxin@chd.edu.cn

Abstract. The introduction of probability model into point set matching has achieved great success. Because the image point set has more outliers and the image has complex transformation, the traditional Gaussian Mixture Model (GMM) does not perform well in image point set matching. In this paper, we propose a Phased Gaussian Mixture Model (PGMM) for point set registration between images. Different from the existing registration algorithms based on GMM, PGMM uses Locality Sensitive Hashing to distinguish the feature reliability of feature points. The initial transformation relation is obtained in the segment with high feature reliability and vector constraint is applied to other segments. The average Hamming distance is used to give different weights to each Gaussian component, which is related to the characteristics between data points and model points. The proposed method has good performance in the case of large image transformation scale.

1. Introduction

Determining the correspondence of feature points between two images is an important task of computer vision. The classical detection method can be divided into three steps, extracting feature points, describing feature points and finding the corresponding relationship between point sets. In order to get a better matching relationship, false matching elimination is needed. In the methods of feature point extraction and description, scale invariant feature transform (SIFT) is widely used because of its excellent stability [1]. This paper focuses on the third step of the matching process, that is, to find a better matching relationship between point sets. In the previous point set matching research, reference [2,3,4] uses the location information of the point set to match the point set. Another matching idea is to use the feature descriptor of point set to match the point set[5,6]. After the probabilistic model is introduced into point set matching, Gaussian Mixture Model (GMM) plays an important role in the process of point set matching. It has good matching effect and is widely used. The core idea is to regard the feature points extracted from two images as template point set and target point set. Each point in the template point set is regarded as the centroid of a Gaussian model, and the points in the target point set are the generated data of the template point set. Through the conversion function of the point set, the template point set moves to the target point set until the point set matches to a certain threshold. However, there are a large number of outliers in the image point set, and the image transformation has strong complexity. Gaussian mixture model is rarely used to find image correspondence. The main reasons are as follows:

- When there are many image feature points, the mixing of interior points and outliers will affect the matching motion of template point set in GMM model, which is often negative.

- When facing images with different rotation angles, the traditional GMM method only uses the position information to align the point set, and can not know the rotation of the image and the scale of the transformation, which often has a very poor effect.

This problem leads us to think about how to fuse the feature information of image point set into probability model to achieve better matching relationship. In the VFC proposed by Ma J [7], after the initial matching of the point set, the probability model is introduced for matching, in order to achieve the purpose of false match elimination. Tao and sun proposed a method called non-uniform Gaussian mixture model (NGMM)[8]. The algorithm assigns different weights to GMM components according to sift descriptor distance, so as to realize the fusion of feature information. Then they proposed a multi-layer feature matching Gaussian mixture model (HGMM)[9]. Using the feature information of the point set, they introduced the local sensitive hash algorithm (LSH)[10]to process the feature points hierarchically, and improved the GMM by quoting non-uniform mixing coefficient.

In our preliminary experiment, for the matching of two-dimensional image point set, we find that the matching points with stable Hamming distance obtained from the feature points after LSH have a high probability of correct matching points, which has high reliability. When the image is in different transformation scales, the number of reliable matching points obtained by LSH is large, so it can be regarded as the correct matching point and make vector constraints on other matching points as a priori. We find that when sift descriptors are used to calculate the mixing coefficient in HGMM, the distance difference between descriptors is not obvious and has little effect on the mixing coefficient of Gaussian mixture model.

To solve the above problems, we propose an image point set matching method based on Phased Gaussian Mixture Model (PGMM). In the PGMM matching framework, we do segmentation processing according to the characteristics of the image point set, and assign appropriate weights to the mixing coefficients of the Gaussian mixture model at different stages. The reasons why we segment the image point set are as follows: a. There are sift descriptors in the point sets of the two images, which have high similarity, that is, the descriptors have high reliability. And the number of this part of the point set is also considerable. b. In GMM, only using the position information of the point set can not get a better matching effect of the image point set. Give greater weight to the points with reliable descriptor similarity, so that these points can correspond better.

2. Phased Gaussian Mixture Model

We need two images and two sets of SIFT feature points, $\mathbf{X}_{N\times D}=(x_1,x_2,\cdots x_N)$ is the data point set and $\mathbf{Y}_{M\times D}=(y_1,y_2,\cdots y_M)$ is the template point set. Where N,M is the number of point sets. D is the dimension of the point set, this paper mainly studies the point set of two-dimensional images, so D=2 in this paper. The SIFT descriptor of the two point sets is $\{descr(x_i)\}_{n=1}^{N}$, $\{descr(y_j)\}_{m=1}^{M}$ .In general $N\neq M$. We use LSH method to segment the feature points. Gaussian mixing coefficient based on Hamming distance is introduced.

2.1. Point set phased registration

In this paper, the SIFT descriptors of the two images are processed by K groups of LSH, and the corresponding points closest to each group are recorded. If this corresponding point appears K times in Group K, that is, there is a unique correspondence between the target point and the template point, it is reasonable to believe that the feature reliability of these matching pairs is very high. Merge these points into the first stage. And through these reliable point sets, the preliminary transformation relationship of the overall point set of the image can be confirmed. Make vector constraints on the registration of other stages. The specific process is shown in Figure 1. The blue points in the third step are considered as reliable matching points, and the initial transformation relation of the point set is determined by these matching points. In the fourth step, it is obvious that the pre-processed point set has better effect.

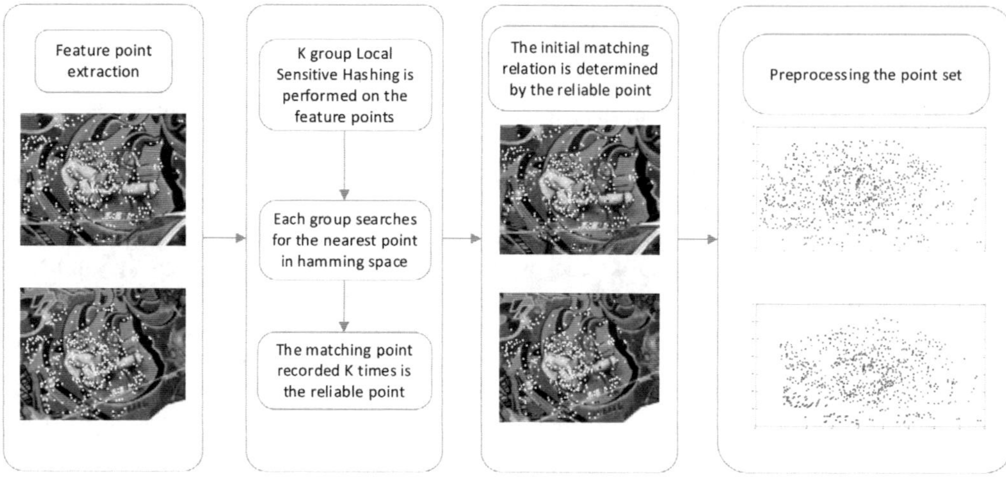

Figure 1. Flow chart of point set differentiation stage.

In the point set processed by K groups of LSH, the template points of the nearest Hamming distance matching with the target points appear more than $K/2$ times, which means that the reliability of features is relatively high, so treat these points as phase two. The rest of the points we will deal with in the third stage. This paper uses the mixing coefficient constrained by the average Hamming distance in the matching process of Gaussian Mixture Model.

2.2. Problem formulation

We refer to the Gaussian Mixture Model framework. $\mathbf{X}_{N\times D}$ is a data point set; $\mathbf{Y}_{M\times D}$ is a template point set. The probability density of Gaussian mixture model is

$$p(x_n)=\sum_{m=1}^{M+1} p(x_n \mid y_m) \tag{1}$$

Considering that there are outliers and noise points in the image point set, a uniform distribution $p(x_n \mid M+1)=1/N$ is added to the hybrid model to deal with these situations. Using equal isotropic covariance σ^2, define the uniformly distributed weight as $\omega(0\le\omega\le 1)$. The probability density of Gaussian mixture model is $p(x_n)=\omega/N+(1-\omega)\sum_{m=1}^{M} P(y_m)p(x_n \mid y_m)$. Since the corresponding relationship between the data point set and the template point set is unknown, the implicit variable defining the corresponding probability of a data point and the template point is $P(y_m \mid x_n)=P(y_m)p(x_n|y_m)/p(x_n)$. EM algorithm is introduced to maximize the log likelihood function. Q function is a lower bound of the log likelihood function in EM algorithm. We maximize the log likelihood function by maximizing the Q function.

$$\begin{aligned}Q&=\sum_{n=1}^{N}\sum_{m=1}^{M+1} P^{old}(y_m \mid x_n)\log(P^{new}(y_{mn})p^{new}(x_n \mid y_m))\\ &=-\sum_{n=1}^{N}\sum_{m=1}^{M} P^{old}(y_m \mid x_n)D\log\sigma-\frac{1}{2\sigma^2}\sum_{n=1}^{N}\sum_{m=1}^{M} P^{old}(y_m \mid x_n)\|x_n-\tau(y_m,\theta)\|^2\end{aligned} \tag{2}$$

Where τ is the transformation matrix and θ is the argument to the transformation matrix. $P^{old}(y_m \mid x_n)$ is a posteriori probability, and its practical significance is the responsiveness of sample points to template points. Therefore, the solution formula of a posteriori probability is

$$P^{old}(y_m \mid x_n) = \frac{P(y_m)\exp-\frac{\|x_n-\tau(y_m,\theta)\|^2}{2\sigma^2}}{\frac{1}{N}\frac{\omega}{1-\omega}(2\pi\sigma^2)^{D/2}+\sum_{k=1}^{M}P(y_k)\exp-\frac{\|x_n-\tau(y_k,\theta)\|^2}{2\sigma^2}} \tag{3}$$

Where $P(y_{mn})$ is the mixing coefficient of the Gaussian mixture model, and the same weight $P(y_{mn}) = 1/M$ is used in the earlier Gaussian Mixture Model method. In the later improved algorithm, SIFT feature distance is introduced into the mixing coefficient. We propose a new method, which is to use the average Hamming distance (AHD) after K groups of LSH as a reference to redefine $P(y_{mn})$.

$$P(y_{mn}) = \frac{e^{-\alpha\|AHD(x_n-y_m)\|^2}}{\sum_{h=1}^{M} e^{-\alpha\|AHD(x_n-y_h)\|^2}+c} \tag{4}$$

$AHD(x_n - y_m)$ is the average Hamming distance of Group K of x_n to y_m. α and c are two control parameters. In Formulas (2) and (3), $\tau(Y,d)=Y+u(Y)$. In order to effectively deal with the registration of irregular point sets with complex transformations, this paper uses Tikhonov regularization framework to obtain a priori data. $p(d)=\exp^{-\frac{\lambda}{2}\|u\|_{H}^{2}}$, Where $\|u\|_{H}^{2}$ is the norm of regenerated Hilbert space $u(Y)$.The optimal function form of u is the linear combination of specific kernel functions, $u(z)=\sum_{m=1}^{M} w_m \Phi(z, y_m)$. $\Phi_{M\times M}$ is the kernel matrix derived from $\Phi(y_i, y_j) = e^{-\frac{1}{2}\|\frac{y_i-y_j}{\beta}\|^2}$. In this paper, Formula (2) will be expressed as

$$Q = -\sum_{n=1}^{N}\sum_{m=1}^{M} P^{old}(y_m \mid x_n) D\log\sigma - \frac{1}{2\sigma^2}\sum_{n=1}^{N}\sum_{m=1}^{M} P^{old}(y_m \mid x_n)\|x_n-(y_m+u(y_m))\|^2 + \frac{\lambda}{2}\|d\|_{H}^{2} \tag{5}$$

Algorithm: Phased Gaussian mixture model

Input: Sift descriptor of two point sets.
Output: P.
Step1: Processing descriptor

1. Perform the K-group LSH function to obtain the k-group hash code and calculate the Hamming distance
2. Find the Y point closest to the X point in each group and record the number of times as H
3. Calculate the average Hamming distance of group K

Step2: Point set phased processing

1. H = K is recorded as the first stage, and make constraints on the second step
2. K/2<H<K is recorded as the second stage, and make constraints on the third step
3. The remaining point sets are considered the third stage

Step3: EM algorithm calculation

- E-step:

$$P(y_m \mid x_n) = \frac{P(y_m)\exp-\frac{\|x_n-(y_m+\Phi(m,\cdot)W)\|^2}{2\sigma^2}}{\frac{1}{N}\frac{\omega}{1-\omega}(2\pi\sigma^2)^{D/2}+\sum_{k=1}^{M}P(y_k)\exp-\frac{\|x_n-(y_m+\Phi(k,\cdot)W)\|^2}{2\sigma^2}}$$

$P(y_m)$ is obtained from Formula (4)

- M-step:

$(\Phi+\lambda\sigma^2 d(P1)^{-1})W = d(P1)^{-1}PX - Y$,1is the column matrix with all elements of 1

$N_P = 1^T P1, T = Y + \Phi W$

$\sigma^2 = \frac{1}{N_P D}(\mathrm{tr}(X^T d(P^T 1)X) - 2tr((PX)^T T) + tr(T^T d(P1)T))$

EM algorithm converges to obtain a posteriori probability P

3. Experimental results

In this section, we will show the experimental results of our proposed PGMM algorithm applied to some data sets. The main data set used to test the proposed algorithm is the data set of Mikolajczyk [11]. The reason why we choose this data set is that it contains some challenging matching situations, such as image rotation at different angles, illumination change, blur change and so on. At the same time, it provides a reliable standard for matching effect. Each group of VGG[11] dataset contains 6 images, and the first image matches the other 5 images. The homography matrix H between every image pair is provided as ground truth.

We use image pairs (boat1, boat3), (ubc1, ubc4), (leuven1, leuven5) to test the reliability of the proposed method. We randomly select 100 correctly matched feature points on each image so that they form 100 correct matching pairs. Different numbers of outliers are randomly added to the two images. The number of outliers is 30-300, the outer point rate is 30% - 300%. The corresponding relationship is calculated by CPD[2], NGMM[8] and the method proposed in this paper. Precision and recall are used to evaluate performance.

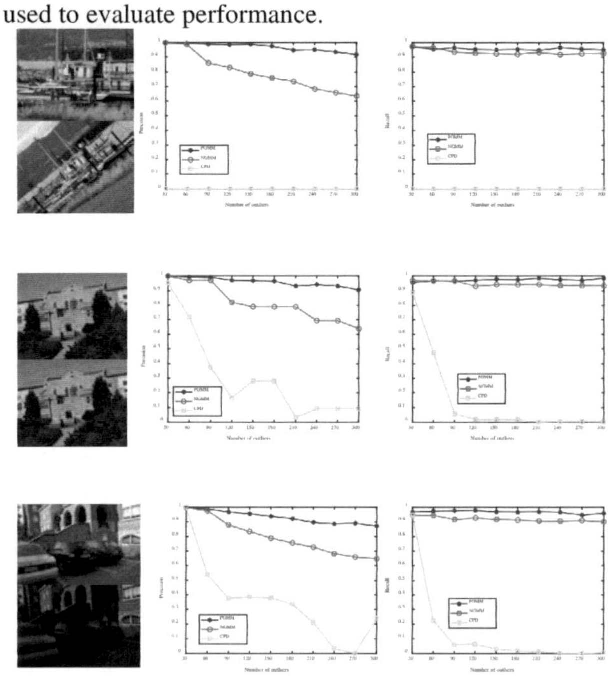

Figure 2. Randomly add 30-300 outliers to 100 true values, and the precision and recall curve.

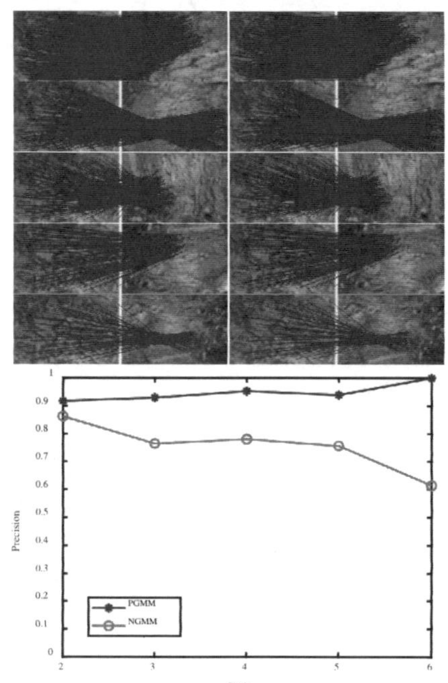

Figure 3. Stability test, in the bark dataset.

In figure 2, (a) the two pictures of boat1 and boat3 have obvious angle of view zoom and rotation. For the zoom and rotation of image point set, the direction proposed in this paper is obviously better than NGMM and CPD. Because the point set has a certain rotation, the performance of CPD is particularly poor, which also shows that CPD can not achieve good results only by using the position information of the point set when rotating the image point set. In (b) ubc1 and ubc4, the picture definition is compressed, and (c) leuven1 and leuven5 are images taken under different lighting conditions. When the image point set is not rotated, CPD can align the point set when there are few interference points, but the effect will be worse and worse with the increase of external points. With the increase of outer points, the matching accuracy of NGMM will gradually decline, and the recall rate is not as good as that of PGMM proposed in this paper. Because we randomly select 100 pairs of correct matching points and different numbers of external points, the experimental results have high reliability. No matter in the face of rotation, zoom, illumination and other changes, PGMM has good performance, which can be concluded that the algorithm in this paper has good robustness.

In figure 3, we use 6 pictures of bark in VGG data set, and use bark1 and bark2-6 for point set registration. The left side of the image is the calculation result of PGMM method, and the right side is the calculation result of NGMM. We have analyzed the bark transform and the matching of five groups of images with different scale. The blue line indicates the correct match and the red line indicates the wrong match. It can be seen that when the image scene transformation scale is large, PGMM shows good stability and accuracy.

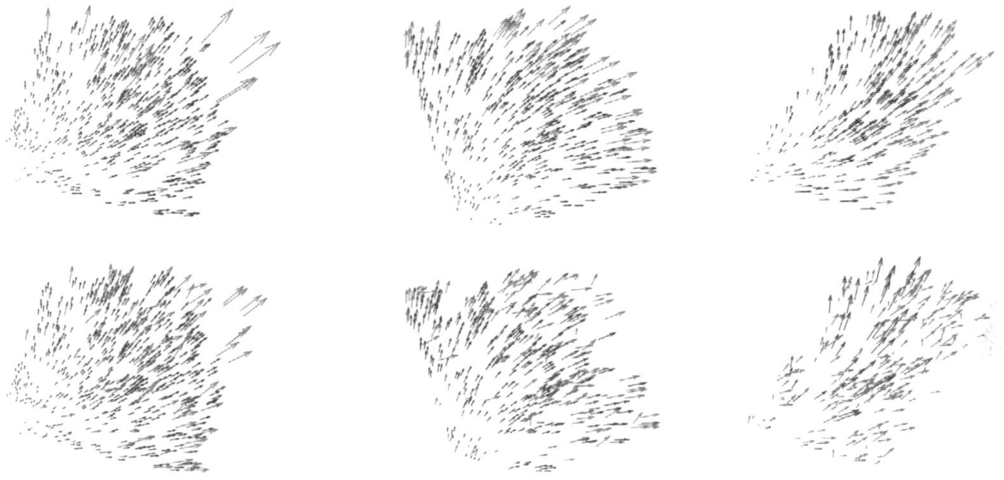

Figure 4. Mixing coefficient test, graf data set, graf2-4 were matched with graf1 respectively using different mixing coefficients. Blue is the true positive vector, red is the false positive vector.

We test the mixing coefficient $P(y_{mn})$ and compare the effects of different mixing coefficients on point set registration. Cancel the piecewise operation of the algorithm proposed in this paper and only use the mixing coefficient proposed by Formula 4 to compare with the mixing coefficient of SIFT feature fitting proposed by NGMM. In figure 4, the mixing coefficient constructed by average Hamming distance is used in the first line, and the mixing coefficient constructed by SIFT feature is used in the second line. It can be seen that in the hybrid coefficient method proposed by us, the vector field of the matching pairs is smoother, and the number of correct registration and registration accuracy are better.

4. Conclusion

In this paper, the image point set registration method of PGMM is proposed. The point set is processed in stages according to the feature reliability of image feature points, and the mixing coefficient of Gaussian mixture probability model is constrained by k-group average Hamming distance. The algorithm proposed in this paper has good performance for complex and changeable image point sets, good robustness for images transformed under different conditions such as illumination and blur. The proposed algorithm can also show good accuracy and reliability for images with large change scales. The next step is to test and verify the method in many aspects and optimize the mixing coefficient of Gaussian Mixture Model.

5. References

[1] Lowe D G. Distinctive image features from scale-invariant keypoints[J]. International journal of computer vision, 2004, 60(2): 91-110.

[2] Myronenko A, Song X. Point set registration: Coherent point drift[J]. IEEE transactions on pattern analysis and machine intelligence, 2010, 32(12): 2262-2275.
[3] Khodadad D, Ahmadian A A, Banaem H Y, et al. CT and PET Image Registration: Application to Thorax Area[J]. Journal of Image and Graphics, 2013, 1(4): 171-175.
[4] Tao L, Nguyen T, Nguyen T, et al. An Adaptive Differential Evolution Algorithm with a Point-Based Approach for 3D Point Cloud Registration[J]. Journal of Image and Graphics, 2022, 10(1).
[5] Yan J, Cho M, Zha H, et al. Multi-graph matching via affinity optimization with graduated consistency regularization[J]. IEEE transactions on pattern analysis and machine intelligence, 2015, 38(6): 1228-1242.
[6] Yan J, Li C, Li Y, et al. Adaptive discrete hypergraph matching[J]. IEEE transactions on cybernetics, 2017, 48(2): 765-779.
[7] Ma J, Zhao J, Tian J, et al. Robust point matching via vector field consensus[J]. IEEE Transactions on Image Processing, 2014, 23(4): 1706-1721.
[8] Tao W, Sun K. Robust point sets matching by fusing feature and spatial information using nonuniform Gaussian mixture models[J]. IEEE Transactions on Image Processing, 2015, 24(11): 3754-3767.
[9] Sun K, Tao W, Qian Y. Guide to match: multi-layer feature matching with a hybrid gaussian mixture model[J]. IEEE Transactions on Multimedia, 2019, 22(9): 2246-2261.
[10] Cheng J, Leng C, Wu J, et al. Fast and accurate image matching with cascade hashing for 3d reconstruction[C]//Proceedings of the IEEE conference on computer vision and pattern recognition. 2014: 1-8.
[11] Mikolajczyk K, Schmid C. A performance evaluation of local descriptors[J]. IEEE transactions on pattern analysis and machine intelligence, 2005, 27(10): 1615-1630.

Journal of Physics: Conference Series **2320** (2022) 012028 doi:10.1088/1742-6596/2320/1/012028

Transmission Line Foreign Body Fault Detection Using Multi-Feature Fusion Based on Modified YOLOv5

Qianyue Wang, Gangquan Si *, Kai Qu, Jiahui Gong and Lele Cui

State Key Laboratory of Electrical Insulation and Power Equipment, Shaanxi Key Laboratory of Smart Grid, School of Electrical Engineering, Xi'an Jiaotong University, Xi'an, China

*sigangquan@mail.xjtu.edu.cn

Abstract. The particular work environment of transmission lines makes it liable to malfunction because of the foreign bodies attached to the power equipment. However, traditional detection methods are difficult to identify the exact category of foreign bodies and ignore the relationship between a foreign body and power equipment, which is correlated to determining whether there is a fault or not. In order to detect and distinguish the foreign bodies and give fault warnings automatically in transmission line patrol with high accuracy and efficiency, we improve the original YOLOv5 object detection model and propose a novel fusion detection model using multi-scale appearance and relationship features. The proposed model is capable of transmission line foreign body category detection and fault determination. By comparing with the original YOLOv5, the proposed model shows better accuracy and achieves the strong ability to give the fault determination simultaneously, which makes it convenient for power system operators and protects the safe and stable operation of the power system.

1. Introduction

Transmission lines play an important role in the power system. With the increasing large-scale consumption of electrical energy, transmission grids are distributed more and more widely, and the work of transmission lines becomes heavier. All of these significantly increase the cost of operation and maintenance of power grids [1]. Based on the particular work environment, the components are always exposed to the complex outdoor environment for a long time period and are easily damaged by nature and man. Among them, foreign bodies such as nests, balloons, kites, and plastic films occur frequently and randomly, which often cause electrical accidents in various places. Once there is a fault that cannot be cleaned up in time, it may cause huge damage to the entire transmission line. However, manual inspection has distinct disadvantages for long cycles, incomplete, inefficient, and difficult working conditions, which can't satisfy the normal operation and maintenance requirements of modern large-scale power grids. It brings a huge requirement and challenge to identify these faults automatically with high accuracy.

With the rapid development of artificial intelligence, computer vision has made great progress in recent years. The neural network based on the Convolutional Neural Networks (CNN) was proposed by LeCun et al. in the LeNet [2], which stacks multilayer convolution. After that, CNN became the main algorithm in image processing. Alex Krizhevsky et al. proposed the AlexNet [3], which uses multi-layer convolutional layer to learn the features of different channels, opening the deep network research of computer vision. Simonyan and Zisserman proposed VGGNet [4], using small size convolution kernel and continuing to make the network deeper, which significantly improves the network effect. Meanwhile, different from the depth of network research, Christian Szegedy proposed GoogLeNet [5], which added Inception block to make multiple paths work in parallel to extract information in different levels. Residual Network (ResNet) was proposed by He Kaiming et al. [6]. It trained a few stacked layers to fit a residual mapping, which gained superior performance in previous image processing networks.

Object detection aims to recognize object categories and predict the position of each object by bounding boxes. The good detection algorithm performs well on semantic and spatial information understanding in images. Currently, object detection models can be mainly divided into two families: (i) two-stage detectors, such as Region-based CNN (R-CNN) and its variants Faster R-CNN and Mask R-CNN [7-10], which has two cascaded detection steps. Firstly, use a proposal generator to decide a sparse set of proposal regions. Then, extract their features and predict the category of the targets based on the proposed region; (ii) one-stage detectors, such as YOLO and its variants YOLOv3, YOLOv4, and YOLOv5 without the cascaded region classification step, directly give object prediction on each position of the feature maps [11-13]. Two-stage detectors are generally believed having better prediction accuracy of detection, while one-stage detectors show significant time-efficient and performance. However, most object detection networks are committed to detecting objects individually, without exploiting their relations, which may miss a lot of connotative and useful information. Visual Relationship Detection (VRD) involves object detection and understanding the interactions between them. Initial research attempted to learn object co-occurrence relationship by vector quantizing feature space like CoLA (for Cooccurrence, Location and Appearance) [14]. Then, with the great development of deep learning techniques in CNN and object detection, many researchers attempt to train a CNN to classify all kinds of relationships after getting the bounding boxes of the two participating objects [15-16].

Foreign body detection in transmission line images has been regarded as a typical target recognition task for a long time [17]. Combined with the large-scale use of UAVs in line patrol, we can get transmission line images more easily and timely. Therefore, it has become an inevitable trend to apply computer vision technology to transmission line fault detection. There have been many applications of object detection algorithms used in this domain. However, most studies are stuck in detecting the foreign bodies independently, without considering the relationship between foreign bodies and transmission lines. In this case, the phenomena of misdeclaration in fault diagnosis are easy to occur. For example, one image containing a bird nest in the tree next to the transmission line will be judged as a failure.
Intuitively, the visual relationship is closely correlated with fault determination. Combining object detection and relationship detection techniques, we propose a novel transmission line foreign body fault detection using multi-feature fusion based on modified YOLOv5. Firstly, we use YOLOv5 with bi-directional feature pyramid network (BiFPN) and Complete Intersection over Union (CIoU) to extract the bounding boxes and features of the target object pairs, including foreign bodies (e.g., nest, kite, balloon, plastic film) and power equipment (e.g., tower, line). Then, use *dual spatial masks* to represent the spatial relationship and leverage an EfficientDet to analyze the multi-scale features of object pairs [18]. All of these object, appearance, and spatial relationship features will be fed to the joint recognition module. Finally, our model will produce a *triplet label* as the output to show the information of object pair and fault condition from each transmission line image.

In summary, the contributions of our fusional model consist of two aspects: (i) modified YOLOv5, a high-precision and effective object detection model, which becomes more suitable for foreign body category detection; (ii) an effective fault determination method using multi-feature fusion, which combines object pairs and relationship features information.

2. Model Architecture

In Figure 1, the architecture of the transmission line foreign body fault detection is shown. There are four main modules: Object Detection, Feature Concatenation, Relationship Detection, and Joint Prediction. Firstly, Object Detection uses a modified YOLOv5 framework to detect the target objects of transmission line images. Then, we can get a set of target objects with labels and bounding boxes. All of these will be divided into two groups: foreign bodies and power equipment. Then, pair them together to form the *object pairs*. Feature Concatenation concatenates the object feature maps and sends them to the later model for further study. For Relationship Detection, on the one hand, use *dual spatial masks* to represent the spatial relationship, one for the foreign body and the other for the power equipment, which will be compressed via three convolution layers. On the other hand, generate *enclosing boxes*, which frame the object pairs with a minimal size, and leverage an EfficientDet network [18] to extract and fuse the multi-scale features. Then, Joint Prediction subsequently concatenates the features from the former detectors and further compressed via two fully-connected layers and outputs a label (*fault* or *normal*). Consequently, our model gives a *triplet label* in the form of (*f*, *e*, *l*). Here, *f*, *e*, and *l* respectively denote the foreign body, the power equipment, and the label as shown in Figure 1.

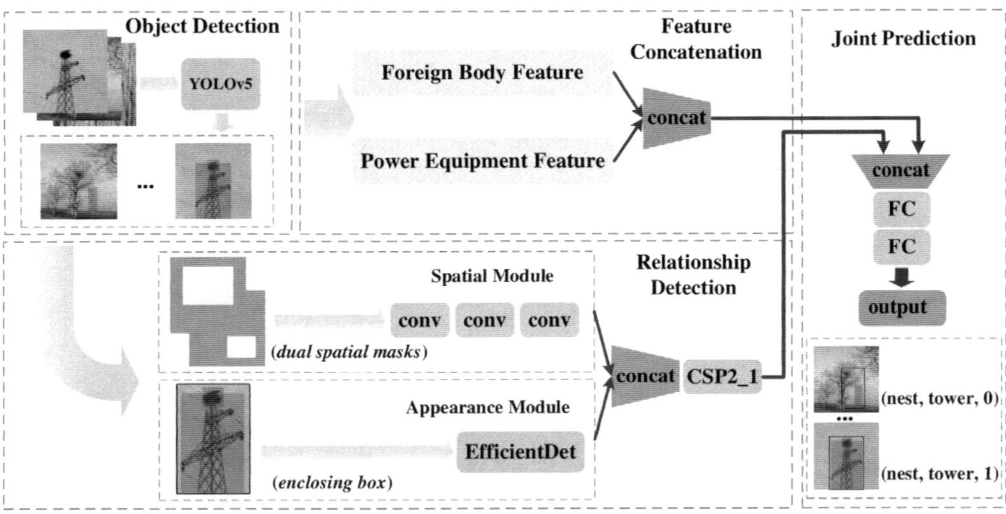

Figure 1. Architecture of foreign body fault detection based on modified YOLOv5.

2.1. Object Detection

Different from two-stage networks, YOLOv5 can directly predict both the bounding box and the classification of the input image without generating a series of candidate regions of interest for bounding box regression, which makes it have great performance in detection speed. Compared to other versions, YOLOv5 is faster and more flexible, which has more data augmentation methods, alternative network sizes of depth and width, and stronger multi-scale feature extraction ability. The accuracy of image positioning can be evaluated by Intersection over Union (IoU) in Figure 2. YOLOv5 uses GIoU regression loss to choose the most suitable bounding box. Unlike traditional Mean Square Error (MSE), which does not consider the integrity of the object itself, GIoU is proposed to find the smallest box covering both the predicted and ground truth bounding box to be the denominator. However, it can't work when the predicted bounding box is inside the ground truth bounding box. Specifically, CIoU loss [19] additionally considers the distance of the center points and the box aspect ratio. The comparison of the loss function can be defined as following:

$$IoU(A, B) = \frac{A \cap B}{A \cup B} \tag{1}$$

$$GIoU = IOU - \frac{C - (A \cup B)}{c} \tag{2}$$

$$CIoU = 1 - IoU + \frac{\rho^2(A, B)}{c^2} + \alpha v \tag{3}$$

where C is the smallest enclosing box covering A and B, c is the diagonal length of the enclosing box, $\rho(A,B)$ is the Euclidean distance between the center of A and B, α is the positive trade-off parameter, and v is the consistency of aspect ratio. Power equipment and foreign bodies in transmission line images sometimes have a big difference in size, which means that the predicted bounding box will be expected to overlap easily. Hence, we choose CIoU loss which is verified to achieve better convergence speed and accuracy on the bounding box regression problem.

The Neck of YOLOv5 consists of Feature Pyramid Network (FPN) and Path Aggregation Network (PAN). FPN builds the bottom-up path and top-down path, and PAN adds an extra bottom-up path aggregation network on top of FPN to fuse different features across stage hierarchies. Unlike the PAN layer, BiFPN layer in Figure 3 adds an extra edge from the original input to the output node if they are at the same level several to optimize the cross-scale connections, which allows scale-wise level re-weighting to fuse more features without adding much cost [20]. Therefore, we use BiFPN layers to replace the FPN + PAN layers.

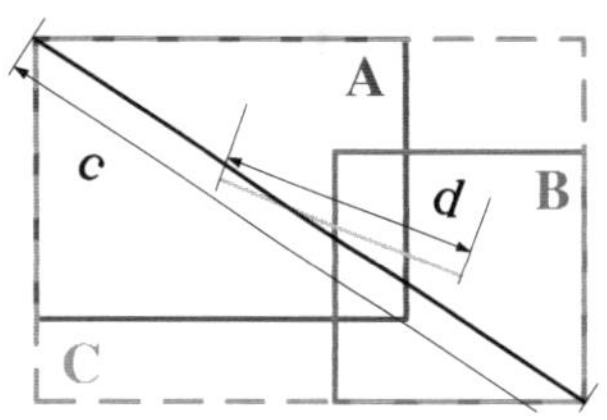

Figure 2. Bounding box of A, B and their smallest enclosing box C.

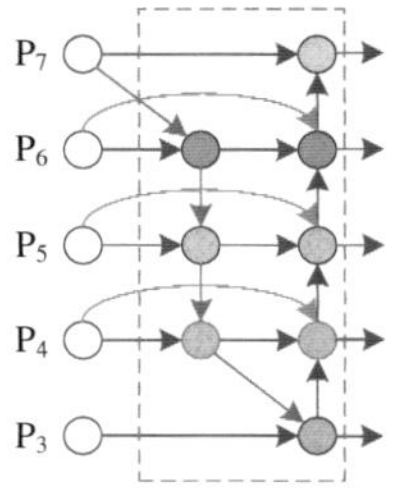

Figure 3. BiFPN layer.

As we discussed previously, our modified YOLOv5 with BiFPN and CIoU network is greatly suitable for transmission line images detection, which contains small foreign bodies compared with the big size of the power equipment in complex backgrounds and may overlap. After locating a set of target objects, including m foreign bodies and n power equipment with their own bounding box and category label, we will combine these to form (mn) *object pairs* and send them to the later modules.

2.2. Feature Concatenation

After the Object Detection module, we can get the object pairs feature maps from our modified YOLOv5 outputs. We choose the $N \times N \times [3 \times (4+1+n)]$ feature map for the $N \times N$ grid points of the feature map, 4 bounding box offsets, 1 objectness prediction, and n class predictions. In this section, we want to combine these individual features containing semantic and spatial information, which may have impact on their relationship detection and fault determination. Therefore, we utilize the concatenation block to combine the object features together, which can extend the dimensions of tensors in different channels. Then, send them to the later module for further study.

2.3. Relationship Detection

The Relationship Detection can be divided into two parts: Spatial Module and Appearance Module. Inspired by Mask R-CNN's great contribution to instance-level recognition, we use two binary masks to form *dual spatial masks* to represent the spatial relationship, one for the foreign body and the other for the power equipment. Unlike Mask R-CNN predicting pixel-accurate of object masks, we simplify the model to mask the pixels in the bounding boxes to focus on the spatial configurations of objects and downsample via three convolution and pooling layers as shown in Figure 4.

Utilize an *enclosing box*, which covers the object pairs with a minimal size. Then, leverage an EfficientDet based on weighted BiFPN for multi-scale feature fusion of the image in the enclosing box. It takes different levels of features from the backbone and repeatedly applies BiFPN layers for feature fusion. Similarly, utilize the concatenation block to concatenate feature maps from Spatial Module and

Appearance Module to get the final relationship feature maps, and transform them into the same dimensions with Feature Concatenation outputs by CSP2-1 block.

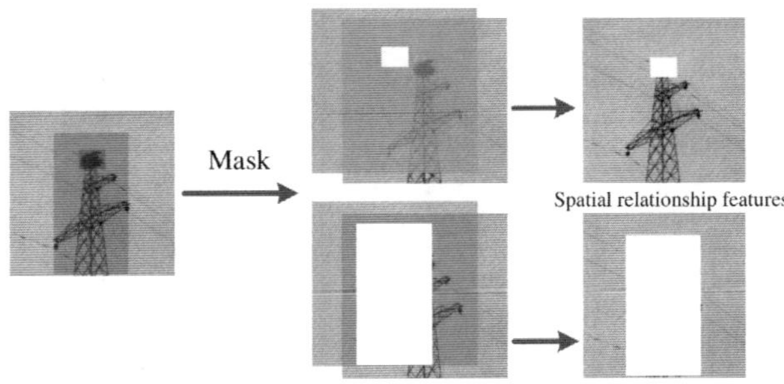

Figure 4. Dual spatial masks of a picture.

2.4. Joint Prediction

Joint Predictor subsequently concatenates the features from the former detectors and further compresses them via two fully-connected layers to output a label (*fault* or *normal*). In this way, our model not only detects the category of the foreign bodies but also determines the fault condition based on both object pairs features and relationship information. Consequently, our model gives a *triplet label* in the form of (*f*, *e*, *l*) of each image.

3. Case study

In this section, we evaluate our method for foreign body fault detection on our transmission line dataset. First, we introduce our dataset and application scenarios. Second, we advocate the evaluation criteria, which will be used in our later analysis of experimental results. Third, we compare our modified YOLOv5 with the original YOLOv5 in Object Detection. Finally, use our full model to determine the fault. Additionally, the transmission line images and statistic data are presented to explain the function and accuracy of our model.

3.1. Dataset

The experiment dataset contains 1706 transmission line images of UAVs in line patrol, including power equipment of tower and line, foreign bodies of the nest, kite, balloon, and plastic film, and the surrounding environment. We label the pictures and give each of them a binary label 0 or 1 representing normal or fault, and classify the pictures based on the object categories. Then, count the number of the pictures as shown in Table 1 (e.g., 0/83 indicates 83 pictures without fault, 1/106 indicates 106 pictures with fault). Then, we use 1300 images in the experiment dataset for training and 406 images for testing. Meanwhile, we use traditional data augmentation methods like Horizontal flip, Random rotating and Cropping method, and also Mosaic data augmentation method, which can mix training images to enhance the training effect.

Table 1. The statistics of foreign bodies and power equipment in transmission line images.

		Foreign Bodies								
		No	Nest		Kite		Balloon		Film	
Power Equipment	No	0/76	0/126		0-76		0/96		0/68	
	Line	0/106	0		0/83	1/106	0/93	1/63	0/71	1/98
	Tower	0/69	0/105	1/201	0/44	1/45	0/49	1/47	0/28	1/56

3.2. Evaluation Criteria

We adopt 4 common types to describe the binary classification: true positive (*TP*), false positive (*FP*), true negative (*TN*) and false negative (*FN*), and four widely used measures to evaluate the performance of our model: Precision (*P*), Recall (*R*), Accuracy (*ACC*) and mean Average Precision (*mAP*), which are defined in formulas as following:

$$P = \frac{TP}{TP + FP} \tag{4}$$

$$R = \frac{TP}{TP + FN} \tag{5}$$

$$ACC = \frac{(TP + TN)}{(TP + FP + TN + FN)} \tag{6}$$

where $TP + FP$ means the total number of located defects, and $TP + FN$ is the total number of actual defects. C is the number of classes, and AP_c is the area of the *P-R* curve of class *c*.

3.3. Object Detection Experiment and Parameter Analysis

In order to research the improvement of the proposed modified YOLOv5 model method on transmission line foreign body detection, we use the training dataset to train 300 epochs based on the pre-trained YOLOv5 network and evaluate it on the testing set. Then, comparing our modified model with the original model, we can find the initial training loss of the original YOLOv5 is lower than ours due to the simple feature fusion method. However, in the subsequent training, our model obtains lower error and better accuracy from *loss*, *mAP*@0.5, and *mAP*@0.5:0.95 for all classes curves of both model in Figure 5.

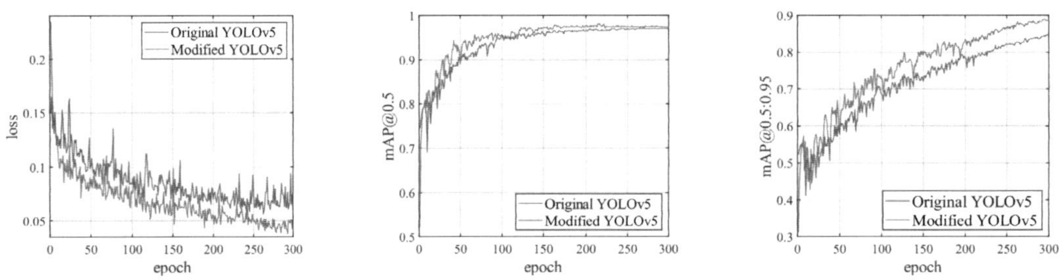

Figure 5. Comparison of the *loss*, *mAP*@0.5, and *mAP*@0.5:0.95 of different models.

Next, we apply the original YOLOv5 model to make a comparison with our modified YOLOv5 model. The data in Table 2 shows the results of *P*, *R*, and *mAP*@0.5 for all classes on testing set. We can see that the *P*, *R*, and *mAP*@0.5 of our model are improved by 7.76%, 9.85% and 10.32% compared with original YOLOv5. Obviously, our modified model gets great improvement of every evaluation indicator. The main reason is that the CIoU and BiFPN methods enhance the abilities of multi-scale feature fusion, which is the basis for subsequent fault detection.

Table 2. Evaluation comparison of original and modified YOLOv5.

	P/%	*R*/%	*mAP*@0.5/%
Original YOLOv5	86.3	80.2	81.4
Modified YOLOv5	93.0	88.1	89.8

3.4. Fault Determination Experimental Results and Discussion

In this section, the effects of our full model on fault determination are discussed. After the Object Detection module, we get target objects including 4 types of foreign bodies and 2 types of power equipment with their own bounding box and category label, which can form 8 kinds of *object pairs*. At this point, we can already determine that some of the images don't have faults, such as only foreign bodies, only power equipment, or neither of them. Then, send the rest of them to the later modules to

Journal of Physics: Conference Series **2320** (2022) 012028 doi:10.1088/1742-6596/2320/1/012028

analyse their relationships. Finally, output a binary label 0 or 1 to represent normal or fault, and combine them into the *triplet label* (*f*, *e*, *l*) to each image. The examples are shown in Figure 6.

Figure 6. The examples for proposed model detection. We output a *triplet label* of each image. (a), (b), (c), and (d) are the 4 kinds of foreign bodies causing faults, labelled by (nest, tower, 1), (kite, line, 1), (balloon, line, 1), and (film, line, 1). Other situations without faults in (e), (f), (g), and (h), labelled by (nest, no, 0), (no, tower, 0), (nest, tower, 0), and (no, no, 0).

Moreover, we also utilize the *P*, *R*, *ACC* to evaluate the fault determination performance in Table 3. The *P* and *ACC* of the Object Detection only model are just 68.0% and 71.8%, while *R* can reach 89.4%. The main reason is labelling the images containing foreign bodies which do not cause fault as label 1. It indicates that if we just use the result of foreign body detection to judge fault may be very confusing because of the surrounding environment of the line. And if we only use the Relationship Detection model, the *ACC* can reach 86.9%. Finally, we combine the two parts together to form our full model, which shows a significant improvement of the *P*, *R* and *ACC* to 92.4%, 94.1% and 91.9%. It also proves that the individual features of object pairs have some impact on fault determination

Table 3. Fault detection evaluation of proposed model.

	P/%	*R*/%	*ACC*/%
Object Detection only	68.0	89.4	71.8
Relationship Detection only	87.4	90.2	86.9
Full model	92.4	94.1	91.9

4. Conclusion

In this paper, we propose a novel fusion detection model using multi-scale appearance and relationship features based on the modified YOLOv5 network for transmission line foreign body detection and fault determination. On the one hand, we modify YOLOv5 for object detection and get the object pairs features. On the other hand, we utilize *dual spatial masks* to represent the spatial relationship and extract multi-scale features of the *enclosing box*. Finally, the *triplet label* (*f*, *e*, *l*) shows not only the category of the foreign body but also the fault condition. Experiments show that: (i) the modified YOLOv5 model achieves high accuracy for object detection, which can help people get the exact category of the foreign body to take precise troubleshooting measures; (ii) the proposed model can give the judgment of fault or normal and reduce the rate of false positives. However, our method can only tell whether there needs troubleshooting automatically without the rank of the fault. We can add fault level as a feature to train. When there are many faults occurring simultaneously, the fault handling plan can be arranged according to the fault level output, so that the damage can be minimized. Research on the priority of faults will be carried out in the future to make the line patrol work more efficient.

5. References

[1] Furse, Cynthia M., et al. "Fault diagnosis for electrical systems and power networks: A review." IEEE Sensors Journal 21.2 (2020): 888-906.

[2] Y. Lecun, L. Bottou, Y. Bengio and P. Haffner, "Gradient-based learning applied to document recognition," in Proceedings of the IEEE, vol. 86, no. 11, pp. 2278-2324, Nov. 1998, doi: 10.1109/5.726791.

[3] Krizhevsky, Alex, Ilya Sutskever, and Geoffrey E. Hinton. "Imagenet classification with deep convolutional neural networks." Advances in neural information processing systems 25 (2012).

[4] Simonyan, Karen, and Andrew Zisserman. "Very deep convolutional networks for large-scale image recognition." arXiv preprint arXiv:1409.1556 (2014).

[5] Szegedy, Christian, et al. "Going deeper with convolutions." Proceedings of the IEEE conference on computer vision and pattern recognition. 2015.

[6] He, Kaiming, et al. "Deep residual learning for image recognition." Proceedings of the IEEE conference on computer vision and pattern recognition. 2016.

[7] Girshick, Ross, et al. "Rich feature hierarchies for accurate object detection and semantic segmentation." Proceedings of the IEEE conference on computer vision and pattern recognition. 2014.

[8] Girshick, Ross. "Fast r-cnn." Proceedings of the IEEE international conference on computer vision. 2015.

[9] Ren, Shaoqing, et al. "Faster r-cnn: Towards real-time object detection with region proposal networks." Advances in neural information processing systems 28 (2015).

[10] He, Kaiming, et al. "Mask r-cnn." Proceedings of the IEEE international conference on computer vision. 2017.

[11] Redmon, Joseph, et al. "You only look once: Unified, real-time object detection." Proceedings of the IEEE conference on computer vision and pattern recognition. 2016.

[12] Redmon, Joseph, and Ali Farhadi. "Yolov3: An incremental improvement." arXiv preprint arXiv:1804.02767 (2018).

[13] Bochkovskiy, Alexey, Chien-Yao Wang, and Hong-Yuan Mark Liao. "Yolov4: Optimal speed and accuracy of object detection." arXiv preprint arXiv:2004.10934 (2020).

[14] Galleguillos, Carolina, Andrew Rabinovich, and Serge Belongie. "Object categorization using co-occurrence, location and appearance." 2008 IEEE Conference on Computer Vision and Pattern Recognition. IEEE, 2008.

[15] Lu, Cewu, et al. "Visual relationship detection with language priors." European conference on computer vision. Springer, Cham, 2016

[16] Dai, Bo, Yuqi Zhang, and Dahua Lin. "Detecting visual relationships with deep relational networks." Proceedings of the IEEE conference on computer vision and Pattern recognition.

[17] Lan, Meng, et al. "Defect detection from UAV images based on region-based CNNs." 2018 IEEE International Conference on Data Mining Workshops (ICDMW). IEEE, 2018

[18] Tan, Mingxing, and Quoc Le. "Efficientnet: Rethinking model scaling for convolutional neural networks." International conference on machine learning. PMLR, 2019.

[19] Zheng, Zhaohui, et al. "Distance-IoU loss: Faster and better learning for bounding box regression." Proceedings of the AAAI Conference on Artificial Intelligence. Vol. 34. No. 07. 2020.

[20] Ying Liu, Luyao Geng, Weidong Zhang, Yanchao Gong, and Zhijie Xu, "Survey of Video Based Small Target Detection," Journal of Image and Graphics, Vol. 9, No. 4, pp. 122-134, December 2021. doi: 10.18178/joig.9.4.122-134.

A Water Gauge Scale Capturing Method in Tidal Well Based on Image Recognition

Hai Guo[1], Yanyan Wang[2] and Yifei Zhang[1]

[1]Key Laboratory of Ocean Observation Technology, MNR, National ocean technology center, Tianjin, China
[2]Tianjin Key Laboratory of Information Sensing and Intelligent Control, Tianjin University of Technology and Education, Tianjin, China

tjdxwyy@163.com

Abstract. Tide level observation data is an important technical support data for the delimitation of warning tide level and disaster prevention. The performance of the tidal well is tested by regularly checking the external water gauge, so as to maintain the tidal follow-up and wave dissipation of the tidal well. The check of water gauge inside and outside the well shall be carried out once a month. It is time-consuming and labour consuming. The paper proposes a real-time borehole water gauge capturing algorithm in tidal well based on image recognition. Experiments have proved that the method can recognize the water gauge scale effectively. The comparison experiments with the other standard measurement method are done. It can be seen from the comparison chart that the error between the water gauge scale and the tide level value is ± 1cm, which meets the national standard.

1. Introduction

Tide level observation is of great significance in the fields of marine disaster warning and reporting, sea level measurement research, marine engineering design and so on. Tide level observation data is an important technical support data for the delimitation of warning tide level and disaster prevention. After years of construction and development, our national ocean observation network has built more than 100 tide stations along the coast to comprehensively monitor the fluctuation of coastal tides in real-time. The acquisition of tidal observation data is realized by installing tide gauge in tide well. In order to ensure the quality of tidal observation data, the tidal gauge system in the vertical shaft is set up. The performance of the tidal well is tested by regularly checking the external water gauge, so as to maintain the tidal follow-up and wave dissipation of the tidal well. The check of water gauge inside and outside the well shall be carried out once a month, and two persons shall observe simultaneously. Obviously, the traditional manual observation mode is time-consuming and labour consuming.

With the progress and development of image recognition technology, computer vision technology based on deep learning has been successfully applied to security monitoring, electronic medical treatment, unmanned driving, intelligent payment and other fields. Based on image recognition technology, some researches are developed to apply in the water gauge scale detection. Zhang proposes an intelligent image water level recognition method for water level identification and monitoring[1]. Niu designs an automatic water level monitoring system based on image gauge[2]. Chen proposes a method of water level data capturing based on video image recognition[3]. These methods obtain good recognition result. So, the image recognition method can be used for water gauge recognition. Nowadays, few research is done for water gauge scale capturing in tidal well, due to the particularity of tidal observation.

This paper proposes a real-time borehole water gauge capturing algorithm in tidal well based on image recognition. The method could carry out long-term and real-time observation of borehole water gauge. It also can reduce the workload of observation staff. The data from the method can be compared with the observation data of tide gauge for a long time, which is convenient for observers to monitor the operation state of tide gauge. At the same time, the observers could find operating obstacles in advance and ensure the quality of the tide data.

2. Characteristic analysis of tidal gauge and the algorithm flow

Figure 1 shows the tidal gauge in well. The tidal gauge will move up and down with the change of tide. There are two types of reading symbols: character marks, long and short scale marks from figure 1. Character marks include three types of marks: meter level character marks (such as 3m, 4m or 3xx, 4xx), decimeter level character marks (such as 30,40 or 3x, 4x), and centimeter level character marks (such as 1,2,3,...,9). The long and short scale marks correspond to the whole centimeter scale, and the minimum unit of scale is 1mm, 2mm or 5mm. By analyzing the image characteristics of the water gauge in the well, the recognition algorithm should obtain the moving area of the water gauge in the well. So, the inter frame difference is developed due to its strong adaptability to dynamic environment. The deep learning algorithms is developed to obtain the character mark, the long and short scale mark. It contains craft, resnet50, canny edge detection and CTC algorithms. Finally, the voting mechanism and continuous reading smoothing processing method is applied to obtain the accurate scale value of the water gauge.

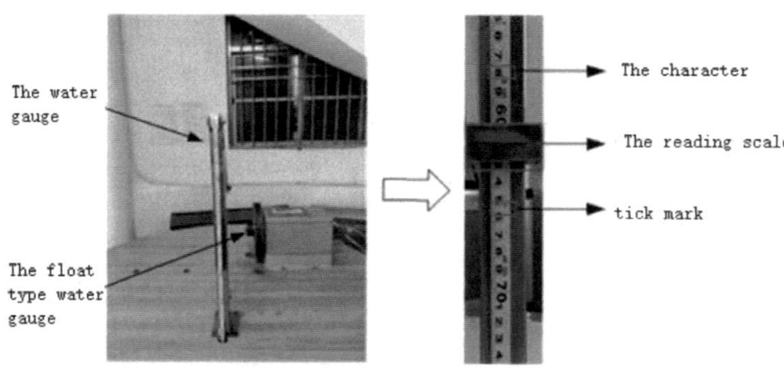

Figure 1. The tidal gauge in well

In this paper, the borehole water gauge scale recognition algorithm is divided into the following steps, as shown in Figure 2.

1. Moving area detection method: To detect and intercept the moving area based on the inter frame difference method. Then the real-time water gauge area in the well can be obtained;
2. Scale mark recognition method: To obtain the position of the scale reading line of the water gauge through the canny edge detection algorithm;
3. Character recognition of the water gauge: A recognition model based on craft single character detection branch and dictionary is proposed. The character marks containing meter, decimeter and centimeter are detected based on craft algorithm. The character marks such as letters and numbers are obtained based on resnet50 and CTC algorithm;
4. The water gauge value output: The accurate value of the water gauge is output through voting mechanism and mathematical morphology filtering.

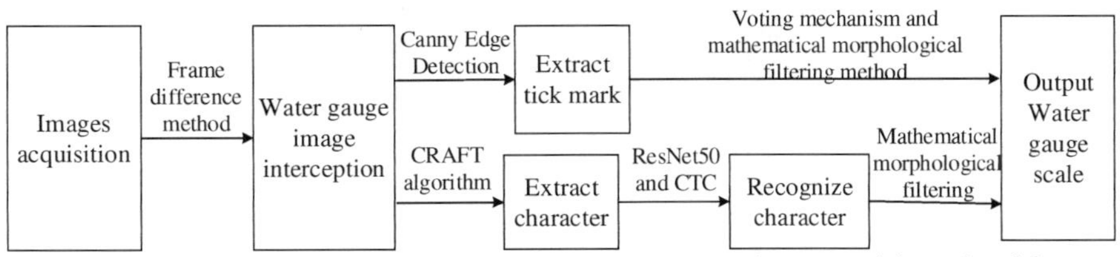

Figure 2. The structure of borehole water gauge scale recognition algorithm

2.1. Borehole water gauge image interception

The principle of inter frame difference method is to obtain the contour of moving target by difference between two adjacent images. It has the advantages of simple calculation, good real-time performance and strong adaptability to dynamic environment[4-6]. It is very effective to obtain moving target in the same scene. In this paper, the frame difference method is used to identify the water gauge images in two adjacent frames. Then the change of the water gauge scale is intercepted due to the tide upgrade. For the adjacent frame images, the water gauge scale is the only motion factor, and the other corresponding pixels are basically the same. When the water gauge appears in the image, it will change for the pixels, even if the motion is slow. Through the comparison of pixels between adjacent frame images, it is found that the changed pixels are represented by the image. The formula is as follows[7]:

$$D(x,y)=\left|I_k(x,y)-I_{k-1}(x,y)\right| \quad (1)$$

Where, $D(x,y)$ is the difference image between the frame image $I_k(x,y)$ and the frame image $I_{k-1}(x,y)$. Threshold segmentation is performed on the difference image to obtain the binary foreground image:

$$T_k(x,y)=\begin{cases}1, D_k(x,y)\geq T\\0, D_k(x,y)<T\end{cases} \quad (2)$$

Where, $T_k(x,y)$ is a binary foreground image; T is the segmentation threshold.

In image recognition, the image changes over time (such as the change of illumination), which will change the pixel value between the front and back frames. By setting the threshold, the target image can be extracted when the environment changes.

2.2. Edge detection

The edge detection algorithm is used to achieve the water gauge scale. The algorithm is needed to identify as many tick marks in the image as possible. The canny operator is selected for edge detection. It is a multistage method. It contains five steps: image graying, noise reduction, finding intensity gradient of the image, non-maximum suppression, double threshold and edge tracking by hysteresis[8-10]. The water gauge image, the grayscale image and the image after canny edge detection are shown in figure 3. The tick mark of the water gauge could be detected clearly.

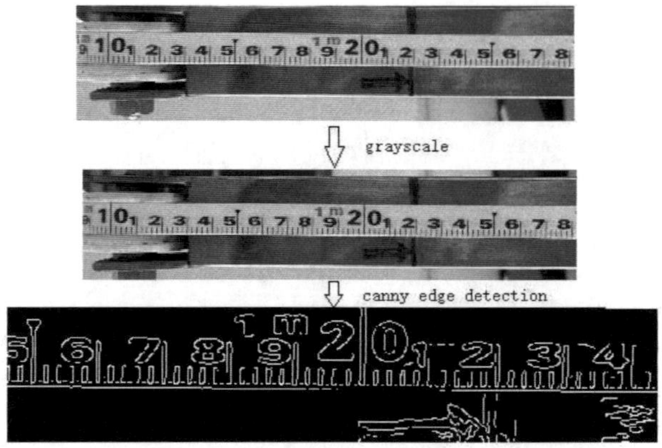

Figure 3. The result image of canny edge detection

2.3. Text detection and recognition of water gauge

Natural scene text detection based on neural network is a widely used text detection method. It is mainly divided into regression based, segmentation based, end-to-end and character level methods. The character mark of the water gauge in the well is mainly composed of Arabic numerals and the length unit “m”. Each character is independently distributed. Based on the character characteristics of the water gauge in the well, the paper selects the craft method based on single character detection. This method belongs to the branch of character level detector. The method could solve the disadvantage existing in the traditional text detection and recognition scheme, such as easily missing single words and can not detect each separated mark on the water gauge separately.

Craft algorithm first detects a single character and the connection relationship between characters. Then the final text line can be detected. The algorithm takes the full convolution neural network based on vgg-16 as the backbone network. The method based on u-net is developed for the decoder part[11-15]. This method adopts a top-down feature aggregation method, which is helpful to aggregate deep and shallow features. Finally, two channels are output to predict the pixels: 1) Region score, which represents the probability that each pixel is the center of the character. 2) Affinity score, which represents the center probability of the space between adjacent characters. According to the text characteristics of water gauge, the text frame produced by craft method is a conventional rectangle. Firstly, there are two probability maps between 0 and 1. The binarization thresholds of the two maps are calculated respectively. The two binarization maps are combined to obtain a mask graph; Secondly, use the connection component label (CCL) for area connection; Finally, opencv’s MinAreaRect (minimum circumscribed rectangle) is used to frame the smallest quadrilateral area. The method is applied in our system. The final recognition accuracy is high from figure 4.

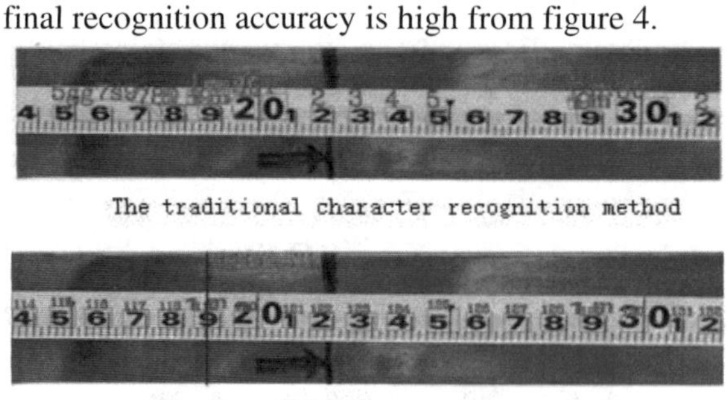

Figure 4. The comparisons between the traditional character recognition method and the resnet50+CTC recognition method

2.4. Algorithm for distinguishing the number mark and the scale mark of the water gauge

The centimeter marks are arranged equidistantly and equidistantly. So the algorithm for distinguishing the number mark and the scale mark of the water gauge is proposed. It is efficient and robust. The principle of the algorithm is to analyze the identified numbers by using the characteristics of continuous growth of numbers and equidistant scale lines. It has a strong correction function for missing and wrong detection of digital marks, missing and wrong detection of scale lines, etc. The algorithm obtains the final true value of water gauge data according to meter level, decimeter level, centimeter level numerical markers and millimeter level estimation. Its flow chart is shown in the figure 5 below.

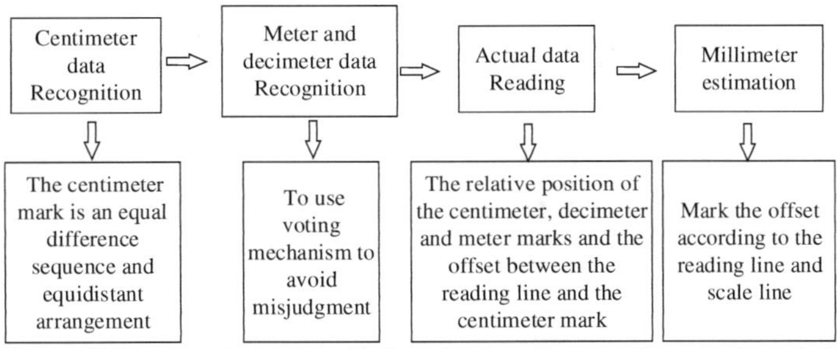

Figure 5. The flow chart of the algorithm for distinguishing the number mark and the scale mark of the water gauge

3. Experiments and Analysis

The verification of water gauge recognition algorithm based on image recognition technology is applied in the current water gauge system and automatic tide level observation equipment. According to the frequency requirements of tidal observation data output by automatic tidal observation equipment per minute, the observation data are obtained synchronously for comparison. The accuracy of the water gauge recognition algorithm is obtained. The diversity of water gauges in the wells of existing tide stations has caused environmental differences to the image recognition technology. At the same time, the poor wave elimination of some tide measuring wells will cause the float to drive the water gauge in the well to fluctuate rapidly up and down under bad weather and sea conditions, resulting in errors in image recognition and output data. The amplitude limiting filtering algorithm is used for data processing.

The physical diagram of the experimental system is shown in the figure 7. It contains a camera, an illumination and CPU. The algorithm is applied in the CPU. The camera and the illumination are combined to collect the images of the water gauge.

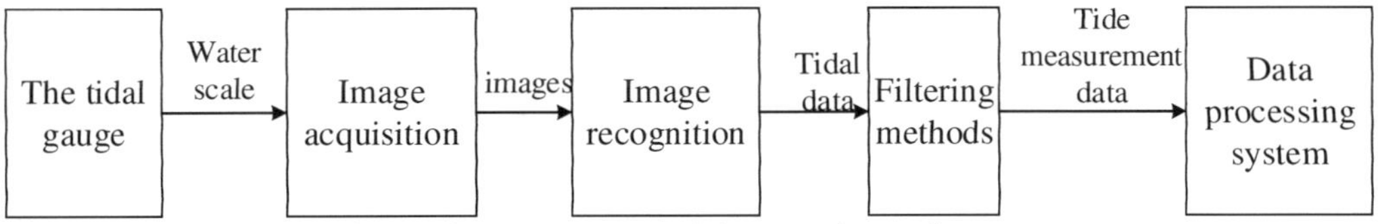

Figure 6. The structure of the experimental system

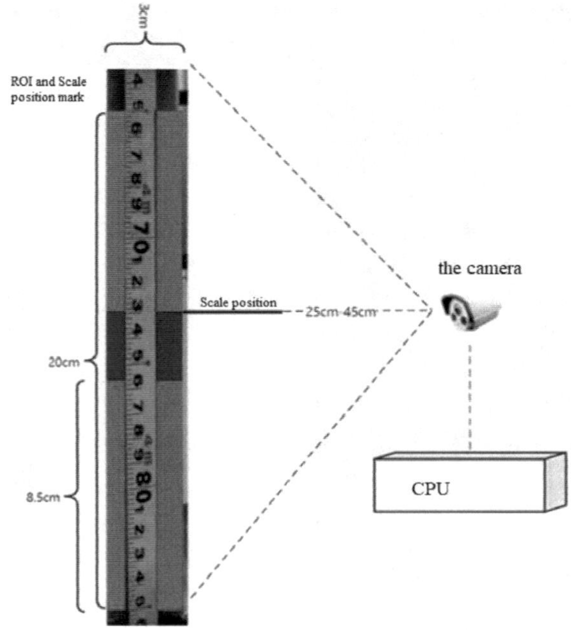

Figure 7. The physical diagram of the experimental system

The water gauge system in the tidal observation well is composed of the water gauge in the well, the float system and the reading pointer. The water gauge in the well usually adopts the ribbon glass fibre flexible ruler. The water gauge in the well will rise or fall with the change of water level in the well through the float. The observer reads the water gauge in the well through the reading pointer to obtain tidal data. The minimum scale of water gauge is 0.1cm, and the cumulative error of gauge length is no more than 0.5cm. The difference between the water gauge and the tide gauge data shall not be greater than 1cm.

The image acquisition module obtains the RGB image of the water gauge system in the well through the camera. The image recognition algorithm is applied in the industrial computer to show the recognition value of the water gauge. The camera type is Hikvision DS-2CD2646F(D)WDV4-LZS network camera, and the industrial computer model is Yanqin OTF-5632. The image quality and system main frequency meet the requirements of image recognition algorithm. The pictures processed during the field test are shown in figure 8.

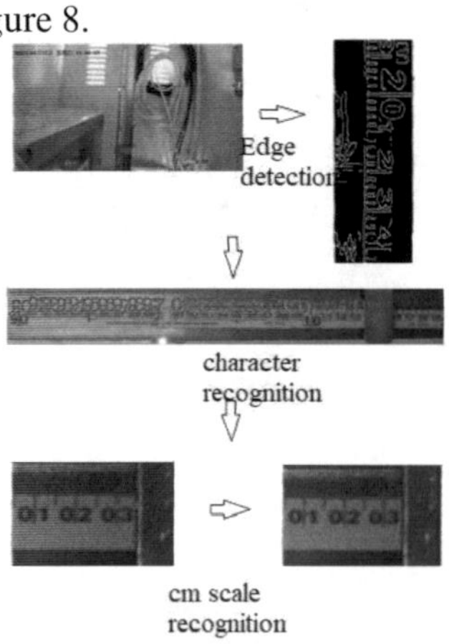

Figure 8. The pictures processed during the field test

In order to verify the correctness and accuracy of the image recognition algorithm, the data extracted by the algorithm are compared with the observation results of the tide gauge. The data sample takes the data of the whole day, and refers to the minute level of the output data frequency of the tide gauge. The number of samples is 1440, as shown in figure 9. The horizontal axis is the time axis, the unit is s, and the vertical axis is the tide level value, the unit is cm.

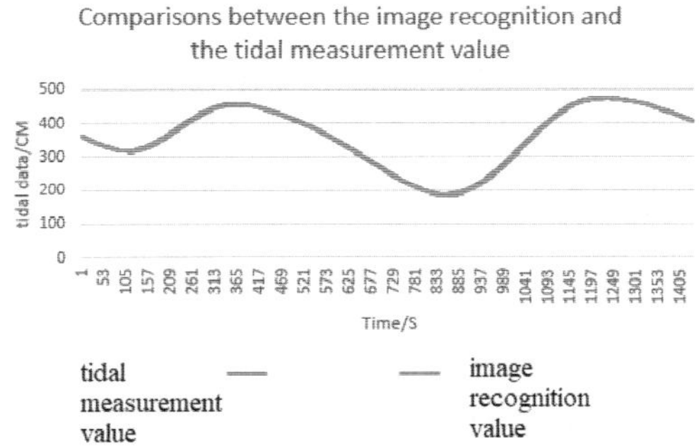

Figure 9. The comparison chart between the tidal measurement value and the image recognition value

It can be seen from figure 9 that the error between tidal measurement value and the image recognition value is ± 1cm, which meets the requirements for manual observation of water gauge in the well in the code for ocean observation part 2: coastal observation (GB / T 14914.2-2019).

4. Conclusions

The tide level extracted by the borehole water gauge capturing algorithm based on image recognition technology has high accuracy. It is enough to replace manual observation. The new method can not only ensure the timeliness and stability of tidal data comparison, but also avoid the observation error caused by the technical level of the observers themselves. At the same time, the application of the new method can greatly reduce the workload of front-line workers, and has popularization value. However, the method in this paper has a silent period of 5 to 10 minutes (i.e. the learning and training time of the algorithm during initialization) to adapt to different environments. It needs to be optimized in the later application process.

5. References

[1] Chen J., Method of Water Level Data Capturing based on Video Image Recognition[J]. 2013 *Water Resources Informatization* **1** 48-57.

[2] Wang Lei, Chen Ming, and et. Al, Research on water level recognition method based on deep learning algorithms[J] 2020 *Water Resources Informatization* **3** 39-56.

[3] Zhang Jingfang, Yin Xianwen and et.al, Solution and application of intelligent image water level recognition[J] 2021 *Water Resources Informatization* **3** 59-65.

[4] Krizhevsky A,Sutskever I,Hinton G E. Imagenet classification with deep convolutional neural networks[J] 2012 *Advances in neural information processing systems* **25** 1097-1105.

[5] He K,Zhang X,Ren S,et al. Deep residual learning for image recognition[C]// *2016 Proceedings of the IEEE conference on computer vision and pattern recognition* 770-778.

[6] Howard A G,Zhu M,Chen B,et al. MobileNetV2: Efficient convolutional neural networks for mobile vision applications[J] 2017 *Neural Networks for Mobile Vision Applications* **1704** 04861.

[7] Sandler M, Howard A, Zhu M, et al. MobileNetV2: Inverted residuals and linear bottlenecks[C] 2018 Proceedings of the IEEE conference on computer vision and pattern recognition 4510-4520.

[8] Rrn S, Hr K, Grishick, et al. Faster R-CNN: Towards real-time object detection with region proposal networks[J]. 2017 *IEEE Transactions on Pattern Analysis and Machine Intelligence* **39** (6) 1137-1149.

[9] Redmon J, Divvala Sivvala S, Grishick R and et al. You only look once: Unified, real-time object detection[C] 2016 IEEE Conference on Computer Vision and Pattern Recognition. Las Vegas: CVPR 779-788.
[10] Zhanf Z, Zhou Y, Liu H Y, et al. Visual measurement of water level under complex illumination conditions[J] 2019 *Sensors* **19** 4141.
[11] Zhang Z, Zhou Y, Liu H Y and et al. In-situ water level measurement using NIR-imaging video camera[J] 2019 F*low Measurement and Instrumentation* **67** 95-106.
[12] Angie M. Ceniza, Tom Kalvin B. Archival, and Kate V. Bongo, Mobile Application for recognizing text in degraded document images using optical character recognition with adaptive document image binarization[J], 2018 *Journal of Image and Graphics*, 6 (1) 44-47
[13] Ryo Hasegawa, Yutaro Iwamoto, and Yen-Wei Chen, Robust Japanese road sign detection and recognition in complex scenes using convolutional neural networks[J] 2020 *Journal of Image and Graphics* 8 (3) 59-66.
[14] Mengting Liu, Guoying Liu, Yongge Liu, and Qingju Jiao, Oracle Bone inscriptions recognition based on deep convolutional neural network[J] 2020 *Journal of Image and Graphics*, 8(4) 114-119
[15] W. U. Wickramaarachchi and S. Vasanthapriyan, Multi-Layer framed offline signature recognition algorithm[J] 2015 Journal of Image and Graphics, 3(1) 11-15.

Acknowledgment
This research was financially supported by the National key R & D plan (No. 2017YFC1405101).

Cloud Resource Scheduling Method based on Markov Process and the Cuckoo Search

Bin Qi[1], Pan Zhang[1], Hong Wu[1] and Miao Yan[2]

1.Beijing Space Information Transmission Center, Beijing, China
2.PLA Unit 32709, Beijing, China

qbfree@foxmail.com

Abstract. With the rapid development of cloud computing, the computing reliability of cloud servers has become an important research area. To improve the load balancing capability of cloud servers, an autonomous adaptive architecture cloud computing framework based on software-defined networking and network function virtualization were proposed. And a cloud resource scheduling method based on the Markov process model and cuckoo search algorithm to realize the load balancing of cloud nodes was designed. The cloud platform verified by the experiment shows that the method can effectively guarantee the service reliability of cloud computing.

1. Introduction

With the development of big data and cloud computing, the ability to autonomously provision service resources have gradually become a constraining bottleneck for cloud services. Due to the complex and changing cloud computing environment with node heterogeneity, changing resources, and uncertainty, it is necessary to strengthen cloud resource management and optimize cloud task scheduling to meet both the computing requirements of users and the benefits of cloud computing service providers.

The core problem of cloud computing is the resource scheduling problem, called the NP-complete problem [1]. The optimization method usually includes static and dynamic methods. Static methods mainly include round-robin scheduling algorithms [2], and the idea is to follow the grid computing scheduling working principle of distributed systems, with long computation time and high complexity, which cannot be adapted to large-scale cloud computing task scheduling and easy imbalance of resource load. Dynamic methods mainly have some swarm intelligence optimization algorithms, whose path solving speed is faster and scheduling scheme is better than static methods, which become the main research direction at present [3]. The genetic algorithms, particle swarm algorithms, and ant colony algorithms all have corresponding advantages and disadvantages [4]. Literature [5] proposed a resource scheduling method for Hadoop Yarn, which integrates the ant colony algorithm and particle swarm algorithm, optimizes the pheromone volatility coefficient of the ant colony algorithm and incorporates the self-recognition capability of the particle swarm algorithm to finally improve the accuracy of the solution. The literature [6] proposes a cloud computing resource scheduling strategy based on the cuckoo search algorithm to improve resource utilization, but there is room for optimization in terms of computational overhead and algorithmic solution space. The literature [7] present a new cluster scheduling system, ROSE, that is based on a multi-layered scheduling architecture with an ability to over-subscribe idle resources to accommodate unfulfilled resource requests. The literature [8] present an efficient resource scheduling technique using the ant colony optimization (ACO) algorithm, to minimize execution cost and time. To verify the effectiveness of OSML and obtain a well-generalized model [9] collect a dataset containing over 2-billion samples from 11 typical microservices running on real servers over 9 months. The literature [10] propose a new YARN scheduler that can effectively reduce the makespan (the total execution time) of a batch of MapReduce jobs in Hadoop YARN clusters by leveraging the information of requested resources, resource capacities, and dependency between tasks.

To improve the efficiency of resource allocation under the cloud framework, we analyse and design a variable-configuration cloud framework based on the "growable software idea" for dynamic adjustment planning. A resource scheduling algorithm based on the Markov process and cuckoo search to improve the reliability of communication cloud services is proposed.

2. Variable-configuration cloud framework research

Benefiting from the unified management and dynamic configuration of global resource optimization by technologies such as SDN (Soft Define Network) and NFV (Network Function Virtualization), the network management capability, resource collaboration level, and service quality are enhanced. The similarities and differences brought about by different networks can be unified through modularization, which makes the convergence of multiple networks easier and provides important prerequisites for collaborative resource provisioning and management. Based on the virtualization capabilities provided by NFV, it can improve the control of virtual network operators and provide users with the required content, applications, and computing services more effectively, thus creating a high-performance, low-latency service environment, accelerating the download and execution of various contents and applications in the network, improving the response speed of user requests, and optimizing network services and user experience.

Based on the theoretical concept of "Growable Software" proposed in the project of "Sustainable Evolution Software Theory, Methodology, and Technology" [11]. It makes cloud services no longer depend on external resources such as hardware and software environment and user service contents, but can be actively sensed, intelligently adapted, and continuously evolved. The system resources can be intelligently deployed according to the task execution queue and execution rate of the whole domain, and the resource service capacity of each node can be estimated based on the task effect evaluation module to guarantee the load balancing of the whole domain resources most effectively.

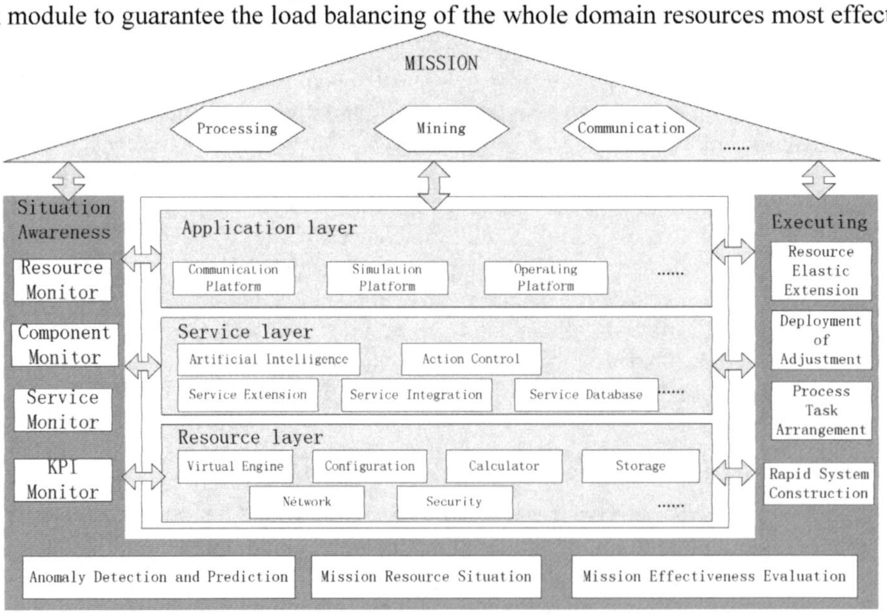

Figure 1. Cloud Variant Framework.

Journal of Physics: Conference Series **2320** (2022) 012030 doi:10.1088/1742-6596/2320/1/012030

The cloud is a cloud-based reconfiguration with the smallest functional module as the plurality of elements, and it realizes the service capability for global planning and various special computing business scenarios. At the technical implementation level, it is designed around three levels of "application-service-resource" and two stages of "situational awareness-decision execution". The core mission-oriented software-defined service and resource assurance framework enables "visibility" of changes in application requirements and operating environment, and "control" of business processes, service interactions, and resource configurations to form a continuous service capability of resources.
As shown in Figure 1, the framework defines the software into three layers: application layer, service layer, and resource layer. It supports the two core functions of environment cognition and decision execution in a deeply software-defined way, which makes all disturbances and changes at each level of the system visible to the control plane on the one hand, and allows the control plane to carry out cross-level global regulation from the perspective of mission assurance on the other.

First, the control plane consisting of environment cognition and decision execution is the core of the framework. Using environment cognitive functions, interfaces such as basic resource monitoring, container status monitoring, service status monitoring, business key index monitoring, and abnormality analysis and prediction, it obtains real-time status information of computing, storage, network, information, and services. Then, it makes intelligent analysis and prediction, forming a "task-resource" situation that reflects the system operation entity's guarantee of tasks in real-time. It evaluates the impact of various abnormalities or predicts the possibility of abnormalities on tasks on this basis. The "task-resource" situation, based on which the impact of various types of anomalies or the possibility of predicting anomalies on the task is evaluated. The decision execution functions and interfaces such as resource elasticity scaling, service deployment adjustment, business process orchestration, rapid system reconfiguration are used to adaptively match or generate various types of anomalies in the resource layer, service layer, application layer, and user-triggered operation level transfer. It adaptively matches or generates optimal policies and sends them to the resource plane, driving the automated execution of adjustment policies covering business processes, service interactions, and environment configurations.
Secondly, the resource plane composed of elements at the application, service, and resource layers, as the monitoring and control object of core mission assurance, and based on its own real-time, reliability, availability, security, and other capability features effectively support the implementation of the policies issued by the control plane, such as in the basic resource layer, switches, routers, switches, and other network resources can realize link updates by changing packet forwarding policies. In the service layer, different services are equipped with request redirection, service migration, rapid deployment, and other capabilities. In the application layer, new systems "assembled" for different special types of tasks provide business process switching, system capacity succession, access rights control, etc. In the application layer, for new systems "assembled" by different special types of tasks, it provides business process switching, system capability succession, access rights control, and other capabilities.

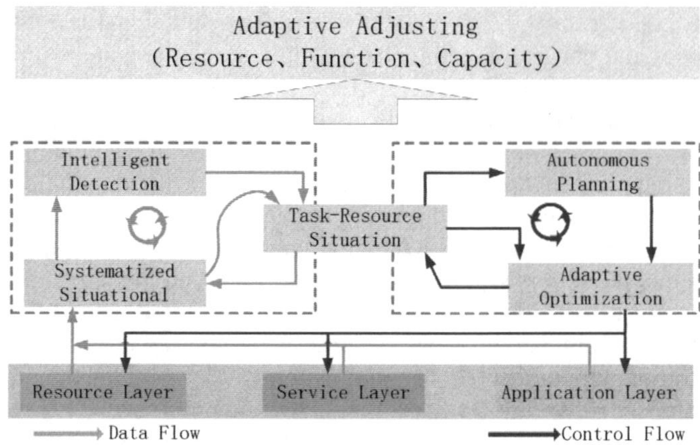

Figure 2. Circumferential autonomous adaptation mechanism.

For the core mission assurance framework defined by deep software, a "double-loop" autonomous adaptation mechanism based on "environment awareness loop - decision execution loop" is established based on the guidance of growable software operation mechanism, as shown in Figure 2. Through active awareness and change control for mission assurance, the system is equipped with the ability to respond to changes in application requirements and the operating environment.

Among them, the environment cognitive loop is equipped with distributed and systematic monitoring for resource layer, service layer, and application layer, as well as intelligent abnormality analysis and prediction. It can also dynamically generate a "task-resource" situation for mission assurance and provide input for the decision execution loop. When the application demand changes or the operating environment changes, the decision execution loop has the ability of autonomous planning and online dynamic adjustment of a "task-resource", and at the same time, it accumulates the experience knowledge such as solution library and strategy library related to autonomous adaptation in the process of system construction or adjustment to improve the system growth capability.

3. Resource scheduling algorithm based on Markov process and cuckoo search

In case of sudden failure of the cloud node, after the end node sends out a computing request for data services, the local computing server will judge whether to accept the request based on its load level and if the threshold value is exceeded, it will obtain computing and storage services from the core server or send out a load collaboration request to the neighboring network servers based on the resource scheduling algorithm until a low-load server capable of carrying the service is found to accept the request and select a suitable load transfer link. The Markov-Cuckoo Search (MarCuS) based processing method realizes the global traffic balancing in the case of business bursts and ensures that the servers are all within a reasonable load range. Based on the autonomous adaptive cloud framework to form a proprietary service network and resources, the cloud service does not degrade in service quality due to uneven load, and effectively improves the reliability of cloud services.

3.1. Markov process-based link weight assignment

Suppose $\mathbf{S} = \{1,2, \dots \dots, N\}$ characterizes the set of all servers, and the three levels of load thresholds high, standard, and low are denoted as MAX(i), NORMAL(i), and MIN(i), respectively, with $i \in \mathbf{S}$. The occupancy metrics of individual services for servers are mainly represented by triadic features, C(t) represents CPU usage, M(t) represents memory occupancy, W(t) represents task waiting time, and the weight parameter is set to λ and its input comes from the hierarchical dynamic weight design method, then the load characterization of task i at time t is

$$L_i(t) = \lambda_1 C_i(t) + \lambda_2 M_i(t) + \lambda_3 W_i(t) \tag{1}$$

In which, $\sum \lambda_i = 1$. Further, the load state of the whole server at moment t can be characterized as

$$\mathrm{LOAD}(i) = \sum_{i=1}^{n} L_i(t) \tag{2}$$

Therefore, when the load LOAD≤MIN(i), it can be judged that the server is in a low load state and can undertake coordination tasks. When LOAD≥MAX(i), it can be judged that the server is overloaded and the user response is lagging, so the required tasks can be forwarded to the low-load server until it converges to NORMAL(i). When LOAD is within the normal load range, the server is in normal operation, and the LOAD in the cloud region can be arranged in ascending order as the backup server for load balancing.

The centralized scheduling policy sets up a module to collect the load situation of each node, adjusts the allocation of dedicated service resources of important single nodes according to the hierarchical task requirements under the cloud framework. It unifies the resource scheduling for multi-task combinations of task sets, transferring the services of high-load servers to low-load servers to ensure the load balancing and business priority of the cloud platform. Therefore, to guarantee a reliable transfer from high-load nodes to low-load nodes, a Markov process model is used to determine the nodes to be migrated.

There are often multiple paths between the nodes in the resource network, and if there are more connectable paths between the nodes, it indicates that the probability of choosing instead of the directly connected path is higher and the corresponding weight is lower. The network topology is represented by an undirected graph G = (V, E), where V denotes the set of nodes in the network and E denotes the set of links, as shown in Figure 3, there are n paths from node i to node j, and only P_1 is a directly connected channel.

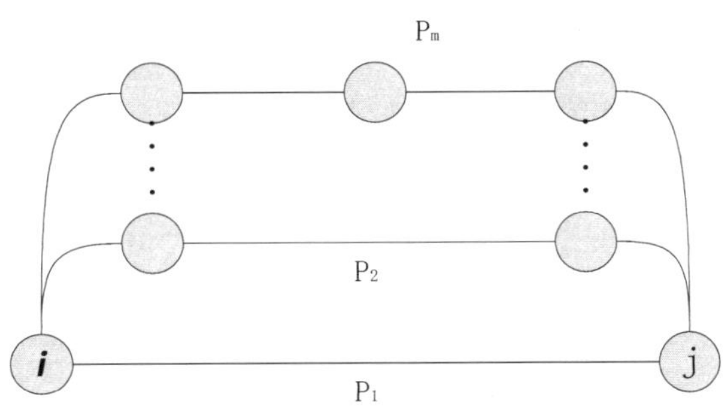

Figure 3. Topology of undirected graph nodes.

To obtain the fully connected graph of the adjacency matrix, the n-step transfer probability expression under the Markov process is

$$P^{(n)} = P^n \tag{3}$$

The one-step transfer matrix of the system state consists of the one-step transfer probability $P_{ij}^{(1)}$. The conditional probability distribution of the state space node X_{n+1} in the Markov process depends only on X_n, then the expression of the transfer probability of arriving at j from state i at moment t is

$$P_{ij}(t) = P\{X_{n+1} = j | X_n = i\} \tag{4}$$

Since the link reachability is proportional to the weights, the sum of the non-one step transfer probabilities of nodes i and j is

$$P_{ij} = {\sum_{t=2}^{n}(P_{ij}^{(t)} + P_{ji}^{(t)})}\Big/_{2} \tag{5}$$

The normalized link weight is

$$\varepsilon_{\text{ij}} = \frac{\frac{1}{P_{ij}}}{\sum \frac{1}{P_{ij}}} \tag{6}$$

After the weights and inter-node transfer probabilities obtained above are used as input to the swarm intelligence optimization algorithm to find the optimal low-load nodes and paths that support load balancing and transfer tasks from high-load nodes to low-load.

3.2. Cuckoo search algorithm for solving

The so-called cuckoo search algorithm [12] is to map the bird's nest location to the solution to be solved by the bird's nest location, the nest location superiority is mainly calculated by the fitness function value, some better nest locations are ranked and saved, the poor nest locations are updated according to the Lévy flight characteristics, and the flight path of the updated cuckoo is determined by combining the exotic bird egg discovery probability pa. Among them, the initial position weights of the nest are the sequence of weights of the Ma's chain, while the bird egg discovery probability relies on the node transfer probability P_{ij} for assignment to ensure that the relevant paths are constrained during the initial search to improve the efficiency of the search direction and speed up the optimal value solving.

Assuming that x_i^{t+1} and x_i^t denote the nesting position at the current moment and the next moment, respectively, the update operation of the nesting position is

$$x_i^{t+1} = x_i^t + \alpha \oplus \text{Lévy}(\lambda) \tag{7}$$

where α is the step adjustment parameter of the cuckoo search algorithm, and the adjustment parameter is determined by the positive correlation of the link weights ε_{ij}.

According to the cuckoo Lévy flight characteristics, the expression of the path (s) is calculated as

$$s = \mu/|v|^{1/\beta} \tag{8}$$

Where u and v are random numbers satisfying $u \sim \text{N}(0, \sigma_u^2)$ and $v \sim \text{N}(0, \sigma_v^2)$, and the expressions of σ_u and σ_v are

$$\begin{cases} \sigma_u = \left\{ \dfrac{\Gamma(1+\beta)sin(\pi\beta/2)}{\Gamma\left[\frac{1+\beta}{2}\right]\beta/2^{(\beta-1)/2}} \right\}^{1/\beta} \\ \sigma_v = 1 \end{cases} \tag{9}$$

4. Experiment and Validation

Before the start of the experiment, a series of Markov sequences are constructed as the initial moments of the Markov chain. The simulation model establishes a Markov chain based on data distribution sequences 100 times under initialization and calculates the probability of one-step transfer between different states in the state space, initializes the weights using equation (9). The optimal path by cuckoo search algorithm and performs load migration is determined. Then it simulates the network of 10 nodes with task load scenarios limited by CLoudSim simulation cloud to satisfy the minimum communication and data transmission task load simulation situation and compared with the mainstream ant colony resource scheduling algorithm (ANA) [13] and particle swarm resource scheduling algorithm (PSO) [14], and the experimental results are shown in Figure 4.

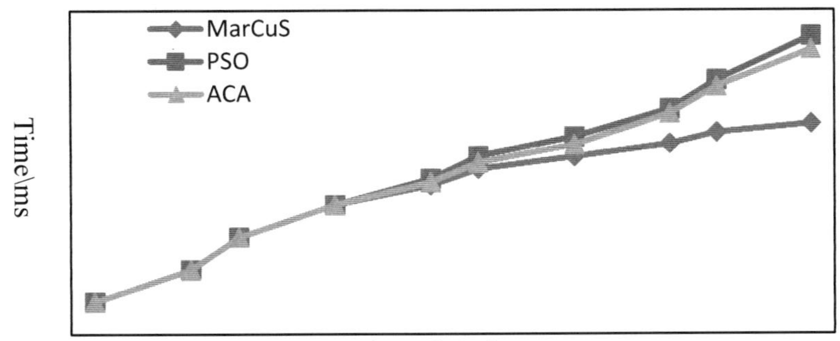

Figure 4. Comparison of task completion time.

The experimental results show that when the number of tasks keeps increasing, the completion time of MarCuS, ANA, and PSO algorithms also rises. When the task size is small, the execution speed of the three algorithms does not differ much, and when the task size satisfies NORMAL and continues to grow larger, the time increase of MarCuS is lower than the other methods, i.e., the computational efficiency is relatively high. The load allocation ratio of cloud resources is listed in Table 1, and the MarCuS significantly outperforms the other compared algorithms in the server resource allocation ratio to achieve the load balancing effect requirement.

Table 1. Resource allocation ratio of tasks on the cloud.

Resource Number / Scheduling Algorithm	1	2	3	4	5
PSO	0.277	0.144	0.356	0.309	0.213
ANA	0.258	0.175	0.332	0.289	0.200
MarCuS	0.257	0.160	0.302	0.276	0.205

5. Conclusion

In this paper, we proposed a variable-configuration cloud framework, the MarCuS resource scheduling algorithm for the resource load balancing problem of cloud computing services under different scenarios, and the experiments show that the resource scheduling algorithm under this framework can meet the requirements of load balancing and cloud service reliability.

6. References

[1] TOPCUOGLU H, HARIRI S, WU M Y. Performance-effective and low-complexity task scheduling for heterogeneous computing[J]. IEEE Transactions on Parallel and Distributed Systems, 2002, 13(3):260-274.

[2] CHENG D Z H, ZHOU X B, LAMA P, et al. Energy efficiency aware task assignment with DVFS in heterogeneous hadoop clusters [J]. IEEE Transactions on Parallel and Distributed Systems, 2018,29(1):70-82

[3] Su Z.F., Wang Guojun, Li R.F.. A multi-dimensional QoS cloud resource scheduling approach based on stakeholder perspective[J]. Journal of Communication.2019,40(6):102-115

[4] Liao W. C.,Xia S. Y.,Jia S. J. Multi-path coverage test data generation based on ant colony algorithm[J]. Journal of Electronics,2020,48(7):1330-1342.

[5] Li Yuanzhen, Yang Qun, Lai Shangqi, Li Bohan. A research on a resource scheduling method for Hadoop Yarn[J]. Journal of Electronics,2016,44(5):1017-1024. (in Chinese)

[6] Li Jia, Xia Yunni. A cuckoo search algorithm for solving cloud computing resource scheduling problem[J]. Control Engineering,2019,26(1):170-174.

[7] X. Sun, et al. ROSE: Cluster Resource Scheduling via Speculative Over-Subscription[C]. 2018 IEEE 38th International Conference on Distributed Computing Systems (ICDCS), IEEE Press, 2018:949-960.
[8] Harvinder Singh, Anshu Bhasin, Parag Kaveri. SECURE: Efficient resource scheduling by swarm in cloud computing[J]. Journal of discrete mathematical sciences and cryptography,2019,3:127-137.
[9] Lei Liu. QoS-Aware Machine Learning-based Multiple Resources Scheduling for Microservices in Cloud Environment. ARXIV. 2019.
[10] Y. Yao, H. Gao, J. Wang, B. Sheng and N. Mi. New Scheduling Algorithms for Improving Performance and Resource Utilization in Hadoop YARN Clusters[J]. IEEE Transactions on Cloud Computing, 2021,3(9):1158-1171.
[11] Xu Chang, Qin Yi, Yu Ping, et al. Theoretical approaches and implementation techniques for growable software: from paradigm to leapfrog [J]. China Science: Information Science,2020,50(11):1595-1611. (in Chinese)
[12] F.B.Ozsoydan, i. Golcuk.Application of Cuckoo Search Algorithm and its Variants[J]. STNIC,2020.6:47-77
[13] DORIGO M, MANIEZZO V, COLORNI A. Ant system: optimization by a colony of cooperating agents[J]. IEEE Transactions on Systems, Man, and Cybernetics, Part B (Cybernetics), 1996, 26(1): 29-41
[14] Hao Liu, Yue Wang, et al. A modified particle swarm optimization for large-scale numerical optimizations and engineering design problems[J]. Journal of Intelligent Manufacturing,2019,30:2407-2433

Acknowledgments

We are grateful for the support from the Center Subject Fund and the experimental help from the project team. Thanks are due to Prof. Yu Wang with the guidness and to Prof. Hongxia Zou for valuable discussion in Space Engineer University.

Hybrid Model Compression for Multi-Task Network

Min Xie[1], Weize Sun[1,2] and Lei Huang[1]

[1]Department of Electrical and Information Engineering, Shenzhen University, Shenzhen, China

[2]corresponding author's e-mail address: proton198601@hotmail.com

Abstract. In recent years, deep neural networks have made significant achievements in various fields in science and engineering. To achieve outstanding performance, neural network models with very deep structures and an extensive number of parameters are usually used, leading to huge memory storage and computational resources requirements of the hardware systems, and thus preventing them from being deployed in devices with hardware limits. In particular, such problem will be further amplified when multiple neural network models for several tasks are to be integrated into one single platform at the same time. In this paper, a hybrid joint-network optimization model is proposed to solve this multi-task multi-model compression problem. The idea of hard parameter sharing in multi-task learning is adopted. Specifically, we turn the sub-network of the same structure of multiple models into a hybrid network model, in which the same parameters are shared for all the tasks. In addition, a unique sparse matrix for each task is added to each task independently. Experimental results show that the proposed compression methods can not only achieve a high compression ratio but also get performance improvement for some tasks.

1. Introduction

With the development of computational resources, deep neural networks (DNN) have achieved significant breakthroughs in many fields such as speech recognition [1], computer vision [2]-[4] and image processing [5][6]. However, deep neural networks usually come with high computational complexity and large memory storage requirement, thus are difficult to be implemented on mobile and/or off-line platforms with constrained hardware resources.

In recent years, Multi-Task Learning (MTL), which can provide multiple outputs for just one input in one run, has received more and more attention. Caruana [7] showed that MTL could be beneficial as it allows for the acquisition of inductive bias through the inclusion of additional related tasks into the training pipeline. In the context of Deep Learning, MTL is typically carried out with either hard or soft parameter sharing of hidden layers. In particular, hard parameter sharing is the most commonly used approach to MTL in neural networks, which is generally applied by sharing the hidden layers between all tasks while keeping several task-specific output layers, as shown in Figure 1 [8].

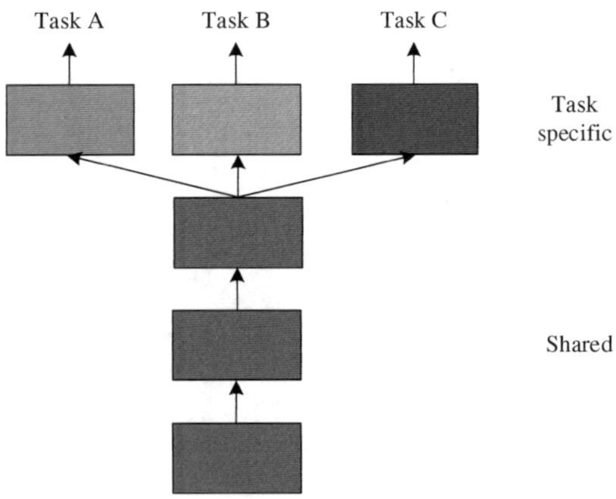

Figure 1. Hard parameter sharing for multi-task learning in deep neural networks.

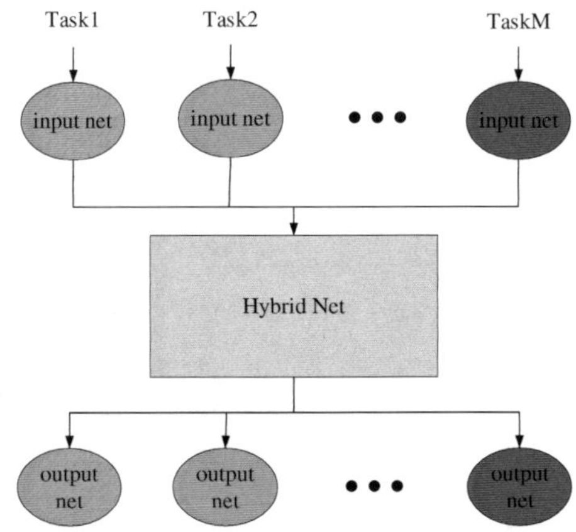

Figure 2. The network architecture of multi-task neural network model.

UberNet [9] was the first hard-parameter sharing model to jointly tackle a large number of low-, mid-, and high-level vision tasks. The model featured a multi-head design across different network layers and scales. Multi-linear relationship networks [10] extended this design by placing tensor normal priors on the parameter set of the fully connected layers. On the other hand, a gradient normalization (GradNorm) algorithm [11], which can automatically train the deep multi-task models by dynamically tuning gradient magnitudes, is proposed to improve accuracy and reduce overfitting effect. Furthermore, MTL is also employed in computer vision tasks, [12] presented an algorithm for simultaneous face detection, landmarks localization, pose estimation and gender recognition using both deep convolutional neural network (CNN) and multi task network (MTN), which fuses the intermediate layers of a CNN using a separate CNN and then followed by a MTL algorithm that operates on the fused features. However, all previous methods use a naive weighted sum of losses, where the loss weights are uniform or crudely and manually tuned. To overcome this problem, [13] proposed to combine multiple loss functions to simultaneously learn multiple objectives using homoscedastic task uncertainly, and can give a good result between these weights.

Compared to single-task models where each task is solved separately by one network individually, the above mentioned multi-task models bring several advantages. First, the resulting memory footprint is substantially reduced due to their inherent layer sharing. Second, in the case of one input corresponding to multiple outputs, the features of the shared layers are calculated only once, and thus the inference speeds for every task are improved. Therefore, these multi-task network models can also be seen as one kind of neural network compression model or lightweight model.

On the other hand, various DNN compression methods, such as transferred/compact convolutional filters [14], low-rank factorization [15]-[17] and parameter pruning and sharing [18]-[21], were proposed in recent years. Usually, when two or more networks from several tasks are implemented in one platform, we can train one backbone network for one task, and then transfer the whole or one part of the network to the other tasks by transfer learning techniques [22][23], then part of the parameters of networks with the same structure are shared, and thus the networks are compressed. However, such approach will use one set of parameters for all the tasks and fails to consider their differences, leading to a performance degradation of some of the tasks.

Network pruning, on the other hand, is one the most popular methods used to reduce the network complexity for one single task. Specifically, [24] proposed a simple pruning strategy: all connections with weights below a threshold are removed, the truncated model is fine-tuned to recover the classification accuracy. In [18], a data-free pruning algorithm is presented to remove the parameters based on a designed specific criterion. Sun et al. sparsified the network on a pre-trained dense model by removing the weakly correlated connections in [19], and the work in [21] discards the filters with minimal impact on the following layer output for network compression and acceleration. Furthermore, evolutionary computing and multi-objective optimization techniques can also be used to find the pruned structure of the networks [25][26]. Generally speaking, pruning methods can generate sparse networks with less parameters and computation to substitute the large network of one single task, but these methods had not been applied to multi-task learning so far.

It is worth noting that many real-world systems are multi-modal in nature. For example, in an unmanned platform, the camera may be required to perform high-resolution imaging [27][28] while the radar is used to detect and classify the objects in the same time. In this case, two different networks should be employed. However, most state-of-the-art MTL methods focus on single-input multiple-output multi-task networks, and thus cannot compress such models simultaneously. Furthermore, the sparse network structure that proved to be efficient in network compression had not been utilized in these models, leading to a huge storage requirement of the platform. To overcome such problem, a multi-tasks neural network compression (MTNC) method, integrating transfer learning, multi-task learning, and network sparsity, is proposed in this paper. First, the sub-networks with the same structure of the network models for all tasks are merged to form the hybrid network, and the parameters of all tasks in the sub-networks are shared. Second, the multi-task model with the hybrid network and a few input and output layers for each task is constructed. Finally, in each layer of the hybrid network, an extra sparse matrix is added for each task to reduce the performance loss caused by compression.

The remainder of the paper is organized as follows: Section 2 introduces the theory of multi-task learning. Section 3 describes the detailed description of our multi-task network compression methods. Finally, the baseline model and experimental results are given in Section 4 and conclusions are drawn in Section 5.

2. Multi-tasks network

In real-world applications, an unmanned system is usually required to jointly process multiple tasks for some general multi-application scenarios. Here we propose to integrate various tasks into one network model and use the hard parameter sharing method between the tasks to compress the model. The mean square loss function for the data fidelity term of this MTL problem is given as:

$$L(\mathbf{X},\mathbf{Y},\mathbf{W}) = \sum_{m=1}^{M} ||\mathbf{X}^m \mathbf{w}^m - \mathbf{y}^m||_F^2 \tag{1}$$

Where $\mathbf{X} = \{\mathbf{X}^1, \mathbf{X}^2, \cdots, \mathbf{X}^M\}$ and $\mathbf{Y} = \{\mathbf{y}^1 \cdots \mathbf{y}^M\}$ are the set of input feature matrix and output target matrix respectively. $\mathbf{X}^m \in \mathbb{R}^{N_m \times D}$ denotes the input matrix of the m-th task, while $\mathbf{Y}^m \in \mathbb{R}^{N_m \times 1}$ denotes the corresponding output vector, and $\mathbf{w}^m \in \mathbb{R}^{D\times 1}$ is the weight vector for the m-th task that map $\mathbf{X}^m$ to $\mathbf{y}^m$, e.g., $\mathbf{y}^m \approx \mathbf{X}^m \mathbf{w}^m$. The scalars N_m, D, and M denote the numbers of the samples for the m-th task, the number of features for each input matrix, and the number of tasks, respectively.

Since the data distribution, batch size, and important of each task are often different, it is not appropriate to sum the loss function of all tasks directly and then use gradient-based back propagation to optimize the network. Therefore, we add a weight parameter for the loss function of each task, and the loss function of equation (1) becomes:

$$L(\mathbf{X},\mathbf{Y},\mathbf{W}) = \sum_{m=1}^{M} \alpha_m L(\mathbf{X}^m, \mathbf{y}^m, \mathbf{w}^m) \tag{2}$$

3. Multi-task network compression combined with sparse matrix

The multi-task network compression (MTNC) model is proposed as in Figure 2. Specifically, multiple neural networks corresponding to multiple applications often contain some common or similar basic sub-networks, which often contains common information. Therefore, we can establish a common net to be a connection between multiple models through the same sub-networks in this hybrid network model. Considering that the input matrices in $\{\mathbf{X}^m, m=1,2,\cdots,M\}$ have different dimensionality of features, we add a task-specific input layer for each task before the common net. Accordingly, the

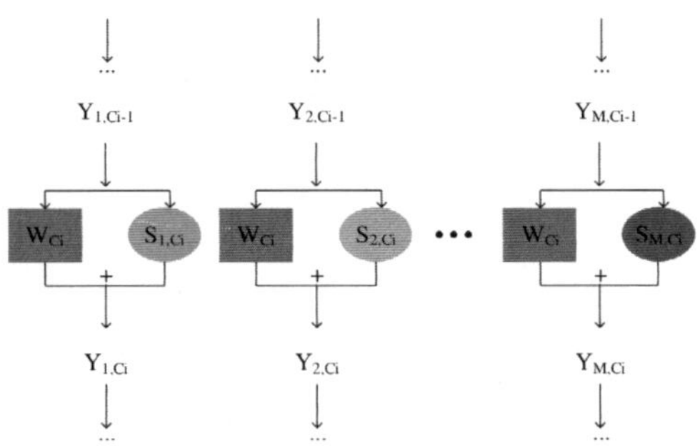

Figure 3. Details of hybrid networks structure.

layer name	output size	ResNet-18
conv1	32 × 32	3 × 3, 64, stride 1
conv2_x	32 × 32	[3 × 3, 64; 3 × 3, 64] × 2
conv3_x	16 × 16	[3 × 3, 128; 3 × 3, 128] × 2
conv4_x	8 × 8	[3 × 3, 256; 3 × 3, 256] × 2
conv5_x	4 × 4	[3 × 3, 512; 3 × 3, 512] × 2
average pool, 10-d fc for CIFAR-10		

Figure 4. Architecture of ResNet-18 for CIFAR-10 dataset.

input layer receives training datasets from all the tasks, and the output layer has M output units with one for each task. Here the outputs of the hidden layer can be viewed as the common feature representation learned for the M tasks. It is worth noting that our proposed MTNC network can reduce the parameters by nearly $(M-1)/M$ comparing to the M networks which train individually. In order to avoid overfitting, L2 regularization is used to reduce the complexity of the model. By this, the training optimization equation becomes:

$$\begin{gathered}\min_{\mathbf{W}} L(\mathbf{X},\mathbf{Y},\mathbf{W}) + \lambda\sum_{i=1}^{n}\mathbf{w}_i^2 \\ s.t. \sum_{i=1}^{n}\mathbf{w}_i^2 \le C\end{gathered} \tag{3}$$

Where λ is the regularization parameter.

For situations where there is little correlation between these different tasks, equation (3) might result in a performance degradation of some of the tasks. To alleviate this effect, a sparse matrix containing extra information is added to the hybrid network. Specifically, for each task, we add an individual sparse matrix with sparsity s to every layer as shown in Figure 2, where the sparsity is defined as the percentage of zero-valued elements over the entire matrix. The details of the *i-th* layer in the hybrid net are shown in Figure 3. The sparse matrix of the *i-th* layer in the hybrid net of the *m-th* task is initialized by:

$$\mathbf{S}_{m,Ci} = \Omega_{c_m} \odot \left(\mathbf{W}_{m,Ci} - \mathbf{W}_{Ci}\right) \tag{4}$$

where $\mathbf{W}_{Ci}$ is the parameter of the *i-th* layer of the common net in MTNC model. And $\mathbf{W}_{m,Ci}$ is the weight parameter obtained by training the *m-th* task separately that corresponds to layer of $\mathbf{W}_{Ci}$. Ω_{C_m} refers to a binary mask matrix reserving only the $c_m = \#(\mathbf{W})(1-s_m)$ largest absolute valued elements in original matrix, where, $\odot$ is the element-wised multiplication operation. Then the output $\mathbf{Y}_{m,Ci}$ can be calculated as:

$$\mathbf{Y}_{m,Ci} = \mathbf{Y}_{m,Ci-1}\left(\mathbf{W}_{m,Ci} + \mathbf{S}_{m,Ci}\right) \tag{5}$$

Consequently, the hybrid network includes both a common part and one additional sparse structure, and the whole multi-task network compress model can be trained by minimizing the loss function:

$$L(\mathbf{X},\mathbf{Y},\mathbf{W}) = \sum_{m=1}^{M} \alpha_m L\left(\mathbf{X}^m, \mathbf{y}^m, \mathbf{w}^m + \mathbf{s}^m\right) \tag{6}$$

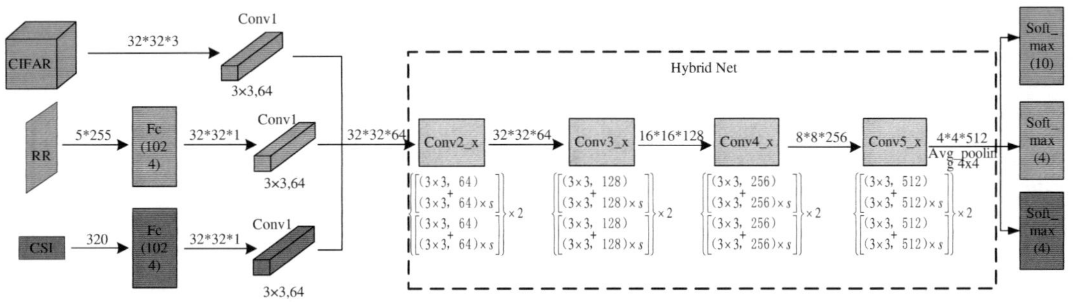

Figure 5. Hybrid model for multi-task network compression based on ResNet-18

4. Experiments

4.1 Baseline model

We evaluate our proposed multi-task network compression methods across the architecture of ResNet-18 [29], using three different datasets, CIFAR-10, CSI and Radar data Recognition (RR) dataset. The CIFAR-10 dataset contains 50000 training and 10000 testing color images of dimensions 32×32 of 10 classes. CSI is a dataset for human activity recognition via wifi, which contains 7273 vectors of dimensions 320 (5823 for training and 1450 for testing)[30]. RR is the echo of road target collected by radar, which contains 8985 training and 3015 testing data of dimensions 5×255 of 4 classes. The ResNet-18 (on CIFAR-10) shown in Figure 4 is slightly modified from the original ResNet-18(on ImageNet), which contains 17 convolutional layers and 1 fully connected layer as well as Batch Normalization[31] after each layer except the output one.

All of the following experiments are based on Tensorflow[32] with Adam optimizer[33]. We train the baseline model with CIFAR-10 dataset to achieve an accuracy of 93.45%. As the input features of the CSI and RR are different from CIFAR-10, we add an extra input layer before the baseline model for RR and CSI dataset respectively so that their input dimensions are changed to 32×32. Likewise, the output is modified to 4, and the other settings remain unchanged. In the CSI and RR model, the parameters of conv2_x, conv3_x, conv4_x and conv5_x are transferred from the trained parameters in the same layer in baseline model, while those of the conv1, input and output layer are initialized by random values from truncated normal distribution. After fine-tuning, the accuracy of CSI and RR are 92.26% and 90.79% respectively.

For convenience, we define the compression ratio (CR) as:

$$CR = \frac{P_m}{P_{CIFAR} + P_{D5} + P_{CSI}} \tag{7}$$

where P_m denotes the number of parameters in the multi-task model, and P_{CIFAR}, P_{CSI} and P_{RR} denote the number of parameters in the original network trained by CIFAR, CSI and RR respectively.

4.2 Result and analysis

In subsection 4.1, we trained and tested the three tasks separately. In this subsection, we merge the three networks for three tasks into one there are three inputs and outputs. The network architecture is shown in Figure 5. The batch size of CIFAR, CSI, and RR is set to 256, 29, 46, respectively, and the parameters in the hybrid network are also transferred from the baseline CIFAR model. The weight parameters of loss function of each task is set to $\alpha_{CIFAR} = 0.1$, $\alpha_{RR} = 0.5$ and $\alpha_{CSI} = 1$. After fine-tuning, the test accuracy of the CIFAR, CSI, and RR datasets are 93.53%, 92.55%, and 92.34%, respectively. It is shown that the proposed multi-tasks network structure can improve the accuracy of all three tasks, with the compression ratio of 36.46%.

In addition, we also carry on a comparative experiment. More specifically, we finetune the original three pre-trained networks respectively after changing the parameter matrix in the convolutional layers

Tabel 1. Test accuracy of the proposed compression method.

method	original	MTNC	MTNC+sparse(s_{CIFAR} / s_{CSI} / s_{RR}) 0.98/0/0	0.97/0/0	0.95/0/0	0.98/0.98/0.98	0.97/0.97/0.97	0.95/0.95/0.95	Original +sparse
CIFAR	93.45%	93.53%	93.76%	93.80%	**93.86%**	93.71%	93.78%	93.79%	92.45%
CSI	92.26%	92.55%	92.97%	93.10%	**93.10%**	92.76%	92.83%	93.03%	91.97%
RR	90.79%	92.34%	93.00%	93.11%	**93.28%**	93.11%	92.91%	93.01%	89.66%
CR	-	36.46%	37.08%	37.40%	38.02%	38.33%	39.27%	41.15%	38.06%

of conv2_x, conv3_x, conv4_x, and conv5_x into sparse matrix. The sparsity of each network is set to 0.66, so that the CR reaches 38.06%. The results of the sparsity network based on the original one are tabulated in the last column of Table 1. It is observed that compared to the proposed MTNC model, the test accuracy of the sparsed original network decreased significantly, even lower than that of the original model.

Finally, the test results of each dataset under hybrid network combined with sparse matrices are exhibited and the results are shown in Table 1. Generally Speaking, adding sparse matrices in the proposed MTNC model can improve the performance of each task. Especially when the sparsity rate of CIFAR is 0.95 and that of CSI and RR are 1, the test accuracy of the three datasets reaches the highest among all the results, and the network's overall performance of the network achieves optimal. It should be noted that sparsity 1 means no additional sparse matrix is required for the task. The main reason is that the amount of CIFAR is much larger than CSI and RR, so the common part of hybrid network is enough to describe CSI and RR well. For CIFAR, additional individual parts are required to improve performance.

5. Conclusion

In this paper, a novel method for multi-task neural network compression, is proposed. A hybrid network structure with both common part and individual parts for each task is divided. Note that for the common part, parameter sharing technique is used, while the individual parts are highly sparse matrices. Experimental results show that the proposed hybrid network can not only compress the multi-tasks network but also improve the performance significantly.

6. References

[1] Huang X, Seide F, Seltzer M, Stolcke A, Yu D, Zweig G, Xiong W and Droppo J. Achieving human parity in conversational speech recognition. IEEE/ACM Transactions on Audio, Speech, and Language Processing,PP(99), 2016

[2] G.Hinton A.Krizhevsky, I.Sutskever. Imagenet classification with deep convolutional neural networks. Advances in neural information processing systems, 25(2), 2012.

[3] D.Trevor M.Jitendra G. Ross, D.Jeff. Rich feature hierarchies for accurate object detection and semantic segmentation. In 2014 IEEE Conference on Computer Vision and Pattern Recognition, pages 580–587, 2014.

[4] S.Sanketh L.Thomas S.Rahul F.F.Li K.Andrej, T.George. Large-scale video classification with convolutional neural networks. In 2014 IEEE Conference on Computer Vision and Pattern Recognition, pages 1725–1732, 2014.

[5] R.Girshick J.Sun S.Ren, K.He. Faster r-cnn: Towards real-time object detection with region proposal networks. IEEE Transactions on Pattern Analysis and Machine Intelligence, 39(6):1137–1149, 2017.

[6] V.Lempitsky D.Ulyanov, A.Vedaldi. Deep image prior. IEEE, 2017.

[7] R.Caruana. Multitask learning. Machine Learning, 28(1):41–75, 1997.

[8] R. A. Caruana. Multitask Learning: A Knowledge-Based Source of Inductive Bias. 1993.

[9] I.Kokkinos. Ubernet: Training a 'universal' convolutional neural networkfor low-, mid-, and high-level vision using diverse datasets and limited memory. In 30th IEEE/CVF Conference on Computer Vision and Pattern Recognition (CVPR), 2017.
[10] J.Wang M.Long and Y.Philip S. Learning multiple tasks with multilinear relationship networks. 2015.
[11] Z. Chen, V. Badrinarayanan, C. Y. Lee, and A. Rabinovich. Gradnorm: Gradient normalization for adaptive loss balancing in deep multitasknetworks. 2017.
[12] R.Chellappa R.Ranjan, V.M.Patel. Hyperface: A deep multi-task learning framework for face detection, landmark localization, pose estimation, and gender recognition. IEEE Transactions on Pattern Analysis and amp; Machine Intelligence, pages 1–1, 2018.
[13] K.Alex C.Roberto, G.Yarin. Multi-task learning using uncertainty to weight losses for scene geometry and semantics. In 2018 IEEE/CVFConference on Computer Vision and Pattern Recognition, pages 7482–7491, 2018.
[14] B.Chen D.Kalenichenko W.Wang T.Weyand M.Andreetto H.Adam A.G.Howard, M.Zhu. Mobilenets: Efficient convolutional neural networks for mobile vision applications. 2017.
[15] W. Sun, S. Chen, L. Huang, H. C. So, and M. Xie. Deep convolutional neural network compression via coupled tensor decomposition. IEEE Journal of Selected Topics in Signal Processing, PP(99):1–1, 2020.
[16] Y.Zhang X.Wang E.Weinan C.Tai, T.Xiao. Convolutional neural networks with low-rank regularization. Computer Science, 2015.
[17] Shaowu Chen, Jihao Zhou, Weize Sun, and Lei Huang. Joint matrix decomposition for deep convolutional neural networks compression, 2021.
[18] R.V.Babu S.Srinivas. Data-free parameter pruning for deep neural networks. Computer Science, pages 2830–2838, 2015.
[19] X.Tang Y.Sun, X.Wang. Sparsifying neural network connections for face recognition. In 2016 IEEE Conference on Computer Vision and Pattern Recognition (CVPR), 2016.
[20] I.Durdanovic H.Samet H.P.Graf H.Li, A.Kadav. Pruning filters for efficient convnets. 2016.
[21] W.Lin J.H.Luo, J.Wu. Thinet: A filter level pruning method for deepneural network compression. In 2017 IEEE International Conference on Computer Vision (ICCV), 2017.
[22] T.Kong W.Zhang C.Yang C.Liu C.Tan, F.Sun. A survey on deep transfer learning. arXiv, 2018.
[23] I. Laptev J. Sivic M. Oquab, L. Bottou. Learning and transferring mid-level image representations using convolutional neural networks. In 2014 IEEE Conference on Computer Vision and Pattern Recognition (CVPR), pages 1717–1724, Los Alamitos, CA, USA, jun 2014. IEEE Computer Society.
[24] J.Tran W.J.Dally S.Han, J.Pool. Learning both weights and connections for efficient neural networks, 2015.
[25] Huang J, Sun W, Huang L. Deep neural networks compression learning based on multiobjective evolutionary algorithms[J]. Neurocomputing, 2020, 378(Feb.22):260-269.
[26] J. Chen, Y. Xu, W. Sun and L. Huang. "Joint sparse neural network compression via multi-application multi-objective optimization." Applied Intelligence, pp. 1-18, 2021.
[27] B.Ghanem J.Zhang. Ista-net: Interpretable optimization-inspired deep network for image compressive sensing. In 2018 IEEE/CVF Conference on Computer Vision and Pattern Recognition (CVPR), 2017.
[28] G.Cen F.Li X.Chen Z.Fu, N.Cai. Image super-resolution via a novel cascaded convolutional neural network framework. Signal Processing: Image Communication, 63:9–18, 2018.
[29] S.Ren J.Sun K.He, X.Zhang. Deep residual learning for image recognition. In 2016 IEEE Conference on Computer Vision and Pattern Recognition (CVPR), 2016.
[30] Zhou N , Sun W , Liang M . Human Activity Recognition based on WiFi Signal Using Deep Neural Network[C]// 2020 IEEE 8th International Conference on Smart City and Informatization (iSCI). IEEE, 2020.

[31] C.Szegedy S.Ioffe. Batch normalization: Accelerating deep network training by reducing internal covariate shift. JMLR.org, 2015.
[32] P.Barham E.Brevdo X.Zheng M.Abadi, A.Agarwal. Tensorflow: Large-scale machine learning on heterogeneous distributed systems.
[33] J.Ba D.Kingma. Adam: A method for stochastic optimization. Computer Science, 2014.

Acknowledgments
The work described in this paper was supported in part by the Foundation of Shenzhen under Grant JCYJ20190808122005605, and in part by the Guangdong Basic and Applied Basic Research Foundation under Grant 2021A1515011706, and in part by the National Natural Science Foundation of China (NSFC) under Grant 62101335 and U1913203.

Photovoltaic Output Prediction Method Based on Weather Forecast and Machine Learning

Kaikai Zhang[1] and Guibin Zou[1,2]

[1] School of Electrical Engineering, Shandong University, Jinan, China

[2] guibinzou@sdu.edu.cn

Abstract. This paper conducts prediction research based on the historical power generation data and meteorological data of a photovoltaic power station. The obtained data is firstly classified by season, and the correlation analysis of meteorological factors is carried out. The data closely related to PV generation is selected for prediction, which improves the prediction efficiency. Then the Kmeans algorithm is used to cluster the historical data, and data are divided into four types: sunny, partly cloudy, cloudy, and rainy, further each type of data is separately trained to obtain four training models. The classified data is decomposed by EEMD, and a set of frequency-distributed IMF components and a monotonic distribution of residual components are obtained. Use support vector machine to predict the dimension-reduced components, and reconstruct the result to obtain the final predicted value. After obtaining the future weather forecast data, select the same type of historical data set for model training. The results show that the prediction accuracy will be significantly improved after signal processing in cloudy and rainy weather.

1. Introduction

In recent years, the number of photovoltaics connected to the grid has grown rapidly, and PV has become a new growth point for renewable energy power generation. Photovoltaic power generation is affected by meteorological and environmental conditions, and has great volatility and randomness. If the photovoltaic power generation cannot be consumed in time, a large amount of photovoltaic surplus power will bring the flow backward in the distribution network lines, and the voltage will exceed the limit value, etc. A series of problems have brought threats to the safety of electricity network. If PV generation can be predicted in a timely and accurate manner, it will provide effective support for the formulation of the day-ahead scheduling plan of the power system. The grid dispatching strategy based on photovoltaic prediction technology will greatly cut back the impact of PV generation on the grid, and provide technical support for better access to the grid with a high proportion of new energy.

At present, there are already international organizations providing photovoltaic power forecasting services, but the core problem is still to improve the accuracy. Some researchers take into account the time series characteristics of photovoltaics, screen the best similar weather on the day to be predicted, and use the power generation amount on that day as an important reference for prediction; or use weather factors as input to analyse the influence of weather on power generation. Existing articles pay more attention to the prediction model, adopt various neural networks, and optimize the parameters to improve the prediction accuracy. Current research is still lacking in mining data features and applying them to model training. For neural networks, the quality of training data has a significant impact on the results. Strengthening data pre-processing and deep mining of data characteristics is an effective way to further improve the prediction accuracy. [1]-[2].

This paper focuses on the pre-processing of historical data. Firstly, the data strongly correlated with PV output are extracted and its data features are mined, and the weather features are clustered by Kmeans algorithm. On this basis, the signal decomposition based on EEMD is performed on each classified signal, and the SVM prediction is performed after the decomposed signal is reduced in dimension. Data pre-processing will greatly reduce the volatility of the input signal, helping to further make predictions more accurate.

In summary, based on the historical data and fully considering data characteristics, this paper pre-processes the initial data and makes predictions. By comparing the prediction accuracy before and after data processing, it is found that the data processing significantly improves the weather forecast results with large fluctuations. The research results can provide improvement ideas for other research on predicting volatility signals, and have certain practical significance.

2. Forecast models and processes

2.1. Data selection

The data used in the forecast comes from a photovoltaic power station in the ashland region of the United States. Since there is no photovoltaic output at night, the historical data from 7:00 to 17:00 is filtered out with a time interval of 15 minutes. There are many weather conditions that affect photovoltaic output. If all environmental factors are used for training, the training time will be greatly increased, and the training efficiency will become very low. Hence, it is critical to implement correlation analysis of meteorological factors and screen out the weather factors closely related to photovoltaic power generation for training. Figure 1 is a graph showing the correlation of photovoltaic power generation in typical weather [3].

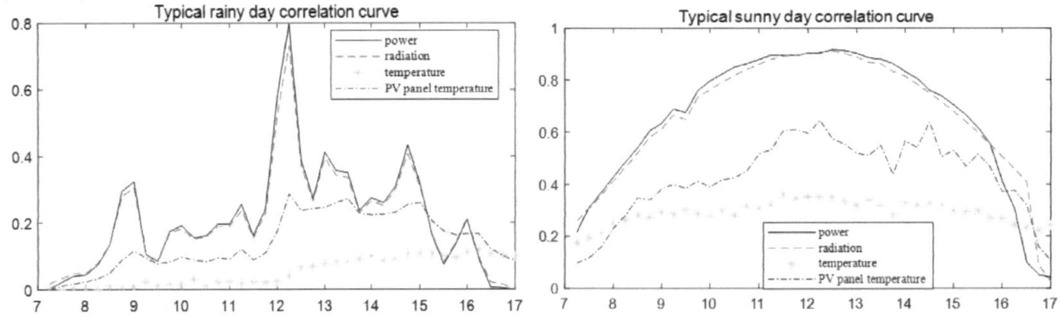

(a)rainy day (b)sunny day

Figure 1. Typical weather factor correlation curve

2.2. Weather type clustering

The photovoltaic output of different weather types is quite different. If the obtained data is directly used for training, the volatility of the training set data will be very large, and the prediction error will also increase. Therefore, the initial data is clustered based on the Kmeans algorithm and divided into four weather types, namely sunny, partly cloudy, cloudy, and rainy. Then group training predictions on data of different weather types to get four training models. The classification results are more intuitive under the two-dimensional scale, but the problem of misclassification is prone to occur at the data classification boundary. This paper select the maximum, minimum and average values of irradiance, temperature and photovoltaic output for a total of nine dimensions for weather clustering [4].

The clustering process is shown in Figure 2 and euclidean distance calculation formula is as follows:

$$d(x,y)=\sqrt{\sum_{i=1}^{n}(x_i-y_i)^2} \tag{1}$$

After clustering, the weather of the same type will be divided into a combination, and subsequent predictions will be made within each combination.

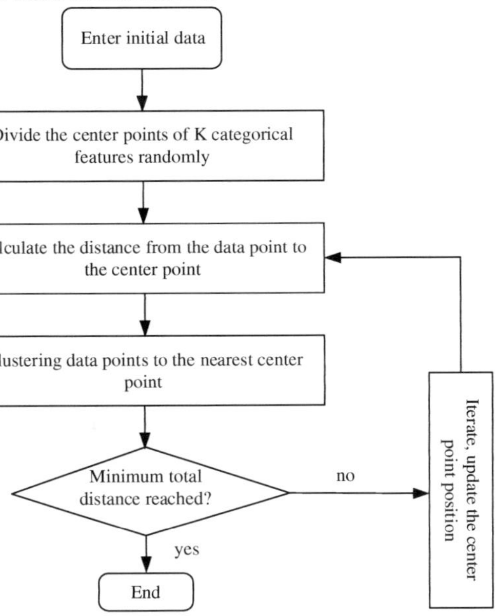

Figure 2. Kmeans clustering flowchart

2.3. Ensemble empirical mode decomposition

Due to the random fluctuation of the obtained original photovoltaic output data, direct use of historical photovoltaic output data for prediction will affect the final prediction accuracy. The EEMD can decompose the signal based on the time scale characteristics of the data, and decompose the original data with strong volatility into a relatively stable sequence. In this paper, the data with large fluctuations are decomposed, and each sequence after decomposition is used for prediction to improve the prediction accuracy. EEMD is an improved method of randomly adding white noise on the basis of traditional EMD in order to solve the problem of modal aliasing after decomposition [5].

The main steps are as follows: (1) determine the number of processing times m for the initial signal; (2) add random white noise to the m initial signals to form a series of new signals; (3) separate the new signals of this series and perform EMD decomposition to obtain a series of IMF components (4) calculate the mean value of the components of the corresponding modes respectively and obtain the decomposition result.

Figure 3 is the result of EEMD decomposition of PV generation. The top one is the original PV output data in October, the time scale is 15min, and the data from 7:00-17:00 every day is filtered. There are 40 data points per day, a total of 1240 points. The graph shows that PV generation will vary greatly within a month. The second to tenth parts show the components with decreasing frequency from high to low, and the bottom is the monotonic residual component after decomposition. It can be seen from the figure 3 that the fluctuation after decomposition is significantly smaller than the initial signal. Using all components for prediction will bring some problems, a better way is to reduce the dimensionality of the components and then predict.

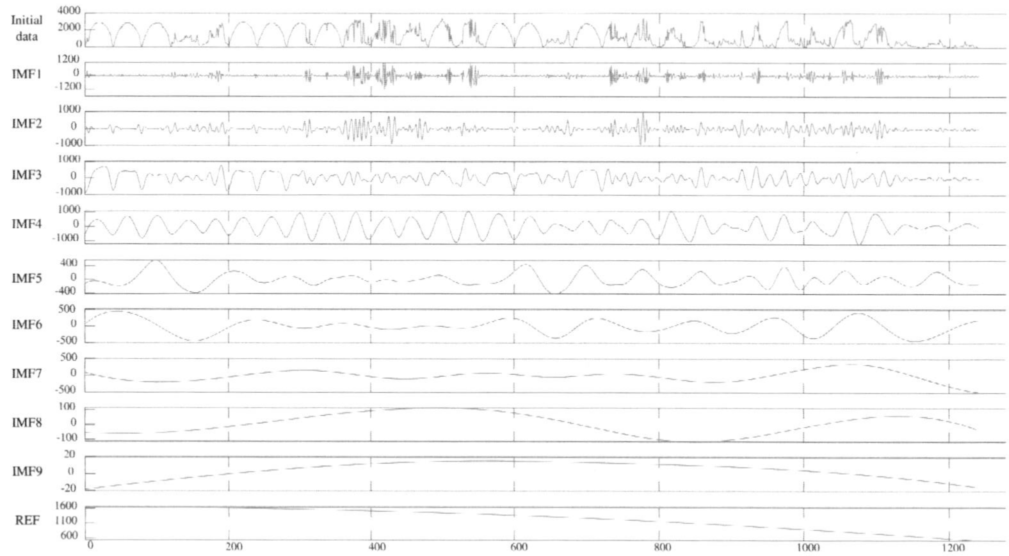

Figure 3. Photovoltaic output EEMD decomposition results

2.4. Support vector machine prediction model

SVM were initially mainly used to cure classification problems such as small samples and nonlinearity, and then gradually extended to research on data prediction such as function fitting. SVM uses a nonlinear function to map the input vector to a high-dimensional feature space, and then uses a linear function to achieve prediction in this space. It is an important tool in the field of regression prediction. Due to the influence of the weather, the change trend of PV generation becomes difficult to predict, the change range is large and the time series is weak. For this kind of data, it is more suitable to use SVM for prediction. In this paper, optimize the parameters C and G of the SVM model by cross-validation, and select RBF kernel function to deal with the non-linear relationship between model inputs and outputs. The detailed mathematical derivation formula of SVM can be seen in [6].

2.5. The basic process of forecasting

Figure 4 is the basic flow chart of photovoltaic forecasting, and PV generation between 7:00-17:00 in the future day is predicted according to the weather forecast. After processing the initial data, the SVM model is classified and trained, and the classification is determined by clustering according to the weather conditions on the day to be predicted [7]-[8]. Use EEMD to decompose the input signal and select the corresponding model for prediction, the predicted value will be obtained after reconstruction. After EEMD, there will be several IMF components. If every component is used for separate prediction, it will have high complexity. Therefore, the high-dimensional component results are reduced in dimension in the prediction. Convert the time domain to the frequency domain, pay attention to its frequency fluctuation and amplitude fluctuation, and reduce its dimension. Use the resulting high frequency and low frequency components for prediction.

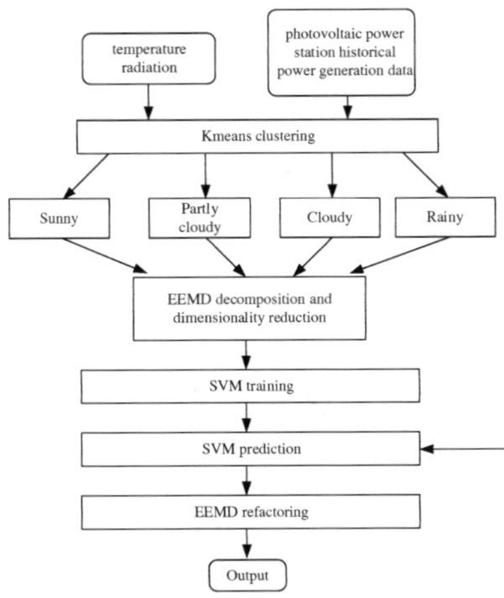

Figure 4. Photovoltaic forecast flow chart

3. Comparison of simulation results

3.1. The evaluation index of prediction effect

The mean absolute error percentage (MAPE) can clearly describe the relative size of the error when the amount of data is large. The root mean square error (RMSE) amplifies the absolute value of the training error. The combination of the two can relatively comprehensively describe the forecast error [9]. In this paper, MAPE and RMSE are used as indicators to evaluate the quality of the results. The specific formulas are as follows, where P_i is the predicted value and $\overline{P_i}$ is the actual value.

$$MAPE = \frac{1}{N}\sum_{i=1}^{N}\frac{\left|P_i - \overline{P_i}\right|}{\overline{P_i}} \tag{2}$$

$$RMSE = \sqrt{\frac{1}{N}\sum_{i=1}^{N}\left(P_i - \overline{P_i}\right)^2} \tag{3}$$

3.2. Comparison of forecast results

PV generation changes drastically in cloudy and rainy days. If the proposed method can have a better prediction effect in cloudy and rainy days, it will also have better prediction accuracy under sunny conditions. After clustering the historical data from June to October, the data set corresponding to the rainy day type is selected for prediction, so as to test the effect of the proposed method. Table 1 lists the actual data and predicted data, and predicted value 1 is the forecast result before data processing, predicted value 2 is the forecast result after data processing.

Calculate the error of the predicted data and compare its accuracy. Draw a comparison chart of its predicted value and actual value according to the data in Table 1. According to predicted value 1, it is judged to be a cloudy day. The MAPE value is 8.54%, and the RMSE is 10.9378. Predicted value 2 is the forecast result after initial data processing. The days to be predicted are clustered into cloudy days through the Kmeans algorithm, and the historical data of rainy days are decomposed for training and prediction. After treatment, MAPE becomes 2.8300%, a decrease of 67.18%; RMSE becomes 5.5564, a decrease of 49.20%. It can be seen that the two evaluation indicators after clustering have been greatly improved. EEMD combined with dimensionality reduction can reduce data fluctuation and improve prediction. But using all components for prediction, the training time will be longer and the prediction accuracy will likely be reduced. Because the correlation between some components and PV generation is very low, if data with low correlation is used for prediction, the model will have difficulty fitting the functional relationship between the input and output, which leads to a large prediction deviation. Figure 5 shows the prediction results of different methods. Figure shows that the bp neural network is difficult to accurately track the photovoltaic output under fluctuating weather. On the contrary, the SVM can more easily adapt to the weather fluctuation and its effect becomes better after data processing.

Table 1. actual and predicted values

time	actual value/W	predicted value 1/W	predicted value 2/W	time	actual value/W	predicted value 1/W	predicted value 2/W
7:15	46	76.41864418	52.34293622	12:15	177.5	172.1780736	171.0426783
7:30	66	90.65941671	70.25393798	12:30	143	143.3674069	137.7008204
7:45	186	185.1579944	184.4299533	12:45	168.5	166.6687342	164.3538038
8:00	335	328.2508531	336.3061257	13:00	196	187.27333	187.7769567
8:15	399	400.7341076	407.5027584	13:15	253	242.293789	247.7669923
8:30	275	282.855046	290.6058993	13:30	251.5	242.3086349	247.7669923
8:45	115.5	131.8714664	121.2027432	13:45	285.5	273.7051666	280.7582958
9:00	124.5	137.5871847	127.7812895	14:00	208.5	202.7036068	204.5002959
9:15	65	88.63112364	67.21098574	14:15	203.5	199.5008615	201.1579113
9:30	50.5	71.79543304	49.44543074	14:30	131.5	135.6259637	127.7812895
9:45	149	149.1740934	144.341692	14:45	158.5	161.049807	157.67184
10:00	70	83.61097296	67.21098574	15:00	198.5	196.3523973	197.8141835
10:15	129	135.1460474	127.7812895	15:15	265.5	261.0799138	267.5932419
10:30	71	86.94731738	70.25393798	15:30	575	579.5833567	576.0337812
10:45	136.5	140.8836806	134.3881924	15:45	310	312.0554479	320.027981
11:00	87.5	99.85394266	85.76442078	16:00	256	257.854779	264.2955044
11:15	155	158.0771987	154.3344323	16:15	182.5	190.982627	191.1234884
11:30	249.5	242.2300825	247.7669923	16:30	145	156.423212	150.9999704
11:45	268	257.909614	264.2955044	16:45	210	214.8652448	217.8530583
12:00	126.5	129.3057356	121.2027432	17:00	203.5	211.8908461	214.5176818

Figures 6 shows the MAPE values at each moment in the two predictions based on Table 1. It can be seen that the moments with large relative errors are always near the starting point and the ending point, corresponding to 7:00 and 17:00 every day. At sunrise and sunset, the photovoltaic output value is small, so the MAPE value is relatively large, which may be because the particulate matter in the air has a large scattering effect on the radiation, resulting in a certain error between the radiation reaching the ground and the forecast. However, the RMSE value is small at this time, so it is more reasonable to combine the two indicators mentioned in this paper to comprehensively evaluate the prediction results.

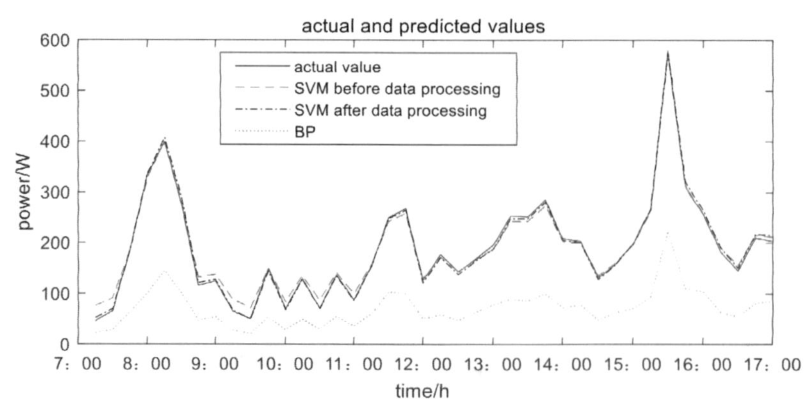

Figure 5. Comparison of results

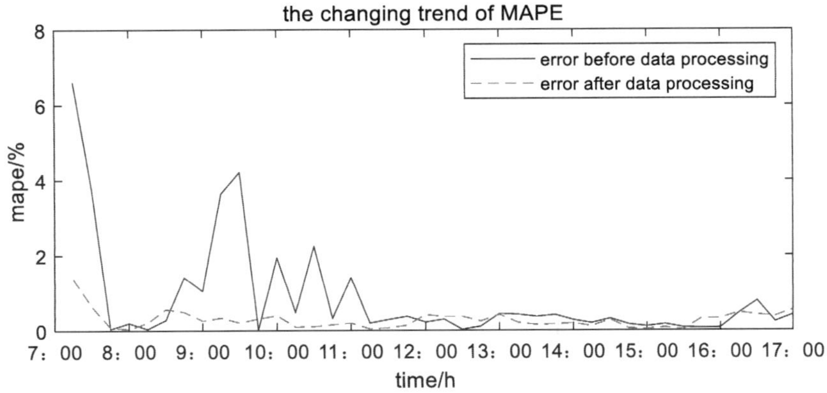

Figure 6. Trend of MAPE

4. Conclusion

Aiming at the problem of low PV forecasting accuracy in weather with large fluctuations, this paper pre-processes the initial weather data, uses Kmeans algorithm to perform weather clustering, and employs EEMD to decompose the initial signal to obtain components with small fluctuations. In MATLAB, the method of SVM is used to predict the data before and after processing, and comparing the prediction results, it is found that both MAPE and RMSE are greatly reduced after data processing. The results verify that data processing can reduce the error and thus improve the prediction effect.

5. References

[1] J. Wang, H. Zhong, X. Lai, Q. Xia, Y. Wang and C. Kang, "Exploring Key Weather Factors From Analytical Modeling Toward Improved Solar Power Forecasting," in IEEE Transactions on Smart Grid, vol. 10, no. 2, pp. 1417-1427, March 2019. doi: 10.1109/TSG.2017.2766022

[2] G. G. Kim et al, "Prediction Model for PV Performance With Correlation Analysis of Environmental Variables," in IEEE Journal of Photovoltaics, vol. 9, no. 3, pp. 832-841, May 2019. doi: 10.1109/JPHOTOV.2019.2898521

[3] X. Yuan, J. Shi and J. Xu, "Short-term Power Forecasting for Photovoltaic Generation Considering Weather Type Index," in Proceedings of the CSEE, vol. 33, no. 34, pp. 57-64, December 2013. doi: 10.13334/j.0258-8013.pcsee.2013.34.011

[4] Y. Zhang et al, "Short-term photovoltaic power forecasting based on Kmeans algorithm and support vector machine," in Power System Protection and Control, vol. 46, no. 21, pp. 118-124, November 2018. doi:10.7667/PSPC171595

[5] G. Zhang et al, "Wind power prediction based on variational mode decomposition multi-frequency combinations," in Journal of Modern Power Systems and Clean Energy, vol. 7, no. 2, pp. 281-288, March 2019. doi: 10.1007/s40565-018-0471-8.

[6] M. Mao et al, "Short-term Photovoltaic Generation Forecasting Based on EEMD-SVM Combined Method," in Proceedings of the CSEE, vol. 33, no. 34, pp. 17-24, December 2013. doi: 0.13334/j.0258-8013.pcsee.2013.34.007

[7] Z. Sun, S. Zhao and J. Zhang, "Short-Term Wind Power Forecasting on Multiple Scales Using VMD Decomposition, K-Means Clustering and LSTM Principal Computing," in IEEE Access, vol. 7, pp. 166917-166929, March 2019. doi: 10.1109/ACCESS.2019.2942040.

[8] L. Ye et al, "Combination Forecasting Method of Short-term Photovoltaic Power Based on Weather Classification," in Automaion of Electric Power Systems, vol. 45, no. 1, pp. 44-54, January 2021. doi: 10. 7500/AEPS20200613003

[9] L. Ye, Z. Chen, Y. Zhao and J. Zhu, "Photovoltaic Power Forecasting Model Based on Genetic Algorithm and Fuzzy Radial Basis Function Neural Network," in Automaion of Electric Power Systems, vol. 39, no. 16, pp. 16-22, August 2015. doi: 10.7500/AEPS20140903004

AUTHOR INDEX

Institute of Physics
Dirac House, Temple Back
Bristol BS1 6BE UK

ISSN: 1742-6588
ISBN 978-1-7138-6040-2

CURRAN ASSOCIATES INC.
proceedings.com